Race to the Stratosphere

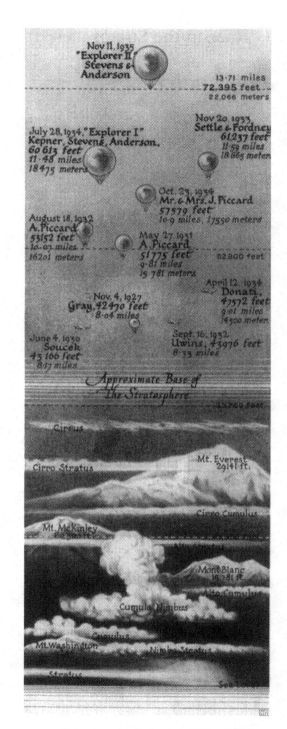

Altitude records in the 1930s. NGS.

David H. DeVorkin

Race to the Stratosphere

Manned Scientific Ballooning in America

With 87 Illustrations

Springer-Verlag
New York Berlin Heidelberg
London Paris Tokyo

David H. DeVorkin
Curator
National Air and Space Museum
Smithsonian Institution
Washington, DC 20560
USA

On the front cover: The National Geographic Society—Army Air Corps gondola is connected to its Explorer II balloon by Goodyear technicians. (NGS/NASM).

Library of Congress Cataloging-in-Publication Data
DeVorkin, David H.
   Race to the stratosphere : manned scientific ballooning in
America / David H. DeVorkin.
      p.   cm.
   Includes index.
   1. Balloons—United States—Scientific applications—History.
I. Title.
TL618.D48   1989
502'.8—dc19   89-4245

Media conversion by Impressions, Inc., Madison, Wisconsin.
Printed and bound by Arcata Graphics/Halliday, West Hanover, Massachusetts.
Printed in the United States of America.

9 8 7 6 5 4 3 2 1

ISBN 0-387-96953-5 Springer-Verlag New York Berlin Heidelberg
ISBN 3-540-96953-5 Springer-Verlag Berlin Heidelberg New York

For Kunie and Hannah,
who are really down to earth

Acknowledgments and Sources

No matter how much research and digging one may do to become familiar with events and issues of the past, and no matter how much planning and framing one may undertake to inform the research agenda, it is not until the time comes to put the results into rational order and to publish them that the real learning process begins. Preparing this book has been such a process, taking me from a preoccupation with presenting scientific narrative as history, to an appreciation of the fact that the history of space science cannot be fully understood as intellectual history. Until very recently, the history of each scientific discipline that today can be found within the loose amalgam called space science really had little to do with how the capability of spaceflight itself developed.

Many people have helped me undergo this learning process. Some have freely provided unpublished works and advice, some have shared their insight and impressions of my work, and a few forced me to see history in a new light. Michael Dennis and Peter Galison kindly commented on earlier drafts of the book, as did Tom Crouch, Ron Doel, R. Cargill Hall, Robert Multhauf, Allan Needell, and Charles Ziegler. Donald Piccard, son of Jean and Jeannette Piccard, generously provided very useful criticism and commentary at a later stage of this work, and kindly permitted me to publish excerpts from the Piccard Family Papers held at the Library of Congress. He also supplied copies of very useful documents still in his possession. Mary Pavlovitch and Larry Wilson of the National Air and Space Museum Library helped me locate materials. Sophie Mayr, Ben Peirce, Renata Rutledge, and Jose Villela all ably assisted in research and translations. Luis W. Alvarez, E.N. Brandt, Brian O'Brien, Paul Forman, Leonard Grant, Paula Hurwitz, Paul Levaux, John E. Moreau, Edward P. Ney, Peter Trower, and Charles Ziegler were all especially helpful in providing research materials, photographs, and advice.

It is inappropriate to impart present-day values upon the characters and events of the past. For example, the terms "manned" and "unmanned" flight, which were used freely throughout the period covered, are offensive to some readers today. We retain the terms as they were used, because alternatives would be clumsy. We also use the words "science" and "scientific" as they were employed commonly during the period, not distinguishing between possible modifiers such as "pure" and "applied" unless they were explicitly stated. We recognize that a wide range of activities throughout the period were labelled as "scientific" and that interpretations changed with time, but our purpose here will be to show how entrepreneurs for stratospheric ballooning attempted to convince the public that their goals were scientific, independent of the exact meaning of the term. If I have managed to maintain objectivity in this process, it is a credit to the attentiveness of colleagues; if I have transgressed, I take full responsibility.

Oral histories proved to be useful resources for capturing a sense of issues too diffuse to be found even in the available archival literature. Those interviewed specifically for this book were Brian J. O'Brien, William H. Gross, and Ned Dyer. All other interviews were part of archival programs at the American Institute of Physics (AIP) Center for History of Physics, and from one of several recent programs in the Space Science and Exploration Department of the National Air and Space Museum (NASM). AIP interviews included Leo Goldberg, John Naugle, Nancy Roman, and Martin Schwarzschild, while NASM interviews were with Phyllis Freier, Herbert Friedman, Robert Gilruth, Leo Goldberg, A.O. Nier, Gordon Newkirk, E.P. Ney, John D. Strong, and James Van Allen. I am indebted to all who freely permitted me to cite from these interviews.

The heart of the research of which this book is but one product is the documentary material collected and preserved in the many archives and libraries consulted in the past four years. Those cited here, with their abbreviations used in the notes, and persons who aided in locating and making available useful material, include: National Geographic Society: Leonard Grant and Barbara Shattuck (NGS); Piccard Family Papers, Library of Congress Manuscript Division, including portions still unorganized and located at the Library's Lanham Maryland storage facility: Ronald Wilkinson (PFP/LC); Lloyd Berkner Papers, Library of Congress Manuscript Division (Lloyd Berkner Papers/LC); Office of Naval Research Historical Files (ONR); Naval Research Laboratory Records, National Archives and Records Administration (NRL); Robert A. Millikan Papers microfilm edition, California Institute of Technology Archives (RAM/CIT); National Air and Space Museum Library and Technical Information Division (NASM); U.S. Naval Observatory Library: Brenda Corbin (USNO); W.F.G. Swann Papers, American Philosophical Society Library: Martin Leavitt and Elizabeth Carroll-Horrocks (Swann Papers, APS); Maxwell Air Force Base Historical Research Center Records, microfilm copies at Air Force History Research Library, Bolling Field: Sgt. Gernagin (MAFB); Registrar's Records, Museum of Science and Industry: Terri Sinnott and Linda Weeks (RR/MSI); National Academy of Sciences Archives: David Saumweber (NAS); Century of Progress Papers, Special

Collections, University of Illinois Chicago Circle: Maryanne Bamberger (COP/UIC); Dow Chemical Company Historical Files: E.N. Brandt (Dow); Records of the National Bureau of Standards, RG 167, National Archives and Records Administration: Marjorie Ciarlanti (NBS/NARA); General Mills Corporate Archives: Jean Toll, corporate archivist; Otto Winzen Collection, University of Minnesota Special Collections and Rare Books: Austin J. McLean, Curator (OW/UM); NASA History Office Files: Lee Saegesser (NHO); American Institute of Physics Sources for History of Modern Astrophysics (SHMA/AIP); National Air and Space Museum Space Astronomy Oral History Project (SAOHP/NASM); National Air and Space Museum Glennan-Webb-Seamans Space History Project (GWS/NASM).

I must acknowledge the fine support and guidance provided by the staff of Springer-Verlag, and the excellent services of Vicki McIntyre, free-lance editor. They, along with Patricia Graboske, Dale Hrabak and Marc Avino of NASM and Joseph Shealy of the Smithsonian attended to technical matters during the production of the book. Finally, I owe the greatest debt to my wife Kunie and daughter Hannah, who generally put up with my fiddling and fidgeting with the past.

David H. DeVorkin

Contents

Introduction and Contentions

The explorers and their craft face a watching world: Albert Stevens, William Kepner, Orvil Anderson, and the Explorer I gondola before the camera, Dow Chemical Company, Midland, Michigan, 1934. Dow.

The stratosphere today is a realm touched by anyone who flies long distance. Astronauts and cosmonauts pass through it in a brief moment. Military jet fighters and reconnaissance craft crew routinely prowl its middle and upper regions. Travelers on Concorde taste the thrill of it for an hour or two, sensing the nearness of space as they witness the curvature of the earth. And sometimes we enter its lowest regions flying west in transcontinental hops or during intercontinental over-the-pole flights. So the stratosphere today is no longer a mysterious, foreboding frontier, a distant place in the atmosphere discovered by the inventive French meteorologist Tisserenc de Bort between 1898 and 1904 where thermal lapse rates seemed at first to cease. It is now an accessible place, a place known for exceptional calmness in its lower regions, a place where the air is no longer turbulent because lighter and warmer layers rest comfortably above cooler and denser strata. Constant monitoring from the ground, and numerous flights through it using an ever-widening array of meteorological devices in the past half century have turned its unknowns into knowns. Very few people wonder about the stratosphere anymore.

In 1930, no human had traveled in the stratosphere and returned to say anything useful about it. Although flights into it of small balloon systems called "balloonsondes" equipped with automatic devices for recording its characteristics were growing in frequency and sophistication, the few intrepid aeronauts who dared to reach it usually lost their lives in the trying. It remained a human challenge, like scaling the highest mountain peak or sounding the deepest ocean abyss. It also remained a difficult place to understand, a frontier for science as well as for exploration.

The stratospheric flights of Auguste Piccard in 1931 and 1932 changed the way people thought about conquering the stratosphere. Although he learned very little about this strange new realm, Piccard demonstrated how this atmospheric barrier could be overcome. His first attempt and then spectacular success stimulated an explosion of flights in the United States, Europe, and the Soviet Union. In America, the first to fly were Navy balloonists sponsored by the "A Century of Progress" World's Fair and Exposition in Chicago. Then came Army aeronauts flying on *Explorer*, backed by the National Geographic Society and the Army Air Corps. By mid-decade, Piccard's twin brother Jean and his American wife Jeannette had flown into the stratosphere, and both pushed for means to return once again after World War II.

The flights in the United States, which form the central discussion in this book, certainly came about because Auguste Piccard solved the chief technical problems that had eluded earlier balloonists. But they were also a reflection of the motives and drives of American aeronauts such as T. G. W. Settle and his patrons, as well as those of Wright Field photogrammetrist and ardent flyer Capt. Albert W. Stevens. Auguste Piccard's deepest influence lay, however, upon Jean, who was filled with an all-consuming resolve to better his brother's achievement in America.

As we follow the hopes and fortunes of these competitors for stratospheric flight, we want to identify their motives and examine the social pressures they faced when they searched for support for high flight. What arguments did they choose to rationalize the great material cost of building vast balloons and sealed pressurized chambers? How did they justify the risk of flying into such a hostile realm? We will find that the forces that led Americans to promote manned balloon flights in the 1930s were not unlike those responsible for the National Aeronautics and Space Administration's (NASA) Project Apollo in the early 1960s. International competition between the United States and the Soviet Union played an important role in each period, a natural consequence of over a century of sporting competition between nations for speed, duration, and altitude that drove balloon technology.[1] Aviation and ballooning were among the numerous arenas in which Americans chose to demonstrate the power of the corporate state, just as Stalin sought to validate his regime to the world through aeronautical achievements.[2] The race for altitude records in the 1930s was not unlike the 1960s race to the moon as the rivalry shifted from the ocean of air to the depths of space, and as the theater changed from sport to the Cold War. We show here that promoters of manned ballooning in the 1930s insisted upon the scientific usefulness of their explorations of the upper atmosphere in much the same way that NASA and advocates of spaceflight in the scientific community "sold" Apollo as a scientific mission. In both eras, the scientist's laboratory smock draped what was a politically motivated demonstration of technical prowess: It was the mystery of the cosmic ray that held the key to the limitless power of the atom in the 1930s, and the moon the key to the origin of the solar system in the 1960s.

This last parallel between the ballooning of the 1930s and the moon race of the 1960s will find fuller elaboration here. Deeply embedded in American culture is the idea that scientific goals provide a higher justification for any enterprise; one could rationalize a large-scale engineering project or justify the expense and risk of exploration if the goal was somehow scientific. Consequently, what was considered to be scientific became clouded over a broad landscape of entrepreneurial activity. At the turn of the century, spokesmen for science proselytized that "America has become a nation of science." Perceived as a shaper of the path to prosperity, the pursuit of science represented the ideal in social behavior: "there is no law on our statutes, no motive in our conduct, that has not been made juster by the straightforward and unselfish habit of thought fostered by scientific methods."[3]

Rhetoric of this flavor, fuelled by the many evident successes of science in service to society, created the feeling that science and its methods demanded and deserved a unique, elevated station. Science had to be served, but it could serve as well. In this synergistic social conditioning process, science inevitably became a marketing tool: In its name and image, every kind of snake oil was sold, arguments won, and new and daring projects undertaken in the service of national interest and all mankind. These included the manned balloon flights of the 1930s, as well as the Apollo Program of the 1960s. Both exploited what Michael Smith has called the "display value" of science because both argued that it was their scientific nature that justified their existence.[4] Moon rocks and facilities to handle them, conferences and symposia on selenology, astronomy and geophysics—all legitimized President Kennedy's famous goal. Apollo was a public spectacle in the name of science, but it was symptomatic of the same motives and drives that created the great world's fairs of the nineteenth and early twentieth centuries, culminating in the Chicago spectacular characterized by Robert Rydell as a "fan dance of science."[5] The Apollo program emerged from a complex interplay of domestic politics and pride, international rivalries, personal ambitions and national purpose that constitutes the profile of contemporary American cultural and political history.[6] But these factors applied as well to the manned stratospheric balloon ascents of the 1930s and 1940s, and so their comparative analysis offers a fresh means of appreciating the consistently perceived place of scientific pursuit in American culture.

The role of humans in the scientific exploration of the stratosphere forms another key theme here; then as now it is the fault line that casts in relief the complex interrelationships between popular patronage, international competition, scientific need, and engineering aspiration in air and space flight. Piloted flights, whether by balloon or rocket-propelled spacecraft, were visible, public symbols of immense prestige and pride of national accomplishment. They were also compelling symbols of romantic journeys of the human spirit with which each and every citizen could identify and vicariously experience, as indelible as the opening of the American West in the popular imagination.[7] But as terrestrial frontiers vanished, and the stratosphere beckoned, the image of the explorer and scientist merged. Just prior to the 1930s, as Paul Forman and others have argued, the image of the scientist in American culture shifted to that of "intrepid explorer or pioneer." Herbert Hoover told a 1922 audience that scientific pioneers would "penetrate the frontier in the quest for new worlds to conquer," and J. J. Carty argued in 1928 that "The pure scientists are the advance guard of civilization."[8] The public stage was thus set for the spectacle of scientists as human vanguards exploring the new frontier of the stratosphere.

Problems created by modern space exploration by humans, as A. Hunter Dupree has found, resemble aspects of the earliest years of western expansion and global exploration, which are so basic a part of the American experience. Although there was much to discover, explore, and to conquer in either era, "the delivery systems, the ships, were so near the limits of their capabilities that the technology of just getting there took up nearly all of resources allotted to the exploration."[9] Problems with guidance, position finding and propulsion, the constraints imposed by military command, and the preservation of mind and body in hostile realms all resurface in stratospheric and space exploration. Piloted flights were enormously expensive; keeping passengers alive on shipboard, in the stratosphere, or in space required expenditures of manpower and equipment far beyond anything needed for flights of automatic instruments. The preoccupation with human survival technology threatened to divert effort from launching an instrument for scientific observation to launching and sustaining pilots and observers. Costs alone, however, did not determine attitudes toward investigative style. The major proponents of piloted flights in the United States—Capt. Albert W. Stevens and the Piccard brothers, Auguste and Jean—enrolled such diverse groups as Chicago businessmen and scientists, the executives of the National Geographic Society, National Bureau of Standards scientists and administrators, and high officials of the armed services in their promotion of manned ballooning. Appealing to specific institutional interests, these advocates and their colleagues tirelessly predicted the scientific value of their proposed flights, prophesied what their efforts would mean for humanity, and promoted the advantages of a human observer and operator in the stratosphere.

This book is also a study in how technical choices are made in science. Only in the 1930s did scientists acquire a clear choice in how they could gain access to the high atmosphere. Lightweight self-recording devices and small rubber balloons had existed since the 1890s, and saw slow but steady improvement over the next four decades.[10] But good reliable rubber balloons were hard to find, retrieving data from balloonsondes was often thwarted by fate, and the data that did survive intact were often compromised by the harsh conditions of the stratosphere. These problems continued into the 1930s, but were somewhat alleviated by improved rubber skins from a newly expanded domestic source, by ingenious methods for isolating airborne instruments from thermal and pressure effects, and by radio telemetry. These improvements did not exist when Auguste Piccard first planned flights in a pressurized gondola, but emerged just as he flew in 1931 and 1932. Thus advances in balloonsonde technology came simultaneously with Piccard's flights in Europe, and those of his followers in the United States and the Soviet Union. The question then became: Should means be found to allow humans better access to these hostile realms, or should instruments be modified in a manner that would allow them to work automatically to record the data as a surrogate for the human observer? Here we explore how advocates for both manned and unmanned flight faced this question in both the ballooning and Apollo eras. We will look not only at what motivated each scientist, but at what they perceived was the most pragmatic way to carry out the observations they felt were essential.

Allied with the question of choosing a mode of transportation into the stratosphere will be how science benefited from military patronage, and if the goals of science were changed somehow in the process. The traditional tie between science and the military, in America since the age of Jefferson, is most evident in government funded scientific exploration through military agencies.[11] Usually military in organization and command, these "contests for empire," made it possible for scientists to collect data and to discuss the results of the voyages.[12] The Wilkes Expedition (1838–1842) mapped coastal waters for safe harbors and sought more efficient sea routes to gain strategic advantage. But it also allowed accompanying naturalists to bring home data and specimens that revolutionized scientific knowledge in the emerging domain of natural history. Expeditions such as the Wilkes naturally were labeled "scientific," but, as David Knight has pointed out, it is sometimes difficult to determine how "scientific" such expeditions were: "Often there had been frustration among scientists on voyages because the captain's instructions, or his interpretation of them, did not let him stay at interesting places as long as they would have liked."[13] In short, it was not unusual that primary expedition directives ran against the interests of the scientific observer, anxious to make the most of what would probably be a unique opportunity.

Did this traditional relationship between military pilots and civilian scientists survive in the stratosphere of the 1930s? Adapting criteria discussed by Knight, we will look at the design of each flight, specifically at the site for launch, the size and nature of the crew, the character of the scientific instrument array, the planned and actual flight profile, and the means used to organize, analyze, and publish the data.[14] We will also look carefully at how these patterns changed after World War II. Whereas the Navy participated passively in the first Chicago flight during "A Century of Progress," becoming active only when things went poorly, the Army took a strong advocacy role from the start in the National Geographic Society's Explorer flights. But after the war, with the growing awareness of the need to support all forms of scientific research in the military's interest, and newly endowed with a mechanism to foster such research, the Navy became the primary patron and controller of Project Helios. Even though the project was conceived and planned by Jean Piccard, the Navy and its contractors soon took it over, but were unable to make it fly. The failure of Project Helios, and what it taught the Navy about manned ballooning, reveals much about the way in which the American experience in World War II deepened the Navy's interest in gaining access to the upper atmosphere, and how it ultimately chose to do so.[15]

Linking the majority of the stated themes and questions posed here are narratives that provide an appreciation of how scientific observation became a part of high altitude flight. It would be anachronistic to call this a study in the origins of space science, as we know that loosely defined amalgam of interests and activities today. Yet there are aspects of problem choice, instrument design, experiment integration, and data retrieval alive in the 1930s that survive in the modern era. The players and institutions change, as do the patterns of patronage. And even though these changes are rather abrupt in the wake of World War II, appreciating how science gained access to the high atmosphere before the war provides needed insight into the contemporary flavor of scientific research in space.

Notes

1. Crouch (1983), chap. 12–13; Stehling and Beller (1962). On early military interests, see also Dupree (1957), p. 127ff.
2. On the Soviet rationale for competition, see Bailes (1976), pp. 55–81. According to Bailes, Stalin's preoccupation with spectacular races and records seriously weakened Soviet air power in the late 1930s. He argues that Stalin needed to cover over the negative effects of his purges in order to achieve political legitimacy and security. See pp. 59; 79–80.
3. W. J. McGee, quoted in Burnham (1971), p. 253. McGee, acting president of the American Association for the Advancement of Science, spoke mainly of practical applications of science, whereas others at the time bemoaned the inferiority of American basic science to that of Europe. See Dupree (1965), p. 3.
4. Smith (1983), pp. 177–209.
5. Rydell (1985), pp. 525–542. See also Rydell (1984).
6. See Van Dyke (1964); Logsdon (1970); McDougall (1985).
7. See Goetzmann (1986).
8. Forman (1988), p. 4.
9. Dupree (1965), p. 8.
10. See Middleton (1969a), pp. 298ff.
11. Dupree (1957), p. 28; pp. 97–100; Goetzmann (1979), pp. 21–34; Goetzmann (1986), pp. ix; 15; 178–181.
12. Dupree (1957), p. 25.
13. Knight (1986), p. 145.
14. On the last indicator identified here, see Dupree (1957), p. 27.
15. See Dupree (1972), pp. 443–467.

CHAPTER 1

Auguste Piccard's Flights
of the *FNRS*

FIGURE 1.1. Max Cosyns (l) and Auguste Piccard (r) ascending into the strato-sphere, n.d., circa 1932. NASM.

One year after Auguste Piccard entered the stratosphere for the first time in May 1931, officials planning for Chicago's "A Century of Progress" exposition asked if they could put his gondola on exhibit. They had already secured William Beebe's bathysphere for their planned Hall of Science and wanted both a visual and symbolic balance for what they argued would be "a spectacular science exhibit." Together the two spherical chambers—one had taken Beebe 430 meters below the waves of the Atlantic and the other had taken Piccard and Paul Kipfer 16,000 meters into the stratosphere—would show how "man has recently overcome physical factors of the universe to explore regions never before penetrated."[1]

Two days after Piccard and Max Cosyns flew into the stratosphere again on August 18, 1932, "A Century of Progress" stepped up its efforts to secure the gondola. Fair officials, convinced that "Piccard's ascent has aroused world wide interest," pushed their embassy ever harder to increase pressure on the Belgian government to participate in the fair as an international exhibitor, as well as to lend the gondola.[2] Henry Crew, physicist at Northwestern University and head of the fair's Basic Science Division, encouraged the fair's Foreign Participation Division to redouble its efforts: "The human prowess represented by these two spheres is certain to attract and hold the attention of multitudes of visitors." Speaking on behalf of scientists underwriting the fair, he assured officials that scientists were "ardent admirers of [Piccard's] courageous and scholarly experiment, and that we regard this exhibit as one having an important international character, symbolizing the fact that science recognizes no geographical boundaries."[3] Eventually, the American Embassy was able to convince the Belgians, and by December 1932, the gondola was pledged for the Chicago fair. A public notice attested that "Together the stratosphere gondola and the Beebe bathysphere will illustrate man's conquest of the natural forces that limit human life to a range of a few scarce degrees and a few pounds of atmospheric pressure of what is on the earth's surface."[4]

What exactly had Auguste Piccard accomplished that drew such attention from fair planners, and, as we shall see, from the world press? As historian Tom Crouch has observed, in that day, the word 'stratosphere' "conjured up images similar to those that 'darkest Africa' had evoked in the nineteenth century or the names Arctic and Antarctic in the early twentieth."[5] The stratosphere was an inaccessible and unknown territory, devoid of life yet barely 16 kilometers away. Auguste Piccard had succeeded in conquering it where intrepid aeronauts like Capt. Hawthorne C. Gray had perished in his 1927 attempt. But unlike Gray, or the late nineteenth century balloonists who met their death or barely escaped death in the high atmosphere, Piccard found a way to carry himself, a colleague, and an array of physical instruments into the stratosphere in comparative safety.

Why Piccard performed this daring feat is the subject of this chapter, and forms the basis as well for understanding what motivated "A Century of Progress" to first celebrate and then emulate his stratospheric flight. Chicago would first display his gondola—the symbol of man conquering the cosmos—but ultimately Chicago would want to fly into the stratosphere as well.

Auguste Piccard (1884–1962) held a doctorate in mechanical engineering and taught general physics at the University of Brussels as a member of the applied sciences faculty when, in the mid-1920s, he began to plan for a balloon flight into the stratosphere in a sealed gondola. Already an experienced balloonist with the Swiss Army Observation Balloon Corps before and during the war, Piccard had contributed to the study of balloon design and was an ardent student of flight.[6]

Auguste was aware that a debate was going on among physicists over the origin and nature of an enigmatic penetrating radiation that kept the atmosphere slightly ionized. Victor Hess, a 28-year-old lecturer at the University of Vienna, had demonstrated in 1912 that the source of the penetrating radiation was beyond the atmosphere, but this finding was the subject of controversy for over a decade. Hess made his discovery by flying with his carefully improved electrometers in open-basket balloons showing that atmospheric ionization increased with altitude. During the next decade, Hess and his supporters Werner Kohlhörster and M. Kofler defended their observations against criticisms that what Hess had detected was the presence of radioactive materials in the upper atmosphere, or instrument error caused by leakage aggravated by low temperature and pressure.[7] As historian Charles Ziegler has shown, much of the criticism directed at Hess arose from a general disbelief that the universe contained a sufficiently energetic source of ionizing radiation. In defense, Kohlhörster tested, evaluated, and refined their Wulf electrometers further, flying them to altitudes in excess of 9 kilometers, vindicating Hess's observations. Nonetheless, they failed to convince others of the extraterrestrial nature of the radiation. Their chief critic in the United States was Robert A. Millikan, who after the war decided to add unpiloted automatic instruments flown on balloons ("balloonsondes") to his scientific arsenal to examine this mysterious radiation at great heights. Millikan, drawing upon his wartime Signal Corps balloon experience, developed balloonsondes with 190-gram electrometers in 1921 and 1922, in collaboration with Ira S. Bowen, and flew them from Kelly Field, Texas. He then developed larger detectors for aircraft, and found that the increase of ionization with altitude was different from what Hess and Kohlhörster had obtained from their flights. Millikan therefore favored a non-celestial origin for the radiation.[8]

The study of penetrating radiation fit reasonably well into Piccard's previous pursuits, which included the study of high-energy radiation and magnetic phenomena. His 1913/1914 Zurich dissertation was on the mag-

FIGURE 1.2. Robert A. Millikan examining tiny Millikan–Bowen cosmic-ray electrometers used for their San Antonio, Texas flights in 1922. Photograph permission of the California Institute of Technology Archives.

netic properties of oxygen and hydrogen and his publications starting in 1921 included ground-based and mountain-based observations of ambient radioactivity, laboratory studies of the radioactive elements, gamma-ray absorption in the atmosphere, attempts at refining knowledge of the velocity of light, and a search for evidence of the "ether" using Michelson interferometers on the ground, on mountain tops, and, as he once proposed, even from piloted balloons.[9] He also followed the work of Millikan, Hess, and Kohlhörster, and therefore was aware that they had established different altitude profiles for the variation in radiation intensity. He knew that many physicists still doubted the extraterrestrial origin of this penetrating radiation, and that there was no agreement on what the radiation was even if it was extraterrestrial.[10]

From his early record, Piccard appears to have been a steady although unspectacular contributor to the exploration of radioactive and electrical phenomena in the atmosphere who followed no consistent research agenda. However, his idea of using a piloted balloon to address the provocative question of the existence of an ether indicates his enthusiasm for ballooning, and his desire to study problems in physics from the platform of a balloon. He was an ardent member of the Swiss Aero Club and patron of a growing group of flight enthusiasts in Zurich, often lecturing them on the many problems ballooning could address in meteorology, aerology, thunderstorm formation and air electricity.[11] Thus he entered cosmic-ray studies in the context of building a stratospheric balloon system. And, as we shall see, it was to be the only context in which he conducted cosmic-ray research after he initiated the balloon project.

Although Piccard may have initially been drawn to ballooning as a means of studying the mystery of cosmic rays, ballooning soon took center stage. His efforts and energies were directed more toward making a balloon and gondola that would carry him into the stratosphere, than toward improving his instruments or technique of observation. Unlike Hess and Kohlhörster, who left the ballooning to balloonists, Piccard assumed the dual role of balloonist and scientist. Also, unlike the German experimental physicist Erich Regener of Stuttgart, who chose to improve balloonsondes for cosmic-ray studies, Piccard chose only to improve piloted balloons. Piccard was thus unique among physicists in his approach to cosmic-ray studies, and for this reason, he remained only on the periphery of cosmic-ray research while assuming the role of visionary in the lore of aviation and spaceflight.

Purpose and Patronage

Surviving correspondence between Auguste and his twin brother Jean Piccard indicates that Auguste had been planning his balloon enterprise well before October 1926, but this was as much for his ether drift experiments as for his study of cosmic rays.[12] By then he probably knew that Millikan, after a series of observations from mountain lakes, had been converted into a believer of the celestial nature of the radiation, and had himself coined the term "cosmic rays."[13] Still, according to his later testimony, Auguste's desire to build a piloted stratospheric craft was undiminished: The unambiguous observation of the variation of cosmic-ray intensity with altitude had still not been accomplished. Auguste Piccard felt that if scientists such as Hess, Millikan, and Kohlhörster could not devise a reliable self-recording device, he would not be able to either, and so the only hope lay with a new approach: to build a pressurized chamber to carry scientists and their familiar manually operated laboratory instruments aloft together as one integrated device.[14]

Auguste Piccard knew that building a sealed gondola and a balloon large enough to carry it aloft required funding beyond the financial capabilities of a university professor, or his university. He applied, therefore, to King Albert I of Belgium, who had just established a fund to support projects useful to science and industry that were beyond the capabilities of normal university-sponsored research. Albert proposed the Fonds National de la Récherche Scientifique (FNRS) in 1927 to foster the intellectual development of the country now that its material restoration after World War I had been accomplished. In a speech on October 1, 1927, Albert hoped that the FNRS would make Belgium competitive in science and that this in turn would make Belgium a power in the new world market. At a time when Belgium was enjoying economic prosperity, having stabilized the franc and showing a national surplus for the second year, the popular Albert had no trouble securing subscriptions to the fund from banks, industry, commerce, and private donors. These amounted to over 125 million francs by the end of the year.[15] As Albert was also an enthusiast for ballooning, since Belgium had just won the Gordon Bennett Trophy for the third year in a row, Piccard knew that his bold plan would probably fly.[16] But more to the point, Piccard was also a member of one of the scientific commissions of the FNRS, that devoted to chemical physics, electrochemistry and radioactivity. It was this commission, along with one devoted to physics and astrophysics, that considered his proposal.[17] Even though he stepped down at the time his application was reviewed, his membership and his familiarity with the goals of the FNRS ensured that his application would be given full consideration. Piccard later justified his 1929 proposal to the Belgian National Scientific Research Fund (FNRS):

The equipment for a manned balloon destined for the direct observation of cosmic rays at high altitudes presented a great difficulty, which constituted for the majority of researchers an insurmountable obstacle. The funds occasioned by the construction of a giant balloon and by the organization of its ascent surpassed, by far, the normal means at the disposition of the research institutions of our continent.[18]

Auguste thought that he might need as much as 500,000 Belgian francs to build a sealed cabin and balloon large enough to carry him into the stratosphere. Throughout 1929 and 1930, he negotiated with the FNRS and finally, on 31 May 1929, obtained the approval of the Steering Committee for the use of 400,000 francs.[19] This grant was a considerable sum: The capital endowment of the entire fund remained at 120 million Belgian francs after its first popular year, and typical grants to both industrial research centers and engineering and trade associations for materials research in 1929 and 1930 averaged only 65,000 francs, the highest being 330,000 for research into the industrial properties of mercury.[20]

Auguste and Jean Piccard corresponded continuously about the design of the balloon and gondola, and funding. Jean was always ready and eager to provide advice and assistance. He argued that the gondola should be made of aluminum, thinking that it would shield the passengers from harmful radiation, and was also quick to offer advice about breathing systems, control lines and ways to reduce dead weight, all the while urging Auguste, if the FNRS did not provide adequate funding, to consider flying from America, "where it would probably be easy to raise the money." Jean added: "Too bad we can't make the trip together, as originally planned. You could most likely make a lot of money if you could commit yourself not to talk about the flight before you give a report to the Associated Press."[21]

Auguste did worry somewhat about funding from the FNRS. Through April 1929, while he was still negotiating, he gave his anxious brother permission to approach American sources such as the Associated Press for additional funds, but he always felt that FNRS funding was virtually assured.[22] There was good reason for Auguste's optimism. Besides being a member of one of its commissions, such a proposal would attract considerable attention in a very new field, and could stand as a highly visible example of Belgium's revival in the field of pure science. The successful fabrication of the gondola would demonstrate the vitality of Belgian industry and the strong connection between science and industry in Belgium that the FNRS wished to foster. The exotic aluminum gondola would represent an advancing capability in materials research, which the FNRS was also intent upon supporting.[23] Auguste preferred funding from the FNRS because he wished to establish a continuing relationship that would also strengthen his university research program. However, he also realized, as he admitted to Jean, that the prospect of American funding, even if only three to five thousand dollars from the Associated Press, "could be of importance during negotiations for the granting of the credit" from the FNRS.[24] Clearly, Auguste Piccard was keenly attuned to methods for raising both interest and funding, and was not above playing one source off against another.

Designing the Balloon and Gondola

Auguste Piccard worked out his basic design for the balloon and gondola by the time of his application to the FNRS in the spring of 1929. The rubberized cotton balloon had to have a capacity of at least 14,000 cubic meters to have sufficient lift to carry his spherical gondola into the stratosphere. But Auguste Piccard also knew that every extra pound was an enemy. He replaced the usual heavy netting that would attach the cabin to the balloon with a weight-saving girdle around the gas bag, to be made by Riedinger of Augsberg, Germany.[25] While the great balloon was by far the largest expense for Piccard and the FNRS, it did not pose the same technical challenge as the gondola. Accordingly, the FNRS withheld the bulk of its support until Piccard could provide assurances from respected engineering advisers that the gondola could be built.[26]

Most critical was the material to be used for the gondola ("la cabine"). It had to have neutral electrical and magnetic properties, so that the electrometers inside could sense the cosmic rays as they traveled unhin-

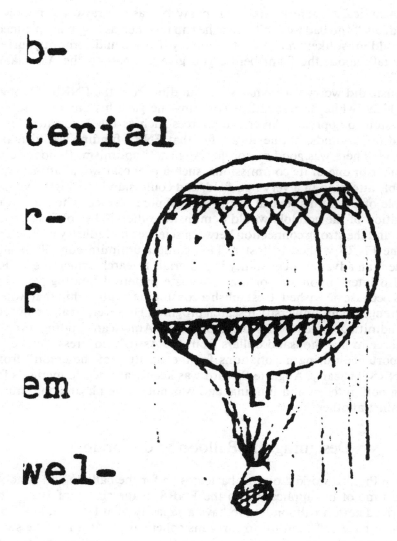

b-

terial

r-

f

em

wel-

FIGURE 1.3. Auguste Piccard's sketch of his balloon showing the girdle used to connect the gondola. From a letter, Auguste to Jean Piccard, March 25, 1929, Piccard Papers, Library of Congress.

dered through the sphere. This requirement, along with low weight and cost, originally led Auguste to consider using thick rubberized canvas as a cabin skin.[27] But aluminum, or something like it, would provide a rigid, safe environment and was overall a better skin material for a pressurized chamber. Moreover, there was no doubt that the future of flight rested with aluminum whereas canvas was a material of the past. The gondola skin had to be lightweight and remain ductile under construction and be amenable to shaping into an airtight hermetically sealed sphere. Still, it had to be strong enough to withstand inadvertent handling on the ground, and at all costs, be reliable against explosive decompression at the low temperatures and pressures expected during flight. At the time, no one alloy could meet these demanding constraints; aluminum alloys were still not well understood by Piccard's advisers, who came mainly from the Liège aluminum industry and from Neuhausen. But some were engineers who had provided the proud Belgian beer brewing industry with very large and reliable welded aluminum vats for brewing their product. These engineers had just found a way to weld pure aluminum sheets into rounded containers that retained their integrity over great temperature ranges. It was just this type of materials research about which the FNRS was so excited.[28] Upon the advice of these aluminum industry specialists, Piccard chose to use pure aluminum over thin steel or refined aluminum materials such as Duralumin, which was known to weaken seriously during the welding process and lose some of its desirable properties.[29] By 1933, after Auguste Piccard had flown in his first and second gondolas, he realized that some alloys then available had slightly better properties, but he later argued that in the autumn of 1929, when he contracted for his first gondola with the firm of L'Hoir in Liège (Société Belge d'Aluminum G. L'Hoir), he thought that pure aluminum was best for its overall properties of ductility, both during manufacture and under flight conditions.[30]

The first gondola was a welded aluminum sphere 2.10 meters in interior diameter fabricated from three large aluminum sections, each rolled and formed and then hand-wrought out of a sheet 3.5 millimeters thick. The three sections were formed into two hemispherical caps connected to a wide equatorial band, and kept the welding to a minimum. The interior had two small internal observing platforms and four shelves for instruments. The lower portion of the gondola had a floor of thin aluminum 1.2 meters in diameter.[31] Piccard wanted to have eight 3-inch viewing portholes fitted with thick glass which were easily accessible from most parts of the interior floor. He also insisted on having two manholes for entry and exit for added safety; no matter how the gondola landed, one manhole would be free from obstruction. The gondola had an empty weight of some 136 kilograms and it had to sustain two passengers for 10 hours at an altitude in excess of 10 kilometers.

Welding was the seam of choice for the simple reason that L'Hoir wished to demonstrate the capabilities of its product and their prowess in shaping it. Piccard later claimed that riveted seams would not hold up against the large pressure differences they were to experience in the stratosphere. On the other hand, the welding procedure for aluminum was not trivial since aluminum weakened somewhat during the process.[32]

The gondola also had to contain devices, both active and passive, for sustaining life. Piccard used a Draeger regenerative valve developed during World War I for German U-boats to regulate the flow of oxygen from pressure tanks carried in the gondola. The Draeger system, donated by its manufacturer, released 0.002-cubic-meters of oxygen per minute while it cycled some 0.08 cubic-meters of cabin air through alkaline filters to absorb carbon dioxide and toxic gases that built up in the sealed environment. Two 8-hour filters were carried in the gondola, which Piccard thought would be sufficient for their planned 10-hour flight. As it turned out, the flight lasted for some 16 hours.[33]

The first gondola was painted black on one side and white on the other to provide a passive system for temperature control. Piccard added small external electric fans that he thought could rotate either side of the gondola toward or away from the sun to aid in controlling the temperature of the interior. These little fans failed on the first flight, and the heating became intolerable. The second gondola was varnished white to reflect heat, but this made the interior too cold. Temperature control remained a problem for all stratospheric ascents in the 1930s; there simply was not

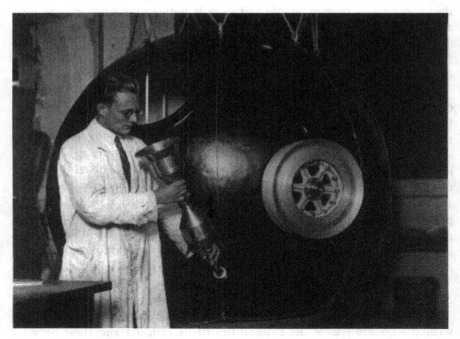

FIGURE 1.4. Paul Kipfer inspecting a valve used in the ballast system in the FNRS gondola, 1931. NASM.

enough electrical power available, nor the technology, for an active temperature control system.

Additional obstacles facing Piccard that complicated the gondola's design were how it was to be connected to the balloon, and how balloon controls, especially the gas release valve, could be operated by the crew. Piccard's first ascent almost ended in failure when this critical valve could not be controlled properly from the gondola. So before his second flight he redesigned the system to employ a series of winches that kept all of the connections between gondola and balloon equally taut. External winches on the first balloon became fouled so that for the second balloon, the control ropes were passed through mercury seals in the gondola skin, and then to winches inside the gondola.[34] These improvements, along with an improved ballast ejection system, increased Piccard's faith in his ability to safely control the craft, but he still included parachutes for the gondola, passengers, and for critical equipment in the gondola that would be ejected if the balloon burst.[35] Finally, Auguste constantly toyed with the placement of both the equatorial and lower-hemisphere girdles, trying to find the best design that would allow the balloon to expand so that equal stress would be applied to all parts of its surface. With his brother Jean he continued to worry about this complex problem well into the 1930s.

Through 1931 construction and testing continued at Augsburg and other sites, but Auguste began to find the project more demanding than he had envisioned. He made numerous test flights in open baskets, testing his endurance and trying to determine the greatest altitude at which he could still operate, so that minimum environmental conditions could be set. He found that at an altitude of 4000 meters, his efficiency at performing tasks was markedly reduced.[36] There were delays in finishing and testing the gondola, but most frustrating were the delays due to the weather, as they prevented him from following an efficient schedule of testing various elements of his balloon designs and making the actual flight itself, which the newspapers heralded would be in September, 1930, to Auguste's great frustration.[37]

The September flight proceeded to the point where Auguste and his partner were sealed in the gondola. But at the very last moment possible, the flight was aborted because the weather turned sour. After four weeks of waiting for good weather in Augsberg, Auguste complained to Jean that "the less I talked to the reporters, the more they invented."[38] Still, Auguste and his party were able to amuse themselves while they waited for good weather. They traveled. At one point, while taking a cable car to the Zugspitze, Auguste felt as if he were in a balloon looking down upon the world. Clearly, he relished the romance of high flight as much as he appeared to despise notoriety. In his protestations to Jean, Auguste made clear that reporters flocking about were frustrating but very necessary to the enterprise. They cared little about the facts, and so the news

services that had been given the rights to the story, especially those arranged by Jean outside Europe, had to be carefully handled. Auguste's
protestations were, in fact, a warning to Jean not to say too much.[39] The
delays only increased his need for control over speculation from the press.
Auguste was not shy of publicity; he merely wished to control it, and
Jean. The weather finally improved early in 1931 and with additional
testing done, Auguste was ready to fly in the spring.

The Flights of the *FNRS*

Piccard's two ascents into the stratosphere aboard the *FNRS*, named in
honor of its patron, contained similar experiments but met with very
different fates. The first flight, on May 27, 1931, was from the Augsberg,
Bavaria, site of the Riedinger plant where the balloon was constructed.[40]
The ascent, with Paul Kipfer as assistant, was extremely hectic, and as
a consequence little science was done. Only one cosmic-ray measurement

FIGURE 1.5. Auguste Piccard making final preparations for the first flight, May
27, 1931. E. Tilgenkamp Collection, NASM.

was made during the first flight, using a single standard laboratory electrometer based on a design by Britain's F. A. Lindemann that saw wide use in scientific experiments ranging from the amplification of cosmic-ray and radioactive decay events to the photoelectric measurements of stellar brightness.[41] Piccard wished to extend the laboratory environment into the stratosphere, and so at first shied away from exotic, ruggedized detectors. Indeed, his first stratospheric laboratory was barely distinguished from its terrestrial counterparts. In fact, he paid comparatively little attention to the scientific equipment; most of his effort was concentrated on making the balloon and gondola system work properly, both before and during each flight.

After launch, when Piccard and Kipfer had reached some 4,500 meters, the time came to seal the gondola, but the planned procedure was thwarted by a deformation in the skin of the gondola caused by a bumpy launch. Oxygen continued to leak from the gondola until they plugged a small instrument hole with a mixture of oakum and waxy jelly.[42] Within only half an hour, they had risen to almost 16,000 meters and there had been

FIGURE 1.6. Auguste Piccard and the interior of the first gondola, 1931. His ionization chamber (Lindemann electrometer) is at center. NASM.

no time to make any meaningful observations of cosmic-ray intensities, or of anything else for that matter. Once at their maximum altitude, however, they did have time to gaze out the portholes and into the world of the stratosphere, but the rope mechanism linking their gondola controls to the exhaust valve of the balloon was broken, and worse yet, as we already noted, the little electric fan that was supposed to turn the gondola to maintain temperature was short-circuited.

With problems like this, scientific experiments were out of the question; the crew could only suffer in the intense heat and congestion of the gondola, in excess of 310 degrees Kelvin. With the arrival of nightfall and the cooling of the gas in the balloon, they could finally descend. The gondola and crew landed safely on the Ober–Gürgl glacier in the Bavarian Alps later that evening, and they walked out the following morning. The gondola lay on the glacier until it was retrieved for an exposition in Belgium in 1932, but it could not be repaired, having been heavily dented and then ravaged by curiosity seekers. This may well be one reason why "A Century of Progress" was unable to secure a quick response to its request for the gondola.

Quick to discuss his one measurement, thinking that it might be the only one he would ever make unless he somehow confirmed the scientific nature of the expedition, Piccard told an eager popular press that his observation of ionization at 16,000 meters revealed a cosmic-ray intensity far lower than what Hess and Kohlhörster found at 9000 meters. On the other hand, Piccard's measurement fit the altitude-versus-intensity curve that Millikan had derived from his balloonsonde and airplane flights in the early 1920s. Hess's reaction was to invite Piccard to bring his electrometer to his Innsbruck laboratory so that it might be calibrated against Kohlhörster's detectors. Subsequently they both found that Piccard's measurement had to be revised upward. In fact, the revised measurement came closer to Hess's, even though one could not place much faith in this single measurement nor in the corrections applied after the fact to account for the electrometer's apparently faulty readings.[43] This revision did not fail to arouse some curiosity, but the mysteries surrounding the proper calibration of electrometers admittedly had not disappeared even though Hess and Kohlhörster had improved the design for balloon flight and F. A. Lindemann had refined it for laboratory studies.

Electrometers, or "electroscopes," measure the presence and intensity of ionizing radiation. The form available to Hess, which was based on improved designs by the Dutch physicist Theodor Wulf, commonly contained two fine electrically conducting quartz fibers sealed within a small air-filled "ionization" chamber. When the fibers were charged, they would move away from one another since like charges repel. If ionizing radiation was present, however, the otherwise insulated fibers would discharge and relax to their original positions. This behavior could be observed directly with a microscope eyepiece focused on the fibers.[44] Hess improved the

Wulf electrometer by making the walls thicker and the seals stronger and more resistant to shock and vibration. Kohlhörster further improved the design for balloon flight, while Millikan and Regener found ways in the intervening years to miniaturize and automate such devices.

Auguste Piccard followed Hess's advice in later discussions, but his primary concern was that his first flight failed to demonstrate that one could fly in the stratosphere in comfort and control, and perform useful work. Although Auguste Piccard was widely quoted as saying he would make no attempt to return to the stratosphere because his wife and family worried so much, he at first did not know if support would be forthcoming for the second flight at all.[45] But he lost little time, and had no apparent difficulty, convincing the FNRS to support a second ascent.[46] As harrowing as it may have been, the first flight broke the world record and gave Belgium a new place in aeronautical history, to say nothing of Piccard himself, who became a national celebrity, enjoyed a personal audience with the king and queen, posed with them inside the first gondola, and with Kipfer was knighted for his success. Piccard also became fast friends with the Italian air minister Gen. Italo Balbo and other influential aeronautical pioneers.[47] There was little question then that a second flight would be funded. Thus far, all costs for the flight, including balloon, gondola, manpower, subsistence, and preparations for both the September 1930 attempt and the May flight, came to just over 441,000 francs, a bit more than Piccard was originally provided. The FNRS made up the difference, and on March 4, 1932, provided an additional credit of 30,000 francs for another aluminum gondola, as well as a continuing subsidy to Piccard of 70,000 francs, again a generous sum.[48]

The second gondola was quickly fabricated, and, along with the original refurbished balloon, carried Piccard and the physicist Max Cosyns from Zurich on August 18, 1932, to a new record altitude of just over 16,000 meters. This time almost everything worked right and according to plan. They remained at maximum altitude to conduct observations, and took observations during ascent and descent. The new gondola, as we noted, was outwardly only slightly modified; now it was varnished a highly reflective white to reduce the temperature within. But more controls were placed inside the balloon, with the mercury seals noted before, and the instrumentation was also more elaborate, as well as means to tie them down during flight. Now, Piccard paid more attention to the scientific instruments, having Hess and Kohlhörster watching over his shoulder. For the second flight, the ionization chamber was ruggedized and two Geiger counters were installed along with a modified Kohlhörster electrometer similar to those manufactured by Guenther & Tegetmeyer for balloonsonde experiments.[49] Encouraged by Kohlhörster, Piccard was now emboldened to include new devices. In 1928, with W. Müller, Hans Geiger announced a new type of radioactivity detector. He replaced the phosphor and microscope of older laboratory radiation detectors with a low-pres-

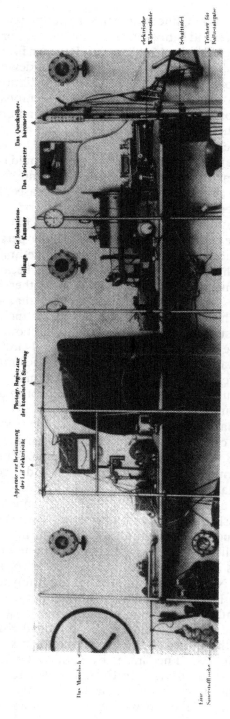

FIGURE 1.7. Interiors of both FNRS gondolas revealing the greater sophistication of the instrumentation for the 1932 flight (shown below). Instruments are more rugged and tied down to more shelving and supports. From Piccard (1933a), *Auf 16,000 Meter.*

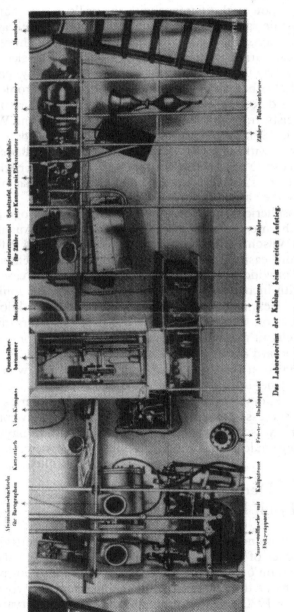

FIGURE 1.7. (*continued*)

sure gas-filled tube that emitted an electrical pulse when an energetic particle passed through the gas, ionizing the electrons. The gas-filled cylindrical chamber had a conducting wire drawn down its axis, which maintained a high potential difference with the walls of the chamber. If a high-energy cosmic ray entered the chamber, colliding with atoms in the gas to produce ions, the ions would accelerate toward the wire and create an electrical current that could trigger an external electroscope hooked to it. The new tubes had greatly increased sensitivity, partly because they could also be equipped with vacuum tube electronic amplification.[50] Geiger's detectors were quickly applied and refined by many cosmic-ray physicists. In 1929, German physicists W. Bothe and W. Kohlhörster, followed in 1930 by the Florentine physicist Bruno Rossi, studied cosmic-ray behavior with sets of newly improved Geiger–Müller counters and found that cosmic rays behaved more like charged particles than Millikan's gamma rays.[51] Everyone agreed with the celestial origin of cosmic rays. Now they argued over what they were.

Piccard's stratospheric laboratory also included several barographs, automatic timers, thermographs, and a radio set. Although Auguste also wanted to use a Wilson cloud chamber, he and Cosyns had difficulties getting one to work easily and reliably.[52] The Kohlhörster chamber on the second flight provided consistent data on the rate of increase of ionization energy with increasing altitude, and agreed with the ionization/altitude curves of Hess and Kohlhörster extrapolated to 16 kilometers, while casting doubt on Millikan's curve. The Geiger counters, arranged in a coincidence circuit to determine directionality, did not provide data that could lead Piccard and Cosyns to any conclusions.[53]

Cosmic-ray data from Piccard's second flight also agreed with the results of Erich Regener's balloonsonde observations, which reached an unprecedented 26 kilometers only a few days prior to Piccard's second ascent. Regener's success in making an unmanned sonde behave properly, at altitudes far in excess of any accessible to Piccard, gave physicists a choice in how they could gain access to the high atmosphere.

Demonstrating a Choice:
Erich Regener's Balloonsondes

One of the most prominent and capable experimental physicists in Germany, Erich Regener of Stuttgart had long been interested in the effect of ultraviolet radiation on atmospheric ozone, the nature of cosmic rays, and determinations of the unit charge on the electron. His cosmic-ray studies began in the 1920s, and at first involved detailed observations on land and under water. But Regener wanted also to produce an extension of the Hess–Kohlhörster curve and decided that unmanned ballooning offered the best combination of high altitude and low cost.

Regener's 1928 ballonelektrometer, influenced originally by Millikan and Bowen's 1921 design, derived mainly from the detectors he had already used for taking measurements under lakes, which had to measure remotely the incident cosmic-ray flux at various water depths.[54] His later photographic electrometers had a small spherical ionization chamber that contained the usual charged quartz fibers. A clock operated a shutter and moving photographic plate holder, which produced an exposure every four minutes. Skylight illuminated both the electrometer fibers and tiny pressure and temperature-sensitive levers; the shadows of all three were superimposed on the photographic emulsion.[55]

When Piccard first flew in 1931, according to an associate, Regener decided to "speed up his efforts" to "enter the competition between manned and pilot balloons."[56] He overcame the temperature problems that plagued earlier balloonsonde observations by placing his instruments inside a clear cellophane shroud that kept the interior protected from winds and from temperature extremes. His observations with this improved sonde confirmed those of Hess and Kohlhörster, but extended them to far greater heights. Added to the data from Auguste Piccard's second flight, Regener argued that the Hess–Kohlhörster curve was correct, in contrast to the one Millikan and his colleagues obtained.

Regener originally believed, with Millikan, that cosmic rays were photons. From his early high-altitude balloon observations, which indicated

FIGURE 1.8. Erich Regener's cellophane-shrouded payload design, just before launch from Stuttgart. From Piccard (1933a), *Auf 16,000 Meter*.

a general flattening of the ionization-versus-altitude curve at the highest elevations, he thought he had entered a realm of primary photons, "which should not ionize until they had produced secondary electrons, and should thus show feeble ionization near the top of the atmosphere."[57] But unlike Millikan, Regener was ultimately converted to the charged particle model when faced with the evidence that he and others had collected.[58] Today we know that the differences between the ionization/altitude curves obtained by Hess, Kohlhörster, Regener, and Piccard, and those of Millikan, were mainly due to the differing magnetic latitudes of observation. But in 1931, there was only the slightest hint that latitude had any bearing on the measurements at all. In fact, it was this dependence on the magnetic field of the earth that provided the most direct demonstration that cosmic rays were charged particles and not photons.

The details of this controversy are of some interest as they shed light on Piccard's *post hoc* motives for balloon ascents and on the reception given to both modes of access to the stratosphere by the physics community. In 1927 the Dutch physicist Jacob Clay found that the intensity of cosmic radiation changed with geomagnetic latitude, which indicated that a significant portion of the radiation might be charged and drawn along magnetic lines of force in the Earth's magnetic field. The charged particle model was strengthened when the Russian D. Skobelzyn found tracks from beta particles in Wilson cloud chamber experiments in 1928; then, in 1929, W. Bothe and Kohlhörster confirmed that cosmic rays had tremendous penetrating power.[59] But Clay's important detection of the latitude effect remained both unconfirmed and obscure. In a story often told, as scientists lined up behind one of the two models, Millikan and his collaborators looked for the latitude effect, and claimed that their expeditions did not show it.[60]

As a number of historians have shown, Millikan was unable to accept the charged particle model partly because of his deep-seated belief that cosmic rays, as photons, provided a remedy for what thermodynamics predicted had to be the ultimate unhappy fate of the universe: a "heat death." Influenced by deeply embedded religious beliefs, Millikan designed a scenario in which cosmic-ray photons were a by-product of the creation of the elements in space, which could counteract the thermodynamic imperative.[61] Millikan's provocative concept was rejected by most physicists, but remained compelling to a few, and to the public. Both Regener and Piccard were momentarily persuaded by it; Piccard in fact was deeply interested in the idea that studying cosmic rays might reveal the processes of creation, as well as a new source of cheap energy.[62] Even when Clay confirmed his detection of the latitude effect in 1932, and convinced University of Chicago physicist Arthur Holly Compton, as well as Regener, of its veracity, Millikan remained unconvinced. Compton, however, had already decided to mount a worldwide cosmic-ray campaign to settle the problem once and for all, and there is some evidence that

Piccard planned to fly for a third time in a far northern latitude to do the same.[63] Compton's campaign was launched in February 1932 and expeditions started out later in the spring. By September the first results were in from some 60 cohorts in the field: There was a 15 percent variation from equator to poles. Compton was confident that he had found a latitude effect, had confirmed Clay's observations, and thus demonstrated that cosmic rays were charged particles.[64] His efforts ultimately settled the question, but in the short run, they only stimulated the already hot debate over the nature of cosmic rays, now rendered far more visible as two of the most prominent physicists in the United States were in pitched battle.

To bolster his argument, Compton cited the results of Regener and Piccard, claiming that "Piccard's 1931 balloon data were far more reliable than those of Millikan in 1921/22, and has more recently been confirmed by the newly announced balloon measurements made by Regener."[65] Thus, by the end of 1932, both methods of access to the high atmosphere— large piloted balloons and tiny balloonsondes—had yielded information of value to the study of cosmic-ray phenomena, and were cited equally at first.

The examples set by Regener and Piccard might lead us to conclude that physicists who wished to study cosmic-ray phenomena in the stratosphere could now feel that they had a choice in how they went about it. How they made this choice in the following several years reveals a great deal not only about the priorities that physicists set for themselves, but also about some of the constraints, expectations, and demands required of them. Compton, like Millikan, later turned to balloonsondes as part of his cosmic-ray research program. But as the University of Chicago's most prominent physicist, he also was involved in planning for Chicago's "A Century of Progress" exposition. These two interests came together in what exposition planners hoped would be a spectacular American assault on the stratosphere, headed by Auguste Piccard, and which would be at least as successful as Piccard's second flight. Millikan was drawn in as well, supposedly to settle the question of the nature of cosmic rays, but really for what were political reasons, as we shall see in Chapter 3.

Auguste Piccard's Rationale for Manned Flight

Regener's continuing series of balloonsonde flights demonstrated that such automata could rise far higher than any conceivable human conveyance and still return useful data. His simple refinements and efficient instrument designs led to success after success, which drew many physicists to his Stuttgart laboratory to learn his techniques and to better appreciate his experimental style. Sensing that what Regener had done might draw interest away from piloted balloons, Auguste Piccard argued in 1933 that manned ascents were still necessary for the study of cosmic rays:

It would never be possible to reach with a manned balloon the altitude at which Regener was able to conduct his measurements. But I believe on the other hand, that the measurements one obtains in a free, manned balloon are more sure and exact than can be obtained with automatic instrumentation [such as is required by balloonsondes].[66]

Piccard admitted that unmanned sondes had major advantages; his problem was how to justify that manned flights were still really needed. He pointed out that on balloonsondes, unattended electrometers had to be charged before flight, and were left to discharge during flight. The discharge would be recorded as a function of time, so that the first derivative of the resulting discharge curve would provide a measure of the flux intensity of the incident radiation, and the second derivative would reveal the energy of the incident radiation at each point. But this was, Piccard argued, only an indirect measurement that depended on many assumptions about what was causing the discharge, and also required that the instrumentation remain true throughout the entire observation period. On the other hand,

In a manned balloon ... one chooses the sensitivity and capacity of the electrometer so the fiber of the instrument reacts rapidly. After each observation, the instrument is recharged to its starting potential. In this way one obtains a large number of precise observations which *directly* indicate the intensity of the rays. If one plots these graphically the first derivative of the curve gives directly the information (which by Regener's method) one obtains only with the second derivative.[67]

Piccard did not have a self-charging electrometer when he first flew. But Millikan was developing one in 1933, although apparently not in time to be mentioned by Piccard.[68] Piccard also argued that unmanned electrometers could only measure intensities, whereas his heavier manned experiments also provided information on penetrating power, since some of the detectors were shielded by the several hundred kilograms of lead shot that were on board as ballast. Quite correctly, Regener's instrument payloads typically weighed only 1.5 kilograms and were dwarfed by the scale of Piccard's manned gondola and the instruments it could carry.[69]

The argument that only piloted flights could handle shielded counters was a consequence of patronage, not of technology; there simply was not enough support for unmanned high-altitude ascensions of extremely heavy payloads, whether they were desired or not. But Piccard never acknowledged that shielding could well be placed on an unmanned sonde if it were made larger, at still far less cost than a manned flight. He did say, however, that many unmanned flights could be carried out for the cost of one manned flight. He was quick to remind his readers and potential patrons that the time was not yet right for such a decision: "in time it may also be possible to conduct balloonsonde observations with great precision. In theory, it is possible to accomplish automatically all of the operations which we performed manually. But up to the present, instruments of this kind have not been invented."[70]

It is not difficult to understand Auguste Piccard's conservatism. Unlike all other cosmic-ray physicists, Piccard thought of cosmic-ray experimentation only in the context of ballooning. He was unwilling to admit that photographic recording was certainly by now an understood technique, and downplayed Regener's success in solving the temperature problems. Of course, radiotelemetry was still in its infancy, and both ionization chambers and Geiger counters were still somewhat unreliable and mysterious, but all showed promise and were constantly being improved. On the other hand, Piccard always freely promoted manned exploration of the atmosphere and the depths of the ocean without any expressed scientific agenda. Thus his conservatism for automata was a strategy for manned flight.

Later in the 1930s, after it had supported Piccard's flights and one by Max Cosyns in 1934, the FNRS prepared a general statement on the relative merits of manned ballooning and balloonsondes. Acknowledging that "there has been much refinement and perfection of sondes by Regener, the British and the United States, still the precision obtained remains well below those that are made by manned balloons."[71] Beyond precision, the statement claimed that the cost per kilogram and per hour of sending instruments aloft was "often ten times higher in an unmanned sonde than in a manned balloon." Nevertheless, the overall conclusion was that manned and unmanned flights "complement each other happily: the precise measures (those of penetration in lead) and those of the deviation of magnetic fields, can only be realized in manned balloons. More qualitative measurements where precision is not as important and the ones that must be made at sea, or in polar or equatorial regions, can be completed by sondes more advantageously."[72] These claims, particularly the one for cost efficiency, represent the views of dedicated partisans for manned flight. The curious cost-efficiency statement was not justified in the statement and therefore cannot be pursued here to any conclusion. Still, Piccard's conservatism for automata was shared by others, and was justified when he began his plans for stratospheric flight. And his efforts did produce a number of significant technical advances in manned flight, including a more efficient balloon design, the pressurized passenger chamber, and, according to the FNRS statement, the technique of partial inflation at launch.

The Image of Auguste Piccard as Explorer and Scientist

Those who introduced Piccard's major review of his first two flights, *Auf 16000 Meter* published in Zurich in 1933, did not have the slightest connection with cosmic-ray physics but were famed explorers of the earth

and of the air. Italian Federal Council member Giuseppe Motta, Adm.
Richard E. Byrd and Italian Air Minister Italo Balbo set the stage for
what was clearly a chronicle of exploration. And just as clearly, Piccard
moved with a certain facility in those circles of fame and glory—always
the shy professor, yet he seemed to relish the attention of crowd and
camera.

Rhetoric urging scientific exploration has always abounded among explorers eager to face great challenges at great personal cost. To Piccard,
exploring the high atmosphere held out the same fascination as sounding
the depths of the oceans: Both were challenges, and meeting both would
benefit society. And he wished to do both. Following an early passion,
possibly as old as his dreams of high flight, Piccard devoted many years
of effort to perfecting a bathyscaphe that could withstand the extreme
pressures of the deep ocean. He long dreamed of oil-filled balloons for
undersea exploration; sounding the ocean's depths was publicly proposed
in 1933 utilizing a "sea balloon" filled with olive oil and free from the
controlling lines of a surface vessel.[73] In 1934, Piccard and Max Cosyns
readied a new gondola for more flights and also encouraged planned
ascensions in Poland and Argentina.[74] And Piccard also obtained Fonds
National de la Récherche Scientifique support for a 7-foot spherical cast
steel bathyscaphe, to be dubbed *FNRS-2*, which he built in the late 1930s
but did not test until after World War II.[75]

Piccard was never silent about his ambitions, nor about his thoughts
on the future of flight. Stratospheric flight was a reality proven by his
exploits, and trips in a stratosphere airplane at 800 kilometers per hour
were not far over the horizon: "Steamers are a thing of the past . . . the
stratosphere is the superhighway of future intercontinental transport."[76]
He was also not shy about speculating on the possibility of a flight to the
moon and planets; according to newspaper accounts, his gondola could
be modified for a flight to the moon or Mars.[77] Piccard was an explorer
and visionary whose dreams and exploits captured the popular imagination. Clearly, his reputation was well deserved as an explorer. By the
end of his life, his achievements in sounding the extremes of height and
depth had brought him fame as an explorer, rather than as a physicist.
Exploration was, after all, his real passion. His contemporary image as
a scientist driven by the pursuit of the unknown was highly marketable
even in a time of economic hardship, for he spoke of ways to travel at
greatly reduced cost, and was not afraid to speculate that his research
into the mystery of the cosmic ray might well yield "unlimited power"
for all mankind.[78]

Thin and looking eternally disheveled, with long and wild hair and a
distant visionary gaze, Auguste was the typical image of the savant. After
his first flight, an awed reporter exclaimed: "If ever there was a typical
'professor,' it is Piccard! A man of medium height, he has the traditional
flowing locks, the high wide, intellectual brow, the thoughtful eyes glinting

FIGURE 1.9. Commemorating the Piccard flights and the FNRS. NASM.

behind spectacles and the courtesy one always associates with the academic."[79] This was America's view of Piccard; he quickly garnered a wide public audience for his exploits as a scientist, even though he remained on the periphery of the scientific discipline of cosmic-ray physics. In contrast, Regener's accomplishments went relatively unnoticed in the popular press, although he became a central figure in cosmic-ray physics and geophysics.[80] It is image that counts in our story.

News accounts of Piccard's exploits said little about the science itself, but made much of how his achievements opened new frontiers for man. What was captivating was the drama of the flight, the man who did it, and what it meant for the future. In dozens of excited accounts, Piccard's feat "proved that the stratosphere is navigable, and that man will be able to master this region of cold."[81] To a scientist at the U.S. Weather Bureau, the safe return of Piccard and Kipfer after the first abortive flight meant that rocket-propelled airplanes could travel in the stratosphere, "provided the passengers could withstand the enormous velocity."[82]

Thus, by 1933, Auguste Piccard's image as a scientific explorer was bright, and his accomplishment was hailed far and wide. Crowned with brilliant scientific success as pioneer explorer of the remote stratosphere, Piccard was just the one to whom America would turn, once he turned to America.

Notes

1. Jay F. W. Pearson to Major Felix J. Streyckmans, 5 May 1932, COP/UIC.
2. See memorandum, Belgian Government folder, 1928–1932, 20 August 1932, COP/UIC.
3. Henry Crew to F. Streyckmans, 4 November 1932; Crew to A. H. Compton, 21 November 1932, COP/UIC.
4. News release on loan of the *FNRS* for display, 13 December 1932 (Streyckmans), folder 1-1201, Auguste Piccard, COP/UIC.
5. Crouch (1983), p. 595.
6. For Piccard's story, see: Poggendorff (1938), p. 2007; Auguste Piccard (1950); McFarland (1974), pp. 597–598; Stehling and Beller (1962), chap. 13, pp. 202ff.; Philp (1937), chaps. 5, 6, 7; Field (1969); De Latil and Rivoire (1962); and Ziegler (1986), pp. 69–92, 85; Naumann and Hohenester (1931).
7. Hess himself drew attention to the fact that observations of his predecessors had been compromised by leaks in the electrometer and by changes in temperature and pressure. He improved the design of the Wulf electrometers, ruggedizing them with better seals. See Ziegler (1986), pp. 69–92. Excellent reviews of Hess's work can be found in Steinmaurer (1985), p. 17; and LePrince-Ringuet (1950), p. 90–91.
8. Kargon (1982), p. 136. See also De Maria and Russo (1987), p. 19, n. 31; and Ziegler (1986), pp. 82–83.
9. Auguste to Jean Piccard, 21 December 1925, PFP/LC. Auguste hoped to travel to between 2000 and 3000 meters in an open, stabilized balloon basket. At that height, according to Piccard's reading of Michelson's conclusions, they would be far enough removed from earth to observe the influence of direction upon the speed of light. Auguste added that Einstein suggested that he make this observation.
10. Poggendorff (1938), p. 2007; See also bibliographical folders, PFP/LC. Auguste Piccard (1933a), pp. 46–48; Auguste Piccard (1933b), pp. 10–22.
11. M. Rikli, in Naumann and Hohenester (1931), pp. 17–28; see p. 27.
12. Jean Piccard to Auguste Piccard, 25 October 1926, 18 January 1927, 22 January 1927, PFP/LC.
13. Kargon (1982), pp. 138–140.
14. Ziegler (1986), p. 85; Auguste Piccard (1933a), p. 47; Honour, (1957), p. 81; Auguste Piccard (1933b), p. 10–22; Auguste Piccard (1933d), p. 353.
15. See FNRS (1953), p. 33; Belgium (1929), pp. 353ff.; Auguste Piccard (1933b), pp. 23–25; Auguste Piccard (1933a), pp. 49–50.

16. Field (1969), p. 36.
17. Paul Levaux (secretary general FNRS) to the author, 10 August 1988, with attachments from the minutes of the Steering Committee of the National Scientific Research Fund (FNRS files). The majority of the 10 members of both commissions held professorships at Louvain, Brussels, Ghent, or Liège universities.
18. Auguste Piccard (1933b), pp. 22–23. Translation courtesy Charles Ziegler.
19. The exact date of the decision by the FNRS was kindly provided by Paul Levaux, present secretary general of the FNRS, 10 August 1988. Auguste did not actually receive the full amount for some time. See Auguste Piccard to Jean Piccard, 13 March 1929; 6 August 1930, PFP/LC. This was equivalent to some $14,000 in 1929 dollars. See [Technical Director, Department of Physics], "Balloon Ascent Into Stratosphere; Flights of Auguste Piccard, 1931–1932," 7 March 1933. Memorandum in Registrarial Files, Museum of Science and Industry, Chicago. It was also mentioned in Auguste Piccard (1933d), p. 357.
20. Beghin (1938), p. 397. Industrial support remained relatively constant through the 1930s, notwithstanding Piccard's contention that the Depression reduced FNRS support for his activities. On the effect of "Wirtschaftskrise," see Auguste Piccard (1933a), p. 49.
21. Jean Piccard to Auguste Piccard, 6 April 1929, in reply to Auguste to Jean Piccard, 28 March 1929, PFP/LC. Translation courtesy Renata Rutledge. The Piccard brothers had contemplated flights together while both were in Europe. Jean held an extraordinary professorship at the University of Lausanne starting in late 1924, but lost the temporary post in the spring of 1926. After a lengthy search, Jean eventually found another temporary post at the Massachusetts Institute of Technology. See Le Chef du Departement to Jean Piccard, 13 October 1925, PFP/LC. Translation courtesy Sophie Mayr.
22. Auguste Piccard to Jean Piccard, 23 April 1929, PFP/LC.
23. See Beghin (1938) sections on "Répartition des subsides par exercice," pp. 397–401; "Fontes à l'aluminium," pp. 33ff.
24. Auguste Piccard to Jean Piccard, 23 April 1929, PFP/LC.
25. Auguste Piccard (1933a), p. 49. The first detailed discussion of his plans appears in Auguste to Jean Piccard, "Mein lieber Hans," 13 March 1929; 25 March 1929, PFP/LC.
26. Approval came only in February 1930 after Piccard had secured assurances from L'Hoir that his gondola could be built. See "Revision de la Decision No. 97/17 du Conseil, du 31 Mai 1929, Relative a l'attribution d'un Credit a Mr. le Professeur Piccard, au titre "pret d'instruments scientifique" (1929–1930)," 7 February 1930, FNRS files, courtesy Paul Levaux to the author 10 August 1988, annex 1.
27. Auguste Piccard to Jean Piccard, 19 April 1929, PFP/LC.
28. Auguste Piccard notes the influence of the European brewing industry in "Ballooning" (1933d), p. 355; but adds, in *Auf 16000* (1933a), p. 87, that the application of welding techniques to aluminum in both the brewing and transportation industries gave his contractors the confidence they required. See "Fontes à l'aluminium," in Beghin (1938), pp. 33ff.
29. See Jean Piccard (1933), pp. 30–31; Auguste Piccard to Jean Piccard, "Mein lieber Hans," 13 March 1929, 25 March 1929, PFP/LC; Auguste Piccard (1933a), pp. 85–88; Auguste Piccard (1933b), pp. 73–77.

30. Auguste Piccard (1933a), pp. 86–87.
31. Ibid., pp. 83–99.
32. See Jean Piccard (1933).
33. "Prisoners of the Air," *Popular Mechanics Magazine 56* (August 1931): 177–178.
34. Auguste Piccard (1933c), p. 30.
35. Auguste Piccard (1933a), p. 100–101.
36. Auguste Piccard to Jean Piccard, 23 April 1929, PFP/LC.
37. Auguste Piccard to Jean Piccard, 21 December 1930, PFP/LC.
38. Ibid.
39. Auguste Piccard to Jean Piccard, 6 August 1930, PFP/LC.
40. Auguste Piccard (1933d), p. 357.
41. DeVorkin (1985), pp. 1205–1220.
42. Auguste Piccard (1933d), pp. 366–370.
43. Ziegler (1986), pp. 87–88. See also Naumann and Hohenester (1931), p. 116; Piccard, Stahel, and Kipfer (1932), pp. 592–593.
44. Ziegler (1986), pp. 70–71; Hess (1940), p. 225.
45. Piccard, in Naumann and Hohenester (1931), p. 116.
46. See "Piccard Now Sees High Rocket Flight," *New York Times*, (31 May 1931), n.p., fragment, NASM Technical Files; "Conquest of the Stratosphere at Hand—Says Piccard," fragment, n.d., Piccard folder, NASM Technical Files.
47. "Ten Miles Up," (3 June 1931), p. 1026, NASM Technical Files.
48. See citations from FNRS records dated 4 March 1932 "Attribution d'un pret d'instrument scientifique..." and "Attribution d'un subside au titre 'chercheur'..." in Annex 1 to Paul Levaux to the author 10 August 1988.
49. Auguste Piccard (1933a), pp. 106–107, and plate 23a. See also Ziegler (1986), p. 88.
50. Geiger and Müller (1928), p. 617. See also Galison (1987), p. 83; Rossi (October 1981), pp. 34–41.
51. Galison (1987), pp. 94–95; De Maria and Russo (1987), p. 23.
52. Auguste Piccard to Jean Piccard, 24 December 1933, 11 March 1934, PFP/LC.
53. Piccard and Cosyns (1932), pp. 604–606, 606.
54. Pfotzer (1985), pp. 85–88.
55. Regener (1932a), p. 695; (1932b), p. 306; (1932c), p. 364. See also Pfotzer (1974) pp. 210–211.
56. Pfotzer (1985), p. 88. "Pilot balloons" here refer to unmanned sondes.
57. For a useful contemporary review, see Compton (1936), p. 1130.
58. Erich Regener pointed out the latitude effect in November 1934 in correspondence with Millikan. See Erich Regener to R. A. Millikan, 10 November 1934; Millikan to Regener, 30 November 1934, RAM/CIT, R45 F42.11.
59. See Braddick (1939), p. 1; Galison (1987), pp. 94–95.
60. See Kargon (1982), pp. 154ff.; Galison (1987), pp. 93–96; De Maria and Russo (1987), pp. 26–66; Rossi (1964), p. 59.
61. On Millikan's "Birth Cries," see Kargon (1981), p. 316. See also Galison (1987), pp. 80–89; De Maria and Russo (1987), pp. 19–21, n. 28–31, and p. 26; Seidel (1978), chap. 7; Millikan (1928), pp. 281–282.
62. Auguste Piccard (1933d), p. 383.

63. Compton (1932a), pp. 681–682; Compton (1932b), pp. 331–333. The evidence for Piccard's plan to fly in the region of the North Magnetic Pole, near Hudson's Bay, Canada, is weak. See "Piccard's Second Ascent" text for photographs, in E. Tilgenkamp collection, Piccard Folder, NASM Technical Files.
64. Compton (1933), pp. 387–403; Kargon (1982); De Maria and Russo (1987), p. 37ff.
65. Compton (1932b), p. 332. Charles Ziegler has commented that Regener soon found that his own measurements were also in error by about 20 percent. This had little effect on his conclusions however.
66. Auguste Piccard (1933b), pp. 19–20. Translation courtesy Charles Ziegler. See also Auguste Piccard (1933a), pp. 46–49.
67. Auguste Piccard (1933b), pp. 20–21.
68. For details of the operation of Millikan's self-charging electrometer, see Millikan to Compton, 3 July 1933, RAM/CIT R23 F22.4. See also Bowen and Millikan (1933), p. 695; Bowen, Millikan, and Neher (1934), p. 641.
69. Auguste Piccard (1933a), p. 48. On the other hand, no evidence has yet been found that any physicist in 1932 desired, or even thought of constructing, heavy automata for balloonsondes. Indeed, it seems that the only ones arguing for heavy payloads were the advocates of piloted flights, save for William Swann, whom we encounter later.
70. Auguste Piccard (1933b), p. 20–21.
71. See "Ascensions stratosphériques," from the FNRS files, in Annex 2 to Paul Levaux to the author 10 August 1988. Translation by Sophie Mayr. The statement claimed that ionization measurements showed up to 10 percent deviations from balloonsondes while only 1 percent deviations were found from manned flights. There was no analysis or documentation.
72. Ibid., p. 2.
73. Lincoln (1933), pp. 34–37.
74. Albert Gilmor, military attaché, U.S. Embassy, Warsaw, Poland, to Chief, M.I.D., WD, Washington, D.C., 9 January 1935, Piccard Folder, NASM Technical Files; Philp (1937), pp. 200–201; Auguste Piccard (1950), p. 107; *Primera Ascensión* (ca. August 1939), NASM Technical Files "Balloons-Science and Technology," file.
75. Noted in Burgess (1975), pp. 184–185, 242. By 1953 Auguste and his son Jacques had reached a depth of over 3,000 meters in an improved bathyscaphe, called *Trieste*, which was used by the U. S. Navy for several years. See Ibid., pp. 186–187; and Field (1969); Honour (1957).
76. Auguste Piccard (1933d), p. 384.
77. "Professor Piccard Here for Lecture, Says It Is Possible for Man to Fly to Moon," *Washington Daily News* (13 January 1933), n.p.; "Future Ventures into the Stratosphere," *Flight 24* (1 December 1932), p. 1151. Fragments in Piccard Folder, NASM Technical Files.
78. Ibid.
79. Maj. Dudley Heathcote, "Conquest of the Stratosphere at Hand, Says Piccard . . ." n.d., fragment in Piccard Folder, NASM Technical Files.
80. In addition to Pfotzer (1974), see Paetzold (1985), pp. 59–63; Pfotzer (1985) pp. 75–90, and other chapters in that volume; Gutbier (1955), pp. 37–48; and DeVorkin (in progress).

81. See "Prisoners of the Air," *Popular Mechanics Magazine, 56* (August 1931), pp. 177–179, 179; see also numerous clippings in Piccard Folder, NASM Technical Files, including "Ten Miles up in a Balloon," *Flight 23* (5 June 1931), p. 511; "Professor Piccard's Balloon Ascent," *Flight 23* (12 June 1931), p. 524; "Prof. Piccard," *Flight 23* (26 June 1931), p. 571; "Professor Piccard's New Venture," *Flight 24* (3 June 1932), p. 498; "Six Hours to Europe in Stratospheric Liner," *Modern Mechanix and Inventions, 10* (May 1933), p. 83. On his scientific observations, see "The Meaning of Piccard's Flight," *The Illustrated London News* (27 August 1932).
82. "Prisoners of the Air," Ibid., p. 179.

CHAPTER 2

Auguste Piccard Comes to the United States

FIGURE 2.1. Jean Piccard and the FNRS gondola on display in the Hall of Science of A Century of Progress. 1933 Science Service collection, NASM.

Piccard's Example

The "A Century of Progress" flight was a direct result of Auguste Piccard's demonstration that man could penetrate the stratosphere. In fact, both Piccard and the exposition authorities worked toward a common goal: to repeat Piccard's demonstration in America. And both eagerly exploited the ongoing, often public, debate between Millikan and Compton over the nature of cosmic rays. By 1933 the debate had taken on a very personal character affecting the public image of these prominent American Nobel laureates, not to mention the sanctity of science itself.[1] Therefore, not only did an ascent into the stratosphere from American soil promise to win the support of American interests, but the prospect of solving a major scientific problem in the process also offered both Piccard and his potential patrons a way of justifying the expensive effort.

The excitement of Piccard's technical achievements in entering the stratosphere in 1931 and 1932 stimulated "a race for supremacy in the stratosphere."[2] Prior to Piccard's flights America's participation in the race was stalled when Capt. Hawthorne C. Gray perished in his 1927 attempt. High-altitude flight, either in powered aircraft or in balloons, had been an obsession in the 1920s. World's altitude records for powered flight were constantly being broken, and Gray finally achieved a record in 1927 when he reached 12,700 meters in an oxygen-equipped open basket under a 1980-cubic-meter single-ply rubberized silk balloon.[3] His record was disqualified, however, because he had to bail out. Although his bailout was a warning, he returned in November for another try, only to meet his death. His fate demonstrated that the existing life-sustaining systems for high-altitude flight were severely limited; if man expected to fly higher, pressure suits or pressurized environments together with reliable respiration systems for maintaining a healthy working environment were required. Two options were available: to develop a pressurized cabin or a pressurized flight suit. Although flight suits did appear—they are best remembered in the celebrated efforts of Wiley Post in the early 1930s— it was the pressurized cabin that first took man into the stratosphere. As a result, it seemed to offer greater potential at the time. The newly established Army Air Service (later called the Army Air Corps) tested pressurized cabins for powered aircraft starting in 1921, but abandoned the effort when they found the cabins difficult to regulate properly.[4]

As Capt. Albert W. Stevens (1886–1949) of the Air Corps noted a few years later, in advocating what would become *Explorer*, all further attempts at high-altitude flight were canceled by "the then Chief of Air Corps [who] disapproved such balloon flights" after the death of Gray.[5] One of the many projects canceled was Lt. Thomas Greenhow Williams ("Tex") Settle's "Flying Coffin." Independent of Gray's heroics, Settle, with C. P. Burgess of the Navy Bureau of Aeronautics, designed a cylindrical pressurized cabin some 2.1 meters long and 0.9 meters in diameter

MEET IN FRIENDLY RIVALRY
Robert A. Millikan, Pasadena (Left), and Arthur H. Compton, Chicago, Who Today Discussed Physical Research Findings

COSMIC RADIATION FOES BATTLE OVER THEORIES OF ORIGIN

P.S-N. Dec.30 '32

Dr. Robert A. Millikan, Dr. Arthur H. Compton Present Opposing Ideas on Whether Cosmos Being Recreated or Disintegrated

What most of the 2000 or more physical scientists, gathered at Atlantic City for the winter meetings of the American Association for the Advancement of Science and associated societies, regarded as the climax of the entire week was reached this morning, when Dr. Robert A. Millikan gave his long anticipated address on cosmic radiation. For many months friendly controversy has been raging between opposing schools of researchers as to whether the cosmic rays originate in stellar space as the result of a continually creative force, or are the result of a similarly ever-siders them the original rays. Dr. Millikan advanced evidences that they are secondary radiation produced in the earth's air by photons smashing into the hearts of air atoms.

Penetrating Radiation

To account for the very penetrating radiations that Dr. Millikan and others have observed in the depths of lakes, Dr. Compton countered with the suggestion that electron cosmic rays produce photons in the earth's atmosphere, just as electrons striking an X-ray tube target produce X-rays.

His argument fell in line with

FIGURE 2.2. Robert A. Millikan and A. H. Compton in the news. *Pasadena Star News* December 30, 1932. Photograph permission of the California Institute of Technology Archives.

that could be placed under a balloon to carry humans in sea-level conditions to great heights.[6] The "Flying Coffin" and other similar projects were dropped because the Navy became sensitive to congressional scrutiny of its racing seaplanes and other sporting activities, but Gray's death no doubt played a role as well.

Auguste Piccard's successes revitalized interest in stratospheric ballooning in the United States; both Stevens and Settle found in his example a challenge and opportunity to convince their military superiors that a safe way had been found into the stratosphere. Piccard's presence in America also helped their cause. In January 1933, Piccard sailed to the United States and with great fanfare, and often wild statements and rhetoric, captured the imagination of the press, the public, and the military services.

Jean Piccard Orchestrates the American Tour

Auguste Piccard came to the United States as much to gain additional support for stratospheric ballooning in Belgium as to help out his brother Jean. Various avenues were open in America, including military support through enthusiasts in the air services, advocates of exploration such as the National Geographic Society, and influential scientists whose research could be augmented by high-altitude flight. Auguste found patronage in a combination of these sources, made possible by the planners of Chicago's "A Century of Progress" Exposition, which had just opened after several years of detailed planning.

Jean Piccard orchestrated his twin brother's American tour. After training and early research in chemistry in Zurich, Munich, and Lausanne, Jean Piccard taught at the University of Chicago between 1916 and 1918. Then, with his new American wife, Jeannette Ridlon, he returned to Lausanne in 1919, eventually finding a post as extraordinary professor of chemistry. The professorship evaporated in 1926 and they moved back to the United States so that Jean could take up a temporary research instructorship in chemistry at the Massachusetts Institute of Technology (MIT). This lasted until 1929, when MIT decided not to continue his position. Jean then went to work for the Hercules Powder Company Experimental Station then located in Kenvil, New Jersey, where he was responsible for the development and testing of a wide range of materials, including explosives.[7]

At Hercules, Jean languished as a frustrated inventor. While Auguste gained world acclaim with his stratospheric ascents, Jean had a stormy time trying to obtain personal patent rights as well as notice from other industries for inventions he had cooked up at home. By late 1932, Jean's employment problems had grown worse; his Hercules position spoiled, partly because of his constant feuding over patent rights, and he had to search for another position, all the while hoping that one of his many inventions might strike pay dirt.[8] Jean tried academe, but even with some good references was unable to find suitable positions at the Universities of Minnesota and Buffalo, among many other places.[9]

Through the Spring of 1932, Jean still had no prospects, and as the year progressed his letters and pleas for support became more stressed. By August 1932, when Auguste flew for the second time, Jean had become a casualty in a world in which scientists in industrial and government laboratories were being furloughed or fired, and meetings of the American Physical Society looked like employment agencies.[10]

But here was an opportunity. His brother's flights in 1931 and 1932 brought worldwide attention, and Auguste had already used Jean as his control point in America for handling the foreign press.[11] Jean Piccard often plied his brother for news of his balloon flight plans. Since 1929, their mother had kept him informed by sending clippings from the *Neue Züricher Zeitung*, but Jean always wanted Auguste's attention. As we have seen, at one point he hoped that Auguste might make his ascension from the United States and that they might fly together.

When the chance came to play an expanded role in furthering his brother's exploits and fortune as well as his own because of the growing interests in ballooning in the United States, Jean Piccard jumped at it. In late August 1932, barely a week after the second flight, Jean was scouting around for American lecture bureaus that could handle a tour for his brother. He happily stepped into the position of his brother's agent, a role that his brother appreciated. William B. Feakins, Inc., of New York City was hired in October to handle arrangements. Feakins had been in touch with American representatives of the Committee for Relief in Belgium, and because of this, the Piccards sought him out. Feakins's office would also negotiate with Auguste's employer, the University of Brussels, to obtain his release, but it was up to Jean to arrange when his brother would come to the United States, for how long, and for how many lectures.[12]

Arrangements were moving along swiftly for what would be an eight-week tour (at $500 to $1000 per lecture) when Feakins became concerned about public reception of an NBC newsreel interview that revealed Auguste's heavy accent and poor command of English. When cancellations started pouring in, Feakins's staff had to go into high gear to keep the commitments they had obtained. He admitted to an associate involved in the planning,

To say that I have been concerned about the whole matter is putting it mildly. Sometimes I have been unable to sleep half the night ... [NBC felt that] The News Reel of the Professor has done him incalculable harm. From the point of view of the lecture buyer it was pretty awful.[13]

Feakins distributed another carefully orchestrated film to counter the NBC disaster, but when Auguste was interviewed in Washington, the newspaper reporters had great fun parroting his accent in print. Nonetheless, the newspapers became Auguste's best friends. In awe of the man and his accomplishments and dreams ("Prof. Piccard is in town, explaining a few simple little matters like the cosmic ray and when man will make a trip to Mars or the moon"), reporters heralded his tour.[14] News accounts told of his every exploit—some reasonably factual and some pure myth—from pulling the teeth of a neighbor's dog in Brussels in order to safeguard his family while he was away, to announcing that he would fly a rocket across the Atlantic, and to gaining an audience with his former Zurich professor, Albert Einstein.[15] While certainly inflated and exaggerated, this was Piccard's image in America. And this would be the image that would gain support for his return to the stratosphere, if he decided to do it from American soil.

Feakins correctly predicted that once Auguste was in the United States, he would become a celebrity. Gradually, Feakins took over primary booking from Jean; his contacts and instincts were better. It was Feakins who suggested that Auguste speak before groups that were specifically interested in aeronautics.[16] Jean's job was to run interference for Feakins with Auguste, and for Auguste with the rest of the world, especially reporters who were impatient for translations of Auguste's colorfully garbled English.

The lecture tour started in Washington, D.C., where, on January 13, Auguste gave a public lecture before the National Geographic Society on cosmic rays and the possibility of traveling to the Moon in one of his pressurized chambers.[17] Thomas Settle, the originator of the "Flying Coffin," recalls that he first met Piccard there. Settle was based at Akron, Ohio, as an inspector of Navy contracts for the production of balloons and dirigibles by the Goodyear Aircraft Corporation. He was an ardent balloonist who had participated in the Gordon Bennett balloon races, and also retained his keen interest in high-altitude balloon flight, having advocated stratospheric flights in 1929.[18] Piccard's lecture rekindled Settle's hopes for his "Flying Coffin" and inspired him to pursue the possibility anew. Ultimately, however, patronage, politics and personalities played a greater role in bringing Settle into the picture as the pilot for the "A Century of Progress" flight.

Feakins and Auguste were well received wherever they traveled. At times, Piccard miscalculated somewhat in trying to determine what his audience was really interested in. At the Institute of the Aeronautical Sciences inaugural meeting, held at Columbia University on January 26, 1933, Piccard presented a sober, carefully designed plan for a stratospheric airplane flight. After James Kimball spoke on advances in long-distance flying and Jimmy Doolittle on "Racer Ramblings," Piccard discussed how his exploits could assist high-speed flight in the stratosphere. He labored hard to excite his distinguished audience, but these aeronautics professionals were not impressed by dreams. During the discussion following his talk, it became apparent that his audience was more interested in hearing about his stratospheric balloon ascents.[19]

The Piccards ran into trouble with almost everyone they dealt with. Sylvestre Dorian, Feakins's operative who attended Auguste and Jean during the tour, arranged that he and Auguste would coauthor an article for the *National Geographic Magazine*. But from the first moment both Auguste and Jean worked to remove Dorian from coauthorship, and after that relations with Feakins's agency deteriorated.[20]

The lecture tour continued as Jean ran interference for Auguste with promoters, autograph hunters, chambers of commerce asking for special appearances, and local scholars inviting the two for lunch. Jean, recently shunned by academics, now rode a heady wave of recognition; the invitations were endless. Karl T. Compton of MIT, Arthur Holly Compton's brother, extended a warm greeting while they were in Cambridge, and many notables scrambled to be photographed with Auguste. Jean, not being the center of attention, often felt slighted. On one occasion, the public relations staff of the Goodyear Tire and Rubber Company photographed Auguste with famous balloon men and Goodyear patrons Tex Settle and W. T. Van Orman. Jean was asked by Goodyear to be sure that his brother's autograph was on the photograph, and was told: "Since the camera also inadvertently caught you in the one picture I'd appreciate also your signature."[21] Slaps like this put a rather heavy chip on Jean's shoulder and left him with an even deeper resolve to repeat his brother's feat in America.

"A Century of Progress" Exhibition

The Piccards' tour took them from the East Coast to the West Coast, and then back into the Midwest.[22] On their way west through Detroit, they stopped in Chicago in mid-February to attend the just-opened "A Century of Progress" Exhibition, which was celebrating the centennial of the founding of Chicago. They returned to Chicago for longer talks in mid-March with numerous fair officials, including Forest Ray Moulton, an emeritus faculty member in astronomy and mathematics at the University of Chicago and both a trustee and local official for the exposition as its "director of concessions." Moulton, according to the Piccards, "thought of the possibility of a stratosphere ascension from the grounds of the Exposition [and] asked [Auguste] Piccard to supervise such a flight."[23] In fact, the idea predated Moulton's meeting with the Piccards and arose from a long series of events surrounding the planning for the fair, specifically the plans for displaying Auguste Piccard's original FNRS gondola. Well before the Piccards reached Chicago in February, the idea of a stratosphere flight was already in the air. The impetus came not from Moulton, but from a Chicago newspaperman anxious to create a story well worth reading.

As noted in Chapter 1, the FNRS gondola was coveted by fair planners. Embassy intermediaries eventually convinced the FNRS to send the second gondola to Chicago so that the fair could celebrate a great achievement: "Man has recently overcome physical factors of the universe to explore regions never before penetrated."[24] Auguste himself had expressed interest in the Chicago initiative, and looked forward to meeting the people who had taken an early interest.

Soon after the FNRS gondola was acquired, Steven Healey of the *Chicago Daily Illustrated Times* proposed that Auguste Piccard fly in it from the fairgrounds and that his newspaper support the venture. "Knowing that Professor Piccard is in need of funds to finance his explorations in the great beyond above the clouds, we felt that the following plan might be of practical use to him in the accomplishment of his desire to aid science." Arguing that "this extravaganza is purely in its embryo state," Healey felt that a flight from Soldier Field, the vast sports stadium on the fairgrounds, would draw enough paid admissions to both reimburse his newspaper completely and fund later flights by Piccard from wherever he chose "to enable him to further realize his ambition along scientific lines." Also, there would be "absolutely no stain of commercialism attached to this entire venture. The Professor is in need of money for science; perhaps we could supply it." Healey knew well that "the public is always looking for a thrill and I am quite sure that they would receive one. In other words, everybody connected with this affair should be satisfied with the particular gain that they receive."[25]

Healey's January 1933 proposal was taken up by Henry Crew, who headed the fair's Basic Science Division. Crew thought Healey's idea was "highly original and generous and if practicable would be one of the most interesting events of the summer."[26] Crew looked beyond Healey's enthusiasm to ask where they could get a suitable balloon and if it would be possible to make use of the old FNRS gondola. Crew knew that Auguste Piccard wished to find support to fly from a more northerly latitude, such as Greenland, and wondered whether Piccard would agree to fly from Soldier Field if prospects of funding for his Greenland trip looked poor. Crew and Charles Fitch asked Moulton to work out the details.

Moulton, as director of concessions, was deeply involved in obtaining patronage for various fair initiatives. He met with Healey to learn more about the idea, and then took it over as a fair initiative independent of Healey. After some preliminary inquiries among industry contacts that he and other fair planners had been courting for three years, he wrote to Auguste Piccard in late January "in regard to a possible ascension from the Exposition grounds during the international scientific convention to be held about June 20th. I think it probable that a plan which will be advantageous to you can be developed."[27] Moulton asked for a meeting, to which Auguste Piccard happily agreed, but Piccard added that a June flight date would not be possible because he already had planned to be back in Brussels then.[28] Auguste was now with his brother in Delaware and was anxious to meet Moulton to discuss plans further. Moulton directed an associate to travel to Delaware to secure Piccard's participation; but if he could not fly the balloon, perhaps he could "consider [the] possibility of having his associate make ascension."[29]

During their visit to Chicago in mid-February, the Piccards assured Moulton that if the fair could find support, Auguste would participate. Traveling west to Des Moines on a fast but bumpy train, Jean wrote Jeannette a letter full of hope, trying to convince himself that he would end up piloting the flight into the stratosphere. He thought that either Moulton or Compton might fly as physicist, but believed "the pilot should be a Belgian or a Swiss." Then he realized that "Compton can not be the physicist because the physicist must spend all of his time preparing the

FIGURE 2.3. A. H. Compton and Luis Alvarez examining Geiger–Müller counters built by Alvarez in 1932. Luis W. Alvarez collection, courtesy Peter Trower.

instruments and he must be responsible for them during the flight." In any event, the fair wanted Auguste to fly for the best "financial result," even though, Jean breathlessly reported, "Compton called up this morning, proposing that I should be the pilot." Images of piloting the craft danced in his head, along with the expected financial returns they would reap. Jean felt it was "very likely that the balloon as well as the gondola could be made free of charge by . . . American industry." Anticipating the gate receipts from a stadium that "seats 100,000 people," Jean exulted: "They would give piles of money to Gusti [Auguste] for himself or for further work."[30]

Moulton turned to two companies that he felt were logical candidates to support the flight: The Dow Chemical Company of Midland, Michigan, and the Goodyear-Zeppelin Company in Akron. Goodyear was already deeply involved in the fair. It was a major exhibitor, having developed an airship dock on the south side of the grounds to give fairgoers short rides in blimps, and it also supplied tires and other rubber products for many of the fair concessions, including "rocketcars" for the 186-meter Skyway. Thus Goodyear already enjoyed high visibility. Feeling the ever growing pinch of the Depression, and being without serious competition in the field of ballooning, Goodyear declined to fund the venture, but was willing to consider building the balloon at cost.[31]

Moulton asked an associate, the bright young chemist Irving Muskat, to contact Dow. Dow had also been thinking about being an exhibitor, but had recently decided to pass it up. However, Muskat's news about the stratospheric flight was too good to miss:

all revenues received from the project beyond operating expenses will be turned over to Professor Piccard for the advancement of his scientific work, a considerable part of which bears upon the nature and origin of the cosmic rays . . . It will be obvious to you that a flight by Professor Piccard or one of his associates . . . would obtain extraordinary publicity. In all this publicity the participating companies would be given full credit for the contribution which they would make in the name of Science. Not only is the balloon ascension a spectacular thing, but one of the objects of the flight would be to make a hitherto unmade test respecting the nature and origin of the cosmic rays. The subject of cosmic rays is, of course, one that has been very much before the public in the last year.[32]

Service to science, the enormous potential publicity associated with solving the mystery of cosmic rays, and the chance for Dow to demonstrate the usefulness of its products for the rigors of high-altitude flight were provocative arguments. Willard H. Dow of the Dow Chemical Company happily offered the firm's light magnesium alloy product, Dowmetal, for the stratospheric gondola, and directed members of his staff to meet with Auguste Piccard in New York to discuss its application. As a nearly pure magnesium alloy, Dowmetal was only 60 percent the weight of aluminum or its popular alloy Duralumin, and so had great potential in the aircraft and transportation industries. The Dow Chemical Company in

1933 was in a generally healthy economic state for the times, but, along with one major competitor, the American Magnesium Corporation, was faced with the task of convincing potential buyers that magnesium was an effective and reliable substitute for aluminum. Dow had made some inroads in the auto industry, which was using Dowmetal for light truck bodies and various aircraft parts, but the industry was by and large wedded to aluminum. If magnesium was to be used more widely, Dow would have to demonstrate improved methods of fabrication for machines and framework structures that proved the virtues of the product. A stratosphere gondola made of Dowmetal instead of an aluminum alloy might just be Dow's ticket. Very much in the spirit of the exposition, where science and industry were working hand in hand to pull America out of the Depression, Willard Dow felt that "if we fabricate the gondola, it will be a credit not only to ourselves but to the 'A Century of Progress' as well."[33] Muskat, carrying the spirit farther, replied that, if the gondola were made of Dowmetal, "it would be an achievement of which every American could well be proud."[34]

National pride, service to science, and science in the service of national needs—all were themes of the fair. Having the old *FNRS* on display while flying a new version would remind fairgoers that the European effort had been bettered on American soil. Moulton contacted Julius Rosenwald's Museum of Science and Industry (soon to be renamed the Chicago Museum of Science and Industry) to prepare for the exhibit of Auguste Piccard's second Belgian gondola in Chicago, along with its instruments, and the museum wished to retain the exhibit after the fair had closed "on account of the scientific importance attached to the expedition and the great public interest displayed therein."[35]

Auguste's fame, as well as his lectures and attendant publicity certainly helped to convince the planners of Chicago's "A Century of Progress" exposition to include a stratospheric balloon flight as a dramatic demonstration of how men of science and men of industry could work together to extend the boundaries of useful knowledge. It would be a captivating and memorable feat, and, if accomplished before the public, would contribute to the "fan dance" atmosphere that the planners of the exposition hoped to create.[36] To understand why this atmosphere was important, we must look at the fair's origins.

The scientific component of the Chicago exposition was the product of several years of planning by members of the National Research Council (NRC) of the National Academy of Sciences, underwritten by corporate interests in Chicago. Reacting to a perceived "revolt against science," the NRC joined with the exposition to "pin popular hopes for national recovery on the positive results expected from the fusion of science and business" at the outset of the Depression.[37] Frank Jewett, vice-president of American Telephone and Telegraph and head of Bell Laboratories, and other "salesmen-scientists" from industry created a Science Advisory

Committee to the exposition, which by March 1930 consisted of 34 sub-committees that brought in 432 scientists and engineers.[38] They hoped that an elaborate public exposition might impress America that science was essential for social and material progress. The best way to convince the public would be to use the art of showmanship to create memorable experiences that would be more ideological than philosophical.[39]

NRC scientists thus became salesmen who, according to Robert Rydell, acted as "intellectual underwriters of the Fair" and realized that simple and dramatic events had to showcase their message. In a 1930 meeting of the physics subcommittee, members agreed that the practical aspects of physics had to be emphasized in dynamic exhibitry that appealed to familiarity, to the spirit of wonderment, to the spirit of curiosity, and to the artistic spirit. Although the physicists claimed that "*truth should never be sacrificed in the interests of showmanship*," they argued that the more theoretical aspects of physics, such as relativity, were not relevant to the mission of the fair and thus should be omitted from the fair's menu. When William Swann presented a case for relativity, the subcommittee opted for demonstrations in classical physics in the areas of optics, electricity, and mechanics. Thus, although committee members remained true to the mission of the fair, they were less true to the mission of physics itself. They looked for stunts that would be wondrous for the 12-year-old mind to behold.[40]

Imagine that the light of a distant star could be captured and transformed into an electrical current that could open the fair. Even specialist journals for the scientifically literate proudly displayed telescopes equipped with newly developed photoelectric devices that accomplished the feat. Thus, even though photoelectric photometry was a young and slowly growing technique in modern observational astronomy, its application at the fair, or in honor of the fair, promised that it would be a commonplace technique in the tool kits of future astronomers. By the time the fair opened, it was hailed as a "huge experimental laboratory."[41]

Another exposition theme was transportation. Both members and visitors to the New York meetings of the NRC's Science Advisory Committee Executive Board argued that dramatic portrayals of long-range, high-altitude flight would have enormous popular appeal. Edward Hungerford, of the New York Central Railroad, promised that simulations of intercontinental air travel and transcontinental high-speed rail travel would be the hit of the fair. He warned that the exposition offered an opportunity that had better not be missed: "If we do not take those people from the Middle West and give them enthusiasm in the thing, why we have lost our opportunity, I think."[42]

A stratospheric flight, in the name of science, was eminently marketable to exposition planners, even though the early NRC scientific committees had not thought of supporting one prior to Piccard's flights and lecture tour in America. As long as such a flight took place in public, captured headlines, and demonstrated what American industry could do when faced with a scientific challenge, it would meet all the requirements of the exposition and its scientific underwriters. Stratospheric ascents, like deep-sea descents, captured the public's imagination, although the former was more visible. "A Century of Progress" planners hoped to have the National Broadcasting Company relay reports from the balloonists during their flight and to simultaneously broadcast reports from Beebe's bathysphere descent in Lake Michigan, to demonstrate "man's supreme achievement to date in projecting himself into an environment which would ordinarily prove fatal."[43] The stratospheric flight, as well as the ocean descent, would extend the experimental laboratory to the very limits of the earth. This was the spectacle Auguste Piccard represented to the exposition, and he used it as a negotiating chip for a trip to the stratosphere for his brother.[44]

Notes

1. On Millikan's image, see Kargon (1982), p. 91; and on the negative effect of the debate, see pp. 156–157.
2. W. G. Quisenberry, United Press correspondent, "News Report," London, 13 August 1932, quoted in Van Orman (1978), p. 167. See also Vaeth (1963), pp. 69–78, and Crouch (1983), p. 609.
3. Crouch (1983), pp. 598–600; pp. 602–603; Emme (1961), p. 11.
4. Armstrong (1939), pp. 366–369, 376; Emme (1961), p. 13; Konecci (1959), p. 175.
5. A. W. Stevens to Chairman, Board of Trustees, National Geographic Society, 25 July 1933, NGS.
6. Crouch (1983), pp. 603–604; Dollfus (1983), pp. 471–473; Vaeth (1963), pp. 68–78; Armstrong (1939), pp. 2–4.
7. *Poggendorff* (1938), pp. 2007–2008. See also "Incoming," 1932 January–June folders, PFP/LC; Donald Piccard to the author, 18 July 1988.
8. See numerous letters in PFP/LC, box 41, "Incoming," 1932 January–June; 1932 July–December folders. See also box 27, "Correspondence, Incoming, Personal," folders for 1932 and 1933, and box 53, especially Jean Piccard to "The Secretary, New Devices Committee, General Motors Corporation," 7 May 1932.
9. Julius Stieglitz to Jean Piccard, 21 April 1932, PFP/LC.
10. Kevles (1978), pp. 250–251. The Piccards were apparently not financially destitute. According to their son, Donald, Jean had savings sufficient to allow the family to live in comfort. Jeannette, however, insisted that he obtain a position and title worthy of their name. She had come from a wealthy, proud, and well-connected Chicago family. Donald Piccard to the author, 18 July 1988.
11. Auguste Piccard to Jean Piccard, 6 August 1930, PFP/LC.
12. Telegram, Auguste to Jean Piccard, 9 October 1932; William B. Feakins to Jean Piccard, 10 October 1932; Feakins to McClenahan, 21 October 1932, and later letters, PFP/LC.

13. Feakins to Sylvestre Dorian, 14 December 1932, clipped to Feakins to Jean Piccard, 15 December 1932, PFP/LC.
14. "Prof. Piccard, Here for Lecture, Says It Is Possible for Man to Fly to Moon," *Washington Daily News* (12 January 1933), NASM Technical Files.
15. "Piccard Would Fly Rocket over Ocean," *New York Times*, 4 January 1933. See also "Piccard to Speak Here January 13," *Washington Daily News*, 5 January 1933, fragments in Piccard Folder, NASM Technical Files.
16. Fragment, Feakins to "Dear Chuck," 21 December 1932, PFP/LC.
17. "Prof. Piccard, Here for Lecture, Says It Is Possible for Man to Fly to Moon," *Washington Daily News*, 13 January 1933. See in general, clipping files, "Auguste & Jean Piccard," NASM Technical Files.
18. Stehling and Beller (1962), pp. 220–221; Settle to H. B. Henricksen, n.d., Spring 1933, box 360, 1933 INA Folder, NBS/NARA RG 167.
19. *Proceedings of the Founder's Meeting, Institute of the Aeronautical Sciences* (New York, 1933).
20. Auguste and Jean Piccard to Gilbert Grosvenor, 1 February 1933, PFP/LC.
21. Hugh Allen to Jean Piccard, 14 March 1935, PFP/LC.
22. See letters and memoranda in box 41, folder 1933 January–June, PFP/LC. See also "Auguste Piccard" folder 1-1201, COP/UIC.
23. "A Century and a Half of Ballooning," box 71, "JF and S&W File," PFP/LC. See also Crouch (1983), p. 606.
24. Pearson to Streyckmans, 5 May 1932, COP/UIC. Auguste Piccard's gondola was the possession of the FNRS and of the University of Brussels. See Belgian Government folder 2-895, 1928–1932, COP/UIC.
25. Steven Healey (*Daily Illustrated Times*) to Charles Walton Fitch (COP Directory of Exhibits), 10 January 1933, COP/UIC.
26. Henry Crew to Charles Fitch, 12 January 1932, folder 1-1201, "Auguste Piccard," COP/UIC.
27. Forest Ray Moulton to Auguste Piccard, 27 January 1933, folder 1-1201, "Auguste Piccard," COP/UIC.
28. Auguste Piccard to Moulton, 29 January 1933, folder 1-1201, "Auguste Piccard," COP/UIC.
29. Telegram, Moulton to M. A. Owings, 29 January 1933, folder 1-1201, "Auguste Piccard," COP/UIC.
30. Jean to Jeannette Piccard, n.d., circa March 1933, black letterbox, Jeannette Piccard, PFP/LC.
31. Moulton to W.C. Young, 22 February 1933; Young to Moulton, 24 February 1933; Moulton to Young, 1 March 1933, folder 1-6463, "Goodyear Tire and Rubber," COP/UIC.
32. Irving Muskat to Willard Dow, 25 February 1933, folder 1-4762, "Dow," COP/UIC.
33. Willard H. Dow to Irving E. Muskat, 9 March 1933, PFP/LC. On the spirit of the exposition, see Rydell (1985), pp. 525, 527. On Dowmetal and the Dow Chemical Company's products in particular, see Titterton (1937), chap. 13, "Magnesium Alloys,"; John A. Gann, "Magnesium in European Aircraft," MSS rept., 1935; "Balance Sheet May 31, 1933." Dow Chemical Company, Dow Archives.
34. Muskat to Dow, 20 March 1933, folder 1-4762, "Dow," COP/UIC.

35. AMM (Technical Director) to LFW, 7 March 1933, RR/MSI. Rosenwald, president of the Sears-Roebuck Company, founded Chicago's Museum of Science and Industry.
36. Sally Rand performed her legendary "fan dance" at the exposition, making it something of a circus. See Rydell (1985), pp. 525–542.
37. Ibid., pp. 525–526. See also Kuznick (1987), pp. 11–16.
38. Rydell (1985), p. 528. On Jewett and the NRC's efforts at restoring science, see Kargon and Hodes (1985), pp. 301–318; Davis and Kevles (1974), pp. 207–220; Kevles (1978), chap. 16, "Revolt against Science," pp. 236–251; Kuznick (1987), pp. 11–37.
39. Rydell (1985), pp. 526–529.
40. K. T. Compton and F. K. Richtmyer, "Physics Subcommittee, Preliminary Report of the Chairman" (n.d., circa 1930), p. 6; National Academy of Sciences National Research Council, "Executive Board Advisory Council to Trustees of Chicago World's Fair, Reports of Subcommittees," NAS.
41. Rydell (1985), p. 531. For specific examples of the fascination with this application of the photoelectric cell, see, for instance, "The Link Between Arcturus and the Fair," *Scientific American 151*, no. 2 (1934), p. 60, and similar notes in *Popular Astronomy* during the period. On the state of photoelectric astronomy at the time, see DeVorkin (1985).
42. "National Research Council Science Advisory Committee to Trustees of the Chicago Century of Progress, 1933," 28 March 1930, p. 28, NAS.
43. Rhetoric used by the Museum of Science and Industry in an attempt to acquire both the stratospheric gondola and the bathysphere for exhibit. See O. T. Kreusser to S. L. Avery, 15 May 1933, RR/MSI.
44. As we shall see in following chapters, before Auguste Piccard left Belgium for his American tour, he was already deeply involved with Max Cosyns's proposal for a third flight, and in June 1933 clearly indicated that the Chicago flight was to be made by Jean and Arthur Holly Compton. See from the FNRS files, "Programme d'une nouvelle Ascension," 30 November 1932 in Annex 2, and 9 June 1933 entry in the FNRS Steering Committee minutes "Examen de l'offre faite par Mr. Le Professeur A. Piccard . . ." in Annex 1 to Paul Levaux to the author, 10 August 1988.

CHAPTER 3

"A Century of Progress" Flights

FIGURE 3.2a. Jean Piccard and the unfinished gondola, showing the shapes of the Dowmetal gores. Dow.

Once back in Chicago in March, Auguste and Jean Piccard met with Moulton, Crew, other exposition officials, and physicist Arthur Holly Compton to work out plans for a stratospheric flight that was to carry Compton's cosmic-ray instruments. Compton, in particular, was a willing and friendly host, calling for the Piccards constantly and seeing to their needs. After their first visit in Chicago, Jean's hopes swelled. Now again in Chicago, he carried with him the heady sensation of possible job offers from Los Angeles ("A professorship at the U. of Southern California. The money to be raised with a balloon trip and I to teach Stratospheric science")[1] and from the new science museum in Chicago. But above all else, he thought of the stratosphere.

After several days in Chicago, as they began to block out what a possible flight might look like, Jean and Auguste went to Akron to meet with Goodyear balloon officials, designers and pilots. Then they returned to New York to await meetings with Willard Dow's engineers to go over plans for the gondola. Back in New York for a final lecture on March 22, Auguste became more distant from the enterprise. He resisted all pleas from Chicago to stay in the United States, and although he continued to advocate the flight, it was more for Jean's sake than his own. Shortly before he left for Europe to help Max Cosyns with his own flight preparations, which had been proposed before Auguste's American tour, Auguste's directions to Moulton were to decide upon a pilot; the stated choices were either Jean or a Zurich associate of Auguste's, possibly Cosyns. Although the latter was a licensed pilot and Jean was not, Jean knew English, was already in America, and was ready to take on the task of coordinating the project planned by Auguste, shuttling between Midland, Michigan, where Dow would construct the gondola, and Akron, where the balloon would be fabricated and tested.[2] Auguste also did not wish their names to be associated with the project until all the details had been worked out.

For the time being, Moulton had no problem with Jean as a pilot. He had more trouble organizing the Chicago end of the project, identifying how gate receipts from the flight would be shared, determining the general logistics of flying from Soldier Field, and trying to convince Auguste to return for the flight. Moulton continued to court Auguste: "It will be announced that this money will be turned over to you to assist you in your scientific researches. No limitations beyond this announcement will be imposed upon you in the use of these funds."[3]

Negotiating for Terms and Expected Profits

The Chicago planners expected a return far in excess of what the balloon flight was expected to cost. Dow quickly donated the gondola, estimated to be worth $4500. The balance, including the construction of the balloon, which was provided at cost by Goodyear, was not expected to exceed $25,000.[4] This was still quite a sum, considerably more than what Compton had available for his world survey.[5] At first Auguste Piccard wanted Ballonfabrik Riedinger of Augsburg, the company that built the FNRS balloon, to provide the gas bag, but when the firm's representatives quoted 34,000 marks and said they would not have it ready for at least five months, Auguste had to accept Goodyear's offer. By the time Auguste returned to Europe to oversee Cosyns's preparations for his flight, all but the financing of the American balloon had been settled.[6]

Moulton and his staff easily found support from many diverse and eager sources. The prospect of a record-breaking flight from American soil was highly marketable. Soon, Dow was joined by the Union Carbide Company which donated the hydrogen for the balloon, and the *Chicago Daily News* and the National Broadcasting Company both backed the flight. In return for their support, each gained exclusive rights to the story for their particular medium, as well as the prospect of full reimbursement from gate receipts. The former restriction bothered Jean continually, who in the following months often ran afoul of "A Century of Progress" backers trying to line up personal patrons in return for personal endorsements.

To ensure good gate receipts for watching the preparations as well as the launch, the planners wanted the flight to take place from Soldier Field in Chicago. Auguste Piccard never objected to this constraint because he was interested in the funding that the event would provide for his own projects in Europe or Greenland, and also wanted a chance to broker the scientific debate between Compton and R. A. Millikan by flying their cosmic-ray instruments together.[7] Solving the great debate would certainly be a coup well worth the possible danger. Auguste well knew from his own experience that flying from Soldier Field might be problematic. Jean must have sensed this too, but Jean was sufficiently inexperienced and Auguste too distracted by the gate receipts from a 100,000-seat stadium not to fully appreciate the problem. Jean was delighted just to be working, if not for a salary, at least for living expenses and potential fame.

Jean returned to Chicago in April to begin sorting out all the details. On the first day of the meetings he reported to Jeannette: "This morning I went to the Fair and was immediately shown a contract, four pages long with the dotted line to sign. There was Dr. Moulton, his people, the men of the press and broadcasting company with the [lawyer] of the press."[8] Jean was quite proud that he had resisted signing, and asked for representation (the NBC lawyer agreed to represent him!) and made other demands that annoyed fair representatives.

Detailed negotiations between Jean Piccard and "A Century of Progress" began to get sticky when they had to face the particulars of the agreement: What was the project going to be called? Who would have title to the balloon and gondola when the flight was over? Who was finally going to fly both as pilot and scientist? And, most difficult of all, to what

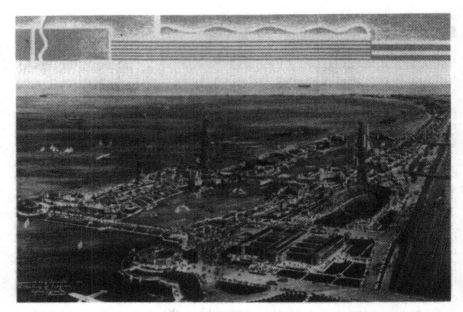

FIGURE 3.1. "A Century of Progress" Exposition, showing Soldier Field with one of the 180-meter piers of the sky ride just beyond. Photograph courtesy of "A Century of Progress" Records, University Library, University of Illinois at Chicago.

extent was Auguste himself going to participate? University of Chicago faculty and administrators acted as liaison throughout these negotiations. Taking over the details from Moulton, Irving Muskat, now a research associate at the university and manager of the exposition's Chemical Section, conducted most of the negotiations.

Trouble started even before Jean returned to Chicago. Moulton told Jean that, upon the advice of both Compton and the Goodyear Company, Jean should not be the pilot since he would probably not secure a license in time.[9] Moulton assured Jean that he would still fly as the scientist; even though the conservative Moulton had reservations about sending a chemist to conduct cosmic-ray experiments, Compton argued that with proper training, Jean should have no difficulty making the measurements. Moulton added that Goodyear, a Navy contractor, strongly suggested that Navy Lt. Comdr. T. G. W. Settle act as pilot. Settle was one of Goodyear's primary patrons within the Navy; certainly he was one of their strongest advocates, even though technically he was the Navy's "inspector for naval aircraft." Jean did not take this suggestion gracefully. In a hasty telegram he argued that Settle was not necessarily qualified because "piloting stratosphere balloon entirely different problem from ordinary free balloon." During this period Jean Piccard began to assume his brother's experience in his pronouncements, using the imperial "we" whenever questions of qualification were addressed.[10]

More trouble ensued during direct negotiations in April at the fair when Jean began to take independent steps to secure patronage and complained that both the balloon and gondola should become the property of the Piccards after the flight.[11] Jean also approached the Du Pont Company independently for support for the balloon, but was turned down flat when he tactlessly stated that the flight from Soldier Field might be dangerous.[12] This blunder eventually got back to Chicago, but nothing came of it at first because funding was no longer a problem. All kinds of financial support, in-kind donations of manpower, and expertise from the military as well as from those who could donate materials flowed in when the news came that a Soviet initiative was well under way to better Piccard's altitude record. And the specter of Max Cosyns's impending flight, aided by Auguste Piccard, also diverted funds quickly into the Chicago coffers.[13]

Sensing the competition and fearing that America might lose out, exposition officials were able to obtain full backing despite Jean's ineptitude and Auguste's duplicity. Both were tolerated for the time being as the Piccard name had immense publicity value, and no agreement had yet been signed.[14] Full funding allowed Jean, along with "Tex" Settle, to work in haste with their patrons to ready their gondola and balloon. Settle, for his part, was delighted that Goodyear wanted him to fly their balloon.

Jean, however, was disgusted with his week in Chicago and still refused to sign anything. Each day's letter to Jeannette grew stronger. Finally, on the eve of his departure to the Dow plant in Midland, Michigan, Jean summarized his relations with the fair planners:

If Gusti [Auguste] can not come for at least two weeks the whole contract is automatically canceled. If I don't sign it they say the navy will make a competition flight here, just like little monkeys. Use all Gusti's plans and then stand on a chair or barrel and cry: Aren't we the grrrrandest country in the world. Sometimes it makes you sick just to be an American.[15]

The Gondola and Balloon

On the surface, the designs for the Chicago balloon and gondola were similar to the craft Auguste Piccard had already constructed, but the United States also had other resources that several parties hoped would play a part in the new venture. Beyond the Dow Company's Dowmetal, balloon skin substances, especially rubberized fabrics, had been extensively developed since the war by the Army Air Corps, the Navy Bureau of Aeronautics, the National Bureau of Standards, and the National Advisory Committee for Aeronautics. American military and government laboratories had improved both rubberized and untreated cotton fabrics, silk cloth, paraffin coatings, rubber cement compounds, and varnishes, testing them for electrostatic properties, comparative strength, gas permeability, and exposure resistance to extreme weather conditions and to the solar ultraviolet.[16] One of the major centers for procurement and testing was the Army Air Corps Engineering Division at Dayton, Ohio. Eighty miles to the east was the site of one of the largest U.S. manufacturers of balloons, the Goodyear-Zeppelin Corporation, where the Navy was a familiar face.

Goodyear-Zeppelin in Akron agreed to build the 16,980-cubic-meter rubberized single-ply cotton balloon in the early spring of 1933, if funds from the primary backers of the Chicago exposition could be found and if Settle was designated the pilot. By now, Settle had moved closer to center stage in most of the negotiations, supported by both the Navy and Goodyear. The basic designs for both the balloon and gondola had been and were awaiting detailed refinements for the scientific instruments and the final testing of the integrated system once it was built. All that remained were the interminable negotiations between the parties involved, orchestrated by Settle.

The Goodyear balloon was similar in design to Piccard's original, except that it was larger and so would provide the greater lift needed to achieve a new record. The Dowmetal gondola, although spherical like Piccard's, was different in both design and construction. At first, no one paid much attention to the gondola except for its weight. Jean and Auguste Piccard, as well as Settle, left the basic gondola design and construction to Dow, and instead worried about details for the life-support system, the gas-release valve system, and the complex connecting system of control ropes between the gondola and balloon.[17] As winter turned to spring, however, the Piccards groused about many design details for the gondola, from the geometry of the entry and escape hatch doors and seals, to the placement of shelving. But overall, the gondola was in Dow's hands.

Auguste Piccard's original three-piece welded aluminum sphere had served him well. But Dow's weldable magnesium alloy Dowmetal was both lighter and stronger; it had distinct structural advantages over aluminum. So when Dow agreed to build and provide the gondola, for the honor of visible patronage and the impact this would have on marketing their product, the choice was easy.[18] Willard Dow was also the most enthusiastic and involved of patrons. Known within his company for supporting engineering research and development, Dow saw the stratospheric flight as a worthy way to test new fabrication methods in his firm's metallurgical laboratory.[19]

The new gondola was the same size as the *FNRS*, some 2.1 meters in diameter with 0.35-centimeter-thick walls, but it weighed only 87.7 kilograms, almost one-third less than its aluminum equivalent, since magnesium is 33 percent lighter than aluminum. According to Dow tests, gas-welded Dowmetal seams also retained 90 percent of their original tensile strength, so Auguste Piccard agreed that it could be shaped from eight gores, as suggested by Dow, instead of the original three. The Dow-metal welds required no hammering; they were simply ground smooth, and carefully x-rayed.[20]

The decision to use eight gores reveals the state of magnesium fabrication at the time. Dow had only begun to fabricate magnesium sheeting in the early 1930s; its rolling mills were put into operation in 1931 and were not fully capable of turning out large thin sheets.[21] On the other hand, Dow engineers had mastered techniques of gas-welding magnesium so that the seams retained most of the strength of the parent material. Furthermore, before it could be rolled, magnesium sheet had to be heated to 500 degrees Kelvin, but this was an advantage: After rolling, the sheet would keep its shape. Thus there was no "spring-back" after fabrication, either after rolling, hammering, or heavy pressing, and the completed gores, retaining their shape, would not stress the welded seals.[22] Dow's metallurgical laboratory, actually a large prototype shop, had no production-scale machines available either to produce or to handle large sheets. Thus, John A. Gann, Dow's chief metallurgist, and A. W. Winston, who ran the laboratory, decided to fabricate the gondola out of a larger number of smaller sections.[23] The eight orange peel-shaped gores were shaped on small hydraulic presses and rolling mills, and then hammered by hand into spherical segments ready for welding. Each was then fitted to its neighbor by tack-welding over a temporary frame, and they were all cropped at the poles to fit .62-meter-diameter octagonal caps.

Dow engineers subjected the gondola skin to many tests. The x-rayed seams were bent and pulled apart time and again, and the completed sphere was filled with water, then compressed air, and finally dry ice to test its resistance to bursting in a low-pressure atmosphere at extremely low temperatures. Winston concluded that the sphere had a bursting pressure of 6900 kilograms-per-square-meter, which provided a safety factor of 10.[24] Before the gondola was painted, its magnesium surface had to be treated by chrome-pickling in a mixture of nitric acid and sodium dichromate. This procedure helped the gondola resist saltwater corrosion and made it less susceptible to oxidation.[25]

Not all of the some 20 people then assigned to the metallurgical laboratory worked on the gondola, but it was by far the most exotic and fascinating project any of them had experienced. William H. Gross, a staff member of the laboratory, recalls vividly the publicity surrounding Auguste Piccard's original flights, and remembers that everyone in the shop was enthusiastic when news arrived that Dow was going to build the gondola for the American attempt. Their experience with forming, joining, testing, treating, and evaluating Dowmetal alloys was now going to be put to a visible and exciting test. Not only would they gain shop experience in meeting exotic technical requirements, but the publicity would be good for the shop and product.[26]

In time the shop became the site for the gondola's first public showing, as well as for those of its successors. The intrepid aeronauts would visit Midland, Michigan, poke around the metallurgical laboratory, and inspect the seams of the stratospheric cabin, smiling bravely before the cameras of the world.

Detailing and Filling the Gondola

Most of the housekeeping devices Auguste Piccard had used in the *FNRS* were to be employed at Chicago; the exposition planners followed his directions to the letter for purchasing liquid oxygen air-conditioning apparatus from Drägerwerke. But Auguste wanted a new and more reliable valving and control system that could be operated reliably from within the cabin. In correspondence through Jean, the two Piccards experimented with designs for self-sealing holes in the gondola skin through which the rip cord and valve lines would pass. At the same time, Settle decided that all the control lines had to be rubberized, and so Jean explored how different types of seals would behave between rubberized ropes and cables drawn through steel tubes fitted through curled-back holes in

FIGURE 3.2b. Auguste (l) and Jean (r) Piccard at the Dow plant, Midland, Michigan. Dow.

test samples of Dowmetal gondola skin.[27] Jean tested the various designs at his home in Delaware, shuttling between Chicago, Akron, and Midland, and sent samples of the material he was using to Settle in Akron, who, as inspector of naval aircraft at Goodyear-Zeppelin, had the technical testing services of that facility at his disposal. Settle and Jean Piccard both seemed to work in harmony during the time that Jean and Auguste were having major contract disputes with Muskat over ownership of the balloon, gondola, and share of gate receipts.[28] Jean also maintained good relations with the Dow engineers and with Willard Dow, to whom he often complained about the exposition agents.

Settle, aided by Ward T. Van Orman, soon became the central figure in the development of the balloon in Akron and also assumed general coordination responsibilities for outfitting the gondola. Meanwhile, Jean's relations with exposition backers deteriorated, which also effectively removed him from a position of responsibility for integrating the scientific experiments. At first he tried to work in uneasy collaboration with Arthur Holly Compton, but Compton, who had extensive experience designing and coordinating large-scale cosmic-ray expeditions, soon took over.

The Cosmic-Ray Competition

Once news of the planned flight was released, Arthur Holly Compton of the University of Chicago emerged as a major player. As an important underwriter of the fair, he was immediately linked to the announced intent of the venture—to study cosmic rays—so he naturally established himself as director of science for the flight. He and his staff ultimately arranged for all the scientific instrumentation that was carried into the stratosphere.

Compton initially accepted the idea of a manned scientific flight; in 1932 he had linked science and exploration in describing his cooperative world survey to study the latitude effect: "As Polo opened up new worlds, so now science opened new worlds."[29] And now, aware of the symbolic value of exploration, Compton saw an opportunity to settle his debate with Millikan through another type of collaboration. Oblivious of either Jean or Auguste Piccard's particular wishes, Compton pushed ahead with his own agenda. In May 1933 Compton wrote to Millikan:

As you may have heard, the Piccard brothers are planning a high altitude balloon flight from Chicago ... It would seem too bad to let an expensive flight of this kind occur without making use of it for some high altitude measurements. The Fair Committee in charge of the flight has, accordingly, asked me to arrange for apparatus for carrying on experiments during the flight. We are planning certain directional and other experiments with and without absorbing screens.[30]

Compton invited Millikan to supply one of his new automatic recording electroscopes already used on aircraft flights, partly to compare their two

FIGURE 3.3. A. H. Compton using a Cenco Hyvac pump driven by a Model A automobile rear wheel to prepare his ionization chambers, Mount Evans, 1931. Photograph by Byron E. Cohn, courtesy AIP Niels Bohr Library and Otto H. Zinke.

types of systems and to reconcile the still elusive intensity-versus-altitude relation. Compton's overture had precedence; in February 1932, with the endorsement of John C. Merriam of the Carnegie Corporation, Compton had asked Millikan to consider collaborating in his world survey "in order that our results would be more closely comparable with each other." Millikan's ionization chambers might be carried along with Compton's apparatus to very high geomagnetic latitudes.[31] Millikan not only declined the earlier offer but he resented the overture and did all he could to discourage cooperation.[32] But in 1933 Compton's invitation was more compelling. Millikan now had to answer to the results of the world survey, as well as his patrons at the Carnegie Institution, and was just then trying to convince John A. Fleming, the director of the Carnegie Institution of Washington's Department of Terrestrial Magnetism (DTM) that the new detector he, Ira S. Bowen, and H. Victor Neher developed was "the best form of instrument" for Carnegie's survey work.[33]

Both Nobel laureates enjoyed support from the Carnegie Corporation of New York, which was funneled through DTM. The DTM wanted the acrimonious debate between the two to be settled through a reconciliation of their results. It had invested in both Millikan's and Compton's detectors, and for the sake of its own geophysical research needed to know which design was superior.[34] Even when Bothe and Kohlhörster introduced the Geiger counter to cosmic-ray physics in 1929, Millikan and Compton continued to use electroscopes along with the newer detectors

FIGURE 3.4a. Cosmic-ray electroscope designed and built by H. Victor Neher and R. A. Millikan and used on the Akron flight of Tex Settle and Chester L. Fordney. The argon-filled spherical ionization chamber is at left, connected to an automatic motion picture camera containing a 31-meter reel of film driven by a 36-hour clock. These detectors were also sent on shipboard cruises and aircraft flights throughout 1933. Photograph permission of the California Institute of Technology Archives.

because they offered superior reliability and stability, and detected ionizing radiation in complementary ways. Millikan employed chambers with internal detectors and Compton preferred external detectors of the Lindemann type and various types of photographic or visual recorders. Each design had its advantages and disadvantages. An internal detector required the ion current to be fed through the walls of the chamber, while an external detector had temperature and humidity problems. In both cases, extremely small electrical currents were involved, and therefore could be influenced by a number of environmental factors. Around 1933, the temperature problems were alleviated to some extent as argon had become available and could replace air in the chambers, but the effect of humidity, pressure, and electrical isolation was still poorly understood.[35]

Few physicists wished to get tangled up in the issue. Thomas H. Johnson, a young experimental physicist at the Bartol Research Foundation of the Franklin Institute and the codiscoverer of the east-west effect (which demonstrated that cosmic rays were positively charged particles) was asked by Fleming in June 1933 to help evaluate the merits of the new Millikan–Neher detector. Johnson had recently commented on the debate, "Never before in the history of science, certainly not in recent times, has there been a subject about which so many able investigators have so strongly

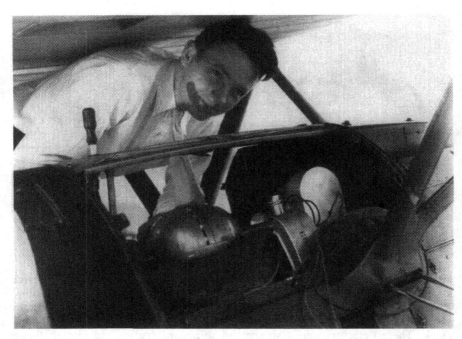

FIGURE 3.4b. H. Victor Neher preparing an electroscope for flight aboard a Boeing P-12 aircraft in June 1933. NASM.

disagreed as they do on questions concerning the cosmic radiation."[36] Therefore, as the recently installed assistant director of the Bartol, he made a good candidate as referee, Fleming hoped. Although Johnson was an excellent experimentalist and a frank commentator, he was naturally reluctant to get involved. Honest disagreement over theory was one thing, but criticism of a detector in which much time, energy, money, and reputation had been sunk by two influential giants was quite another. Johnson hedged: "I greatly hesitate to make any comments on instruments for this type of measurement since I have had no direct experience with them." But as the DTM was also his patron,[37] he had to say something, and so repeated points made already by Millikan and Compton that favored their respective detectors. Millikan's new chamber was compact and electrically stable and possessed excellent magnetic isolation properties, but could be difficult to fix in the field and lacked an internal source of radiation compensation.[38]

One physicist who could not avoid becoming involved was John A. Fleming. Fleming, with John C. Merriam, funded Compton in 1931 to check Millikan's work. "Supplementary investigations" might provide a check "by securing a different view in the field," argued Merriam, or, according to Fleming, Compton could "approach the whole subject from a wider point of view and in a thoroughly systematic way."[39] But by 1933, there still had been no reconciliation, or check that Millikan was publicly willing to accept. The debate heated up and the Carnegie officials grew more concerned. They had to stabilize relations in the physics community if they were to regain credibility for their scientific programs. Numerous editorials had to be answered. In September 1932, the *New York Times* claimed: "Electroscopes and ionization chambers and other cosmic-ray measuring devices seem strangely like wands and totem poles, and Einsteinian equations [are] but incantations that make us believe we know more than we really do."[40]

Fleming had to act to counter this serious attack on the veracity of the Carnegie's DTM programs. The characteristics of detectors had to be reconciled before standardization was possible, and standardization was all-important for Fleming's and DTM's global programs in geophysical research. Thus Merriam, with Fleming's advice, established a Committee for the Coordination of Cosmic Ray Investigations headed by Fleming. It was to address this problem, but, as Robert Kargon has shown, the committee really wanted to "monitor and mediate" between Compton and Millikan.[41] The Carnegie Institution and Corporation both felt they had to demonstrate that they were in the business of supporting precise and continuous geophysical observations worldwide, and not romantic reconnaissance work. Thus its committee had to determine which instrument—Compton's or Millikan's—was the best one to use as a world standard.[42] The Carnegie committee also wished to control the public debate; although some competition was healthy, a better competitive atmosphere had to be fostered.[43] It therefore encouraged cooperation and even collaboration, not only in detector design work, but in testing, evaluation, and operation. A collaborative flight of both instruments on a piloted balloon would demonstrate that all this was beginning to happen.

Thus, as the debate continued in public, privately the two camps struggled to establish their detectors as the best. Compton had the upper hand for the time being in both the debate and the detector competition, but still had to negotiate with Millikan. The "A Century of Progress" flight became his platform for negotiation.[44]

On the same day in May 1933 that Millikan wrote Fleming promoting his ionization chamber design, Compton wrote Millikan inviting him to participate in the Chicago flight.[45] Fleming asked Millikan to justify his statements, but was not heartened by Millikan's defensive reply. Millikan had just received Compton's invitation and felt compelled by the circumstances to participate in the Chicago ascent so that a comparative test of their two detectors could be run. Compton's May 1933 invitation was indeed a challenge. He felt that the manned balloon flight provided "conditions . . . unusually good for making a satisfactory test," and "would make an interesting check on your sounding balloon measurements if you would care to send up with this balloon one of your new recording electroscopes, such as is described in the last number of the Physical Review."[46] Compton added that the eight-hour ascent would stay at well-defined altitudes for specific amounts of time. There would be a naval balloonist to act as pilot, and Jean Piccard could be trained to operate Millikan's apparatus. Finally, if Millikan decided not to participate by providing one of his detectors, Compton was prepared "to build an approximately equivalent one to send with the balloon." Millikan, it seemed, had no recourse but to cooperate.

This "interesting check" was a strategy planned by Compton, no doubt with Fleming's knowledge, to test both detector designs under realistic, controlled conditions. Neither Compton nor Fleming was experienced with balloonsondes, so the piloted Chicago flight offered them a way to better understand how Millikan's devices worked. They already knew that laboratory tests with standard sources could not reproduce the operational conditions found by Millikan's detectors at high altitude, and so the manned balloon seemed to be the best answer to their dilemma. The best result Compton and Fleming could hope for would be that both systems would yield the same answer, and that it would agree with Compton's model.

At the time of Compton's invitation, many in the physics community favored his detection of the latitude effect and thought of cosmic rays as charged particles. Millikan, however, deftly managed to confuse the issue by reconciling the new data with his photon model. The charged particles could well be secondary products of collisions of cosmic-ray photons with the earth's atmosphere. Both physicists knew well that to settle the matter it would be necessary to detect primary cosmic rays before they interacted with the earth's atmosphere, and that the higher they sent their detectors, or the longer their detectors remained at high altitudes, the greater were the chances that primaries would be encountered. In addition, the behavior of both detectors at altitude had to be determined, and this was best done by observing the precise form of the curve describing the variation of cosmic-ray intensity with altitude, which was also very important for discriminating between the two theories and for determining the exact character of the latitude effect at high altitudes.[47] Manned balloon ascents afforded a challenging test to evaluate the performance of both detectors and to interpret their results in terms of either the particle or the photon model.

When Millikan received Compton's and Fleming's letters, he was courting Merriam for improved, cost-efficient unmanned automatic systems that could rise to over 15,000 meters.[48] He had gained considerable support from his many wartime contacts and was developing enormous momentum with colleagues stationed around the world. Even so he still needed the support of the Carnegie Corporation.[49] Thus Millikan could not have relished what appeared to be a forced test of the basic element of his entire program—his detector—at this time. It would have meant a

direct confrontation with Compton, his chief critic, and in a program
controlled largely by Compton. Millikan also had little use for expensive
manned expeditions, either around the world, as Compton proposed to
undertake with Carnegie support, or in balloons. In December 1932 he
noted to a correspondent who was a manned flight buff: "At the moment
I cannot see that more is to be learned by going up in manned balloons
than in unmanned so far as cosmic rays are concerned."[50] To Millikan,
participation in the manned flight would not add anything new, would
take his time and energy, and would tie up some of his valuable manpower
and detectors, both of which always were in short supply. He was also
well aware that the virtues of his compact design could work against him
in a short balloon flight. Millikan had already defended the small volume
of his detector against Fleming's and Compton's queries: "Large volume
of course reduces the fluctuation in the observations over a given time,
but for all the practical purposes of the survey work it is much better in
general to double the time of observation than to double the volume."[51]
But Millikan knew that they were not going to be doing survey work.
Even though Compton had said that the balloon would be under a controlled ascent and would stop for observations, Millikan knew too much
about how balloons behaved to be convinced that enough time would be
available.

But there were political factors to consider. "A Century of Progress"
was a symbol of the scientific progress that Millikan was committed to
endorse. Both senior scientists agreed to act in concert with the goals of
the fair, which were shared by the National Research Council's Science
Advisory Board. In fact, Millikan was a member at large and was keenly
aware that it was his responsibility to promote science, as he was "perhaps
the most famous American scientist of his day."[52] Millikan therefore had
to agree to send along one of his self-recording instruments on what was
by then called the Piccard–Compton flight. He curtly replied to Compton
that "from our point of view, [this would] be desirable since it will yield
data which will not otherwise be obtained."[53] His instrument package
weighed over 11 kilograms with batteries adding another nine kilograms.
Millikan pointed out, however, that

Dr. Piccard will not have to do anything about it whatever, since our new instruments record everything except the barometer, which they could also be adapted
to record if they had access to the outside, but I assume that mercury barometers
will be taken along for the pressure measurements any way.

Millikan agreed to cooperate only to a point. When Compton later
asked to use a Millikan–Neher camera in his own high-pressure ionization
chamber on the flight, Millikan refused, claiming, with impunity, that
"shortness of time makes furnishing recording mechanisms impossible."[54]

To Millikan, a human presence only added another variable that could
alter the results. Accordingly, he sent detailed instructions on how to

install his instrument, and how one might check that it was in operating properly during flight. Here Millikan described all the possible problems he foresaw arising from the combination of an automatic self-recording instrument and the presence of an observer, whether they had to do with the placement of illuminating sources, the operation of the automatic self-charger, or the initiation of the measurement cycle. Above all, Millikan cautioned that the presence of humans would cause "cosmic-ray shadows" that "may change the rate of discharge a good deal."[55] For Millikan, humans were an unnecessary addition to what was already a proven and reliable automatic system.

The press did not fail to appreciate the news potential of the two famous rivals working together. Auguste Piccard had already sensitized the public to the debate and the reasons why manned balloon flights were needed to solve the problem. The Chicago chapter of the Advertising Men's Post of the American Legion pleaded with Jean to give them a status report on the Compton–Millikan debate later in June, and Jean kept a running correspondence with Waldemar Kaempffert, science editor of the *New York Times*.[56] The *Chicago Daily News* proclaimed that the "Piccard Flight May End Compton–Millikan Debate on Cosmic Ray Properties" and noted that the celebrated debate "enjoyed by [the] two Nobel Prize winners may be settled once and for all this summer, the cosmic ray itself acting as referee."[57] Compton sent this news story to Millikan, observing that:

much as we naturally would like to avoid publicity, in connection with these experiments, I suppose in the present case it is only by virtue of some publicity that the cost of the flight can be met. If the News correspondents continue to write in the vein of this article, it will not be so bad.

For their part, fair officials only saw the prestige that the participation of two Nobel laureates provided. Rhetoric, possibly prepared by Henry Crew, for a speech to be given by fair General Manager Lenox Lohr on the eve of the flight in August, waxed eloquent on the secrets of the cosmic ray and capitalized on the Compton–Millikan debate and how science might serve humanity:

Whether in the secrets of the cosmic ray there is or is not the guarantee of the perpetual existence of the universe, the fact remains that these rays do possess a source of energy which may at some future time be harnessed into the service of mankind. The willingness of Dr. Arthur Holly Compton . . . to become scientific director . . . has given the expedition the hallmark of approval in the royal courts of science, and the cooperation of Dr. Robert Millikan . . . has given further assurance as to the lofty scientific purposes of this undertaking.[58]

What's in a Name?

Millikan's provocative but reassuring assertion that cosmic rays were evidence of the permanency of the universe, the impact of the Compton–Millikan debate, and the question of scientific legitimacy and the utility of scientific knowledge all came together in the manufactured image of the "Piccard–Compton" stratospheric flight. But personal feelings were involved as well. Irving Muskat wanted to emphasize scientific advancement in America and that, in America, science came first. In April he told Willard Dow that he favored Settle over Jean Piccard as pilot and general adviser for what was now called the "Piccard–Compton Stratosphere Ascension" and that what they were up to had to be an all-American effort: "As you have no doubt gathered from this letter, the flight will be made by an American pilot with a gondola made by an American manufacturer out of a material developed by an American manufacturer and with a balloon designed and constructed by an American company."[59] Auguste Piccard was now only an adviser; to Muskat, Compton was the critical figure. Because Compton was a Nobel prize winner and the most prominent scientist at the University of Chicago, his name would be a better label than Piccard's. Jean and Auguste Piccard had only caused trouble for Muskat. Later in May, after Jean arbitrarily told the press about "their" flight, Muskat cabled him to cease and desist: "Do not name the flight. We shall name it here. Letter following."[60] The name of the flight soon became an issue as well as a matter of honor.

To reduce the Piccard role, Muskat wanted a "Compton" flight, but promoted the idea of reversing the names to "Compton–Piccard Stratosphere Ascension." His argument was that both Piccards were really playing a very minor role, and they may "fail us" in spite of a [hoped-for] signed contract. Convinced that the main rationales for the flight were the Compton–Millikan cosmic-ray experiments and the solution to the debate, Muskat made his point to a *Chicago Daily News* official:

It may also interest you to know that Dr. Compton has stated that the greater the altitude reached the more valuable the data obtained and this factor will permit us to publicize the flight as a supreme effort to defeat the elements in order to obtain the data so essential for our scientific progress. From this standpoint, the Piccards' stratosphere ascension simply becomes a vehicle for the scientific measurements of Compton and I believe it would be well to regard it as such.[61]

To Muskat's frustration, the fair's director of promotion disagreed about the value of the Piccard name. He argued that it would be a mistake to remove "Piccard" from the name of the flight: " 'Piccard' has far greater publicity value than the name of Compton" and, after all, to him, the flight was largely a 'publicity venture' on the part of the participating companies."[62] Publicity won out, even though, ironically, Auguste Piccard was by now sufficiently annoyed about the negotiations in Chicago so that he too wanted his name placed behind that of Compton.[63]

A subdued Muskat, still angered personally at Jean Piccard, pressured him to convince his illustrious brother to return to the United States and to fly with Settle.[64] If this would happen, he would have less trouble accepting the "Piccard–Compton Stratosphere Ascension." Auguste did return to ensure that the Piccards would meet their contract obligations, but he did not have to worry about flying because the balloon was still under construction during the period Auguste was required by the contract to be in the United States. Thus the name of the flight remained in doubt throughout the summer.

Integrating the Experiments

The construction of the gondola and balloon continued through the late spring as Compton enlisted a number of other experiments for what was now variously called the "Piccard Stratosphere Ascension," and the "Piccard–Compton Stratosphere Ascension." Compton was already far beyond the weight limit of 200 kilograms set by Jean and Auguste Piccard for the cosmic-ray experiments and radio equipment. Much of the weight of the lead shielding for the cosmic-ray experiment lay in the lead shot to be used as ballast (some 250 kilograms), but the overall weight problem remained a supreme concern, as it affected the maximum altitude that the balloon could attain.[65]

In spite of the weight restrictions, Compton arranged with Henry Gale, a colleague at the University of Chicago with close connections to the astronomers there, to procure a simple ultraviolet spectrograph from the Gaertner Corporation. Although the prevailing opinion was that the ozone layer was far higher than the expected 16-kilometer limit of the balloon, Compton explained to Jean Piccard that "No direct test has however been made at altitudes as great as that to be reached in the present flight."[66] Piccard was completely uninterested in ultraviolet observations, and regarded them as a distraction. But he had to pay attention. Compton and Gale's associate, G. A. Monk, told him how to make the observations and also where to place the other cosmic-ray experiments, especially Millikan's, which had to sit above the human passengers on one of the top shelves. Compton provided detailed and exhaustive directions for installing the various thermographs and barographs, as well as his own shielded and unshielded chambers, a double Geiger–Müller counter developed by Louis Alvarez, one of Compton's graduate students who had been co-discoverer with Johnson of the east–west effect and was interested in refining his observations, and a triple coincidence counter developed by D. S. Hsiung, another member of Compton's team.

Compton dealt directly with the Dow Chemical Company to make the changes needed to accommodate the spectrograph in the gondola. Since the instrument had to penetrate the ultraviolet, both the spectrograph optics and the gondola window it peered through had to be made of optical quartz. Compton therefore asked for two small quartz windows in the upper part of the gondola. Jean Piccard, momentarily stationed at

Midland, Michigan to coordinate the detail work, did as Compton wished.[67] Jean Piccard also worked with Settle to develop the flight materiel inventory, which included two minimum-recording barographs from the National Bureau of Standards that were calibrated against temperature-sensitive bimetallic strips. These were all-important to establish the maximum altitude attained during flight. Most of the remaining incidental equipment was for safety and survival.[68]

Jean Piccard, as the observer, had to be trained in the operation of all of this equipment, and Compton's staff took great pains to provide explicit instructions, just as if they were training an uninitiated student. Each of the participating experimenters also had to train Piccard and advise him on all possible details and contingencies. Jean retained the responsibility for securing meteorological devices and photographic equipment for infrared horizon pictures and worked with the National Broadcasting Company on integrating their radio transmitter with all the other equipment in the gondola. He assured NBC that the cosmic-ray instruments were passive detectors and would not interfere with radio transmissions. Jean told his NBC contact that the detectors would emit an audible click when they were discharged by a cosmic ray. There would, however, be intervals of silence when the devices were recharging. But in Piccard's view, the clicking was an asset because he wanted the audio record of cosmic-ray coincidence events to be heard by the radio audience. At one point in his negotiations with NBC, Piccard stated that at higher altitudes the clicking might reach a frequency of once per second, which might interfere with voice transmission. Piccard was willing to turn off the counters at intervals to ensure clear voice transmission, but knew that he would have to consult Compton. To the scientists, even suggesting this was heresy.[69]

During the early part of the summer, Jean Piccard's role in the flight changed. He remained on the roster as scientific observer, but had less say in choosing the scientific complement, and in deciding how the project was to be governed. As already indicated, Jean constantly got into hot water when he overstepped his perceived authority. Throughout the contracting process, both he and his brother failed to establish a common ground with exhibition interests.

Deteriorating Relations in Auguste Piccard's Absence

At least five versions of the contract between the exposition and the Piccards were prepared between March and May. Jean saw their portion of the gate receipts dwindle when Auguste returned to Europe to aid Cosyns, and when he refused to return to, or fly from Chicago. Muskat warned that if the Piccards persisted in behaving this way they could lose their rights to the gondola not to mention revenue, which the fair also stood to lose. But Auguste still refused to take a direct role, preferring to act through Jean, even to the point of refusing to grant any additional interviews to the American press in April.[70]

Auguste's interests lay elsewhere. He and Cosyns were well-funded by the FNRS (Piccard had just received 37,000 francs to cover all remaining costs from his flights and to continue his own research, and Cosyns was granted 87,000 francs for their new flight) and had very real responsibilities that compelled Piccard to remain in Europe. Furthermore, Cosyns's planned flight with Nérée Vander Elst was designed initially to rectify a major problem with previous flight profiles of manned balloons: Since these flights had all started as attempts to achieve maximum altitude, the balloons rose too quickly to allow the crew to take unambiguous cosmic-ray observations at intermediate altitudes. Manned flights were also time-limited by the capacity of their respiration systems. Piccard and Cosyns decided to try to slow their ascent profile for Cosyns's later flight by adding a heavily ballasted military balloon (complete with pilot) that would physically accompany the stratospheric balloon to 9,000 meters when the two would then separate. The second system would both stabilize and slow the ascent. This expensive and radical concept depended upon the reuse of Piccard's original balloon, and a new lightweight gondola that L'Hoir decided to construct out of magnesium, following the example set in America. German industry was a world-leader in magnesium fabrication, and the Belgians hoped to catch up.[71] Still, it was a bold design that offered much to Auguste Piccard that was unavailable in America.

As a result of Auguste's absence, in late April, Muskat, preferring to deal with Settle, attempted to remove Jean from any direct coordinating role. Jean was deeply angered by this treatment and claimed that his brother would never sign the final contract under these conditions. At one point Jean tried to go over Muskat's head and make direct contact with Rufus Dawes, president, and Lenox Lohr, general manager of the exposition. Dawes was a friend of Jeannette Piccard's father, John Ridlon, a prominent Chicago doctor, but the end run failed: They refused to meet with him and left Jean Piccard to a hardened Irving Muskat.

Auguste did not make negotiations any easier, of course. On the one hand, he refused to be present or to fly, but on the other hand he expected full control, through Jean. When Muskat balked, Auguste demanded that his name be removed from the project and that the exposition rename the flight. Moreover, when later versions of the contract failed to describe the flight as a scientific enterprise to Auguste's satisfaction, and Jean's role was reduced, Auguste, according to Jean, refused to sign: "He will not . . . give his name for the sole purpose of attracting a crowd to the Century Fair, even if he were paid for it."[72]

In fact, Auguste never signed the contract. He gave Jean power of attorney and never bothered responding to Muskat's continued pleas that he fly as scientific observer. Jean had to fly, and both Piccards worked under that assumption through June, when Auguste returned to Chicago and Midland to inspect the gondola, after Jean had signed what he thought was the final contract.[73]

Who Flies?

Muskat never trusted Jean Piccard as an adviser or a flyer. He confided in Willard Dow that Jean "has but once ascended in a balloon and at no time did he participate in [its] design." Muskat preferred the company and counsel of Van Orman and Settle; both had a great deal of relevant experience.[74] Muskat also played on Compton's publicly stated desire to fly his instruments as high as possible, along with his privately shared difficulties with Jean Piccard.[75] Thus Jean did not help matters at all when he dealt unsuccessfully and clumsily with Du Pont, or when he attempted to gain an audience with fair managers Lenox Lohr and Rufus Dawes, or especially when he visited the fair to announce that the intended flight was meant to break his brother's record. Using once again the imperial "we," he stated: "We are planning to break our former record. Either myself or my brother will make the flight and our pilot will be Lt. Commander T. G. W. Settle."[76] As this news release came from the "A Century of Progress" Publicity Office, Muskat was clearly fighting an internal image battle as well.

Starting in June, Muskat, in concert with Compton, began to hint that Jean would not fly at all. In an effort to restore Auguste to the roster, Muskat threatened that if Auguste did not fly, then "one of Dr. Compton's men [Luis Alvarez] will be the scientist on the flight," and the Piccards would receive a smaller portion of the gate receipts. Muskat drew up another contract to reflect this latest demand, which took all discretionary powers from Jean. In effect, Muskat reduced Jean's role in an attempt to control Auguste.[77]

Muskat wanted Auguste's presence at the fair, to maximize gate receipts and to squelch Jean. He pushed Jean to intervene, implying that it would be in all of their best interests, but after hurried and acrimonious phone calls, Jean complained to his friends at Dow and asked them to speak to Muskat or his superiors. However, Jean knew that he was in a bind and had to try to convince his brother to cooperate. He told Muskat he would unite with Chicago interests as long as he was restored as alternate scientist for his brother. Thus Jean drafted a letter to Auguste, signed by Muskat and himself, laying out the entire situation, including the name change, the "financial advantage" if Auguste agreed to fly, and a reassurance that Jean would indeed fly with Settle even if Auguste refused, but that this would result in reduced gate receipts.[78]

Auguste arrived in New York on June 21 and stayed for less than two weeks. Jean was then named scientific observer once again, but all control was given over to Settle. Jean, frustrated, still considered Settle an ally and confided in him: "I shall only begin to enjoy life again when we have left this old earth behind us and are floating slowly through the dark sky." Jean told Settle that he saw the stratosphere as a haven away from the frustrations of earth, and it was as if his brother's experience were his own. Unfulfilled and unwanted by the professional world he longed for, Jean turned to the stratosphere as a refuge from politics and economic uncertainty and as a path to fame and freedom. Jean was an unhappy remnant of the romantic explorer, one of *Humboldt's Children*, who hoped for the best: "There is a disagreeable atmosphere of politics and business interference here, but as long as you and I work together, it will go o.k."[79]

To Settle, the stratosphere was not a haven but a means to fame and freedom, and above all a sporting challenge. When Settle's participation was first suggested by Goodyear and then approved by the Navy, he wrote to Jean claiming that he was not interested in "personal glory" and that he "wanted to make the flight solely because of the great scientific and sporting interest that I have had, for a long time, in such a project."[80] Clearly Settle was very anxious to fly; in early 1933 he confided to a friend at the National Bureau of Standards that he had been quietly "agitating" for a Navy high-altitude balloon program since 1929, and was happy to report that "I have finally gotten tentatively lined up for a stratospheric hop out of Chicago in July."[81] Settle's association of science and sport was not an idle remark for this avid balloon racer, romantic adventurer, and military technical officer. He exemplified the "aggressive and competitive individualism" of sport and the playing field, which many thought at the time promoted discipline and moral character.[82] Just as there were special rules on the playing field, a special discipline and esprit accompanied ballooning. To "Tex" Settle, science and sport were inseparable in ballooning: His science was the discipline of making balloons and airships fly higher and faster; he was the true explorer, who was motivated by pride, prestige, progress, and profit. He could not pass up this opportunity and was grateful that forces sympathetic to him were in control.

In mid-July, fair officials grew increasingly nervous about the flight delays and poor attendance at the fair. The flight had missed the highly desired June slot when the fair had hosted a prestigious international conference of scientists held in conjunction with the American Association for the Advancement of Science. Crew, Muskat, and Moulton all hoped that the flight would crown the proceedings of the conference. A scientific flight during a world-class scientific conference was just what the advocates of science had hoped for. The conference turned out to be an unqualified success without the balloon flight,[83] but it was missed by exposition officials. Thus any additional annoyance became intolerable. And Jean Piccard was a major annoyance.

At the same time that Muskat was trying to dump Jean Piccard, Compton was vigorously promoting his graduate student Luis Alvarez as the scientific observer, although Jean believed Alvarez had conspired with Compton to have him removed. Jean also believed that "Settle was afraid he wouldn't have all the glory if he went up with a Piccard," and so as Jean later recalled, one day before the flight on August 4, the exposition planners, NBC, and Settle all demanded that Jean be medically tested before the flight. The test would include a parachute jump. If he failed, then Auguste and Jean would lose title to the balloon and gondola. But if he agreed to bow out peacefully and was willing to cooperate, their financial interests and ultimate title to the balloon and gondola would be preserved. At the time, Jean actually had no claim to the balloon and gondola and, his sense of timing of events was quite far off. But it was a fact that he was being squeezed out, and had to make the best of it. He later confided to Auguste that he wished to take everyone to court, but did not have the funds and evidently the guts for the fight: "Your interest was the best guarantee if I gave in."[84]

Alvarez recalls that Jean did not get along with Settle, was removed from the flight roster well before the flight, and was replaced at first by himself. He recently noted, "I discovered then what it was like to be a celebrity."[85] Alvarez actually trained briefly for the flight with Settle, as he continued with R. J. Stephenson to prepare the cosmic-ray instruments for flight. But on the day of a critical parachute test, which he evidently relished, he too was dropped because lawyers for the fair wished to avoid a legal battle with Jean and Auguste.[86]

By threatening a lawsuit, Jean Piccard actually played right into Muskat's and the fair's hands. Jean decided that if he could not fly, then Settle had to fly solo. The fair officials, to rid themselves of Jean, had already modified their original agreement requiring a Piccard to fly, and now happily endorsed the idea of a solo flight. They all knew that the Russian competition was using a larger balloon, but also that Compton's public statements in the past had more than justified the need to carry the cosmic-ray instruments as high as possible.

Thus, on July 21, Jean and "A Century of Progress" came to terms through a formal agreement prepared by the legal division of the fair. The memorandum of understanding stated that Jean Piccard would remain on the ground, "permitting Commander Settle to go alone. The reduction in weight thereby produced will most assuredly enable Commander Settle to reach a higher altitude than would otherwise be the case, thereby adding to the value of the flight from all standpoints."[87] In return, Jean and Auguste would receive the full expected payment, just as if either one had flown. Strangely, nothing was said about the disposition of the balloon or gondola. This would soon cause great confusion.

Settle, who was in Akron, was informed of Jean Piccard's submission by Lenox Lohr that day and promptly agreed to the solo flight. In a letter penned the next day, Settle endorsed the solo flight: "I appreciate very greatly the fine sportsmanlike proffer of Dr. Piccard to withdraw as flight

FIGURE 3.5. Luis Alvarez (r) and T. J. O'Donnell (l) at Soldier Field, August, 1933. Luis Alvarez collection, courtesy Peter Trower.

aide in the stratosphere project, in the interest of gaining an additional increment of altitude and thereby obtaining more valuable cosmic ray and other scientific data."[88] Sensing a potential problem with Compton, however, Settle agreed to the flight only on the condition that Compton approved of a solo flyer handling both the craft and the cosmic-ray experiments. Even though they were largely automatic devices, the experiments had to be tended from time to time, and various logs had to be maintained as well. But Compton wished above all to be rid of Jean Piccard, who was simply not to be trusted. At the same time, both Settle and Ward T. Van Orman had come to dislike the Piccards. Van Orman argued that neither Piccard, but especially Jean, would be the best companion on a stratospheric flight:

When two are seated nearly in each other's laps for many hours ten miles above the earth, nothing is trifling. Therefore I always had an eye for detail when it came to picking partners for races or matters involving scientific endeavor which, by their nature, would lock us in the air as partners for a long period of time.[89]

Once Muskat had obtained Compton's approval for Settle's solo flight, and Settle had agreed to go alone, there was no hesitation. The rationale, both public and private, was that a solo flight would increase the attainable altitude, which was in the best interests of science. Compton, however, would have preferred to have Alvarez fly. Compton cabled from New York, where he was a visiting professor at Columbia: "Shall want all cosmic ray apparatus sent on flight with Settle if he goes alone. Prefer Alvarez as scientific observer for reasons stated but will accept solo flight in deference Piccards's wishes."[90] A two-man flight certainly had a better chance of managing the scientific apparatus, but Jean insisted that if anyone flew with Settle, it had to be a Piccard. So a solo flight for science was a nice cover for all parties concerned: It would maximize the chance of grabbing the world's altitude record and it gave Jean Piccard an honorable way to back out.

But this still could not have been a comfortable plan. Both Goodyear and "A Century of Progress" officials knew that the site for the launch was going to be troublesome as the Goodyear blimp ground wires kept getting tangled in the skyride and on buildings of the fair as it cruised over the grounds. This frightened riders on the skyride, and created electrical fire hazards. Fair officials complained that Goodyear was being reckless, while Goodyear shot back that the fair had placed too many overhead electrical wires in the vicinity of their concession.[91] Soldier Field was in an even more congested area than the Goodyear dock, and a launch from there was going to require deft control of the craft.

Even though Jean had signed the agreement, he tried to get around it. Settle tried to console Jean, saying that "Your action regarding flight most sportsmanlike and fine and I appreciate your expressions toward me very greatly. I sincerely regret not having you as shipmate."[92] But as the launch date approached, Jean did not feel very sportsmanlike and desperately called on his allies to put him back on board, using Settle's disingenuous remark as evidence. He telegrammed E. H. Perkins at Dow saying that Settle supported his reinstatement: "Dow is giving gondola for scientific flight not for stunt flying. Settle prefers going with me says scientific results would increase. Is pressure by Dow not possible."[93] And on the eve of the flight, Jean called Rufus Dawes, president of the exhibition corporation, pleading for reinstatement. Dawes told legal counsel William Dever to step in, who then, finding Jean Piccard staying at the University of Chicago's Quadrangle Club, stalled for time by pointing out the legal problems: "it was impossible" for Piccard to fly with Settle now on so short a notice, because of the difficulty of modifying the oft-amended contract "at this late hour."[94]

With Dever running interference, Lenox Lohr waxed eloquent in Soldier Field as the balloon was being inflated. In words strangely reminiscent of Settle's, Lohr explained Jean's absence: "The sportsmanship and unselfishness displayed by Dr. Jean Piccard in surrendering his place in the balloon so that a greater altitude may be achieved through the lessened weight of himself and his equipment—is a note of sacrifice that will not be forgotten."[95] The flight sponsors wished to trade the weight of the observer for the weight of the scientific instruments themselves, but most assuredly they wished to rid themselves of Jean Piccard. Certainly, the exposition was devoted to both regional and national propaganda; not the least would be the boast that an all-American expedition held the altitude record for ballooning. Exposition officials knew that they were in a race, and that they were far from being winners in that race. Years later, Jeannette recalled that they wanted the Piccard name for its prestige; other than that, the exposition officials were bent on gaining a world's

FIGURE 3.6. The inflation of the balloon at Soldier Field, on the night of August 4/5, 1933. Note the partially filled stadium. NASM.

record.[96] Jean could only see that he had been outsmarted and betrayed, but he was determined that he would yet fly into the stratosphere.

Settle's First Flight

On August 5, Thomas Settle took off alone at 3:00 a.m. from Soldier Field in Chicago. He knew it was certainly a dangerous and problematic site for the launch, and the newspapers played it up as well. He had to avoid thousands of excited onlookers who had been sitting in the stands waiting for over seven hours for the big event, and once aloft he had to avoid the exposition's Skyride cables and 186-meter towers, "about as dangerous as a rocky reef to a steamer."[97]

The flight was, indeed, a "fan dance of science." Those who pushed Settle into a solo flight from Soldier Field—in full view of what they hoped would be an excited and filled stadium that was anxious to grab a piece of history, even if it was only the souvenir program which hailed "The Piccard–Compton Stratosphere Ascension from Soldier Field . . . A Solo Flight by Lt. Commander Settle, U. S. N.,"—wished to emphasize the display and spectacle of doing science, rather than the science itself. In their drive to restore public faith in scientific and corporate America, the patrons of the flight threw out safety as well as the second aeronaut. A world's record, won before the watching world, would be more memorable than the resolution of the Compton–Millikan debate, which in any case was unlikely. And there was always the need to insure maximum gate receipts.[98] In this race to the stratosphere, misplaced priorities overcame careful preparation and prudent launch procedure.

As Settle rose from the field, clearing the aerial obstacles, the valve in the balloon vent control system which had given trouble just before flight started to act up again. Before the launch, Settle had decided to ignore this problem because he felt pressured to hold to the launch schedule. The malfunctioning valve had almost an immediate effect. Settle lost control after dodging the Skyride and had to abort less than 20 minutes after takeoff. Even though the balloon and gondola landed only slightly bruised but safe in the nearby Burlington railroad yards, near 14th and Canal streets, hundreds of treasure hunters began to tear pieces from the balloon, and, according to a number of observers and commentators, it was only through the quick action of the Marine launch-crew detachment headed by Maj. Chester L. Fordney that anything was salvaged. Combined Army, Navy, and Marine units worked with volunteers from the Museum of Science and Industry to dry and stow the gondola and balloon.[99]

Then the real confusion set in. The question was, who really owned the balloon and the gondola now that they flew, if even for only a few minutes. As already mentioned, exposition officials had agreed that after the initial flight, the Piccards would take possession of the balloon. But according to a recollection by Jean and Jeannette's son Donald Piccard, which was assuredly based upon Jean Piccard's impressions at the time, they were to obtain both the balloon and gondola. In the spring, Willard Dow had given Jean the impression that he would have use of the gondola after the flight, but at the same time assured "A Century of Progress" officials that the title to the gondola remained with the Dow Chemical Company. Both, in fact, were correct: Willard Dow, a relaxed and congenial fellow, had no problem retaining title and letting worthy aeronauts use it in Dow's name.[100] At the time, Dow failed to appreciate the delicacy of dealing with Jean Piccard.

Saving the gondola was not a major problem because Dow was fully committed to it. But the forlorn balloon in the railroad yards also had to be saved somehow. As Donald Piccard recently recalled, the Navy insisted that if it saved the balloon from the yards, it should obtain the title. But neither Jean nor Jeannette would allow the Navy to obtain rights to the gondola and balloon, so Jeannette exploited her considerable Ridlon family ties in Chicago to ask Julius Rosenwald, owner of Sears, Roebuck and Company and founder of the Museum of Science and Industry, for help. The museum was happy to do so, and in payment would take ultimate possession, provided the Piccards had the right to reuse the gondola and balloon whenever they wished.[101] The museum rounded up manpower to save the balloon, and were profusely thanked about one week later by Jean for "the generous and efficient way in which you helped me to take care of my balloon which was wet and needed immediate attention. Mr. Teberg and the men you sent to the Naval Reserve Armory have made possible the preservation of the balloon."[102]

In fact, most accounts cite Fordney and his marines for the salvage effort. Two days after the aborted flight, the balloon bag lay carefully rolled up on a gondola car in the railroad yards of the Chicago, Burlington & Quincy Railroad. At the expense of Dow and the *Chicago Daily News*, Piccard was told, "a continuous guard was put in custody of the balloon so that you might not suffer from having the property left uncared for."[103] But the Piccards had to act fast: "You should make immediate arrangements to take the balloon into your possession, as it is your property and you alone should be responsible for it." Settle estimated that the balloon would begin to deteriorate within 48 hours, so evidently it was at this point that the museum was called in to help dry and store the balloon.

There is no record of the Navy insisting that it should obtain title to the balloon. It seems that after its retrieval, Jean Piccard forgot his pledge to the museum and sold his rights to the balloon to the *Chicago Daily News* when exposition officials decided that they wanted Settle to fly again. Piccard, however, reserved his right to buy the balloon back at some future time for a nominal sum. The *News* had to obtain title to the balloon before Goodyear would agree to refurbish it for another flight.

The Second Flight: Negotiations with Jean Piccard

Irving Muskat thought that Settle deserved another chance to touch the stratosphere. Reviewing the disaster of the moment, he rationalized Settle's decision to fly and also tried to restore Compton's confidence in the overall enterprise:

[Settle] chose to make the ascension in spite of the fact that the valve was not operating properly. He risked his life in order to remove the hydrogen from Soldiers' Field and not to disappoint the large crowd that had waited patiently until 3 a.m. to witness the ascension. His handling of the balloon and bringing it down to a safe landing in the heart of Chicago's business district proves him to be one of the ablest balloonists in the country.[104]

Muskat added that Settle persevered in spite of considerable confusion just before the flight. In the testy hours before launch, Compton's proxy, R. J. Stephenson, was unable to execute his coordinating duties handling final integration and testing when arguments arose because no one was clearly in charge. Next time, urged Muskat, Stephenson should clearly be in charge of all experiments. Compton hoped too that there would be a next time.

When Millikan heard about the fate of the flight, he did not hesitate to ask for his instruments back. Compton replied in a hand-written postscript to a previously prepared letter:

I have just received your note regarding your apparatus in the Piccard gondola. We are naturally much chagrined over the fiasco of the flight, especially after several of our men have spent solid months of work preparing the apparatus. I have not been informed when a second attempt will be made. I am however instructing Dr. Stephenson to return your electroscope, and am hoping you will loan it to us again if there is a "next time."[105]

At first, exposition officials thought there would be no next time. On August 7 they stated that the balloon was the possession of the Piccards, which immediately restored Jean's dreams for a flight.[106] Jean and Jeannette had been making plans for a flight the following summer, but strangely after the aborted flight, Jean told Willard Dow, "As to me this entire episode is now closed and I am again an organic chemist who is looking for a position."[107] When the Piccards obtained the balloon once again, Jean's hopes were rekindled. However, exposition officials soon reversed their decision. They had no choice. They knew that the Soviets were about to fly. The exposition planners, to say nothing of their scientific and corporate patrons and advisers from the National Research Council, could not let this become a national embarrassment, not in the depths of the Depression, nor at any time. The "fiasco" could not be allowed to mark the end of their efforts; another flight under more sane conditions had to be made.

Settle convinced both the exposition sponsors and the Navy's Bureau of Aeronautics that another flight was quite possible since the needed repairs were small. E. H. Perkins of Dow inspected the gondola and found very little damage; the bottom pole segment had been bent and cracked, and Dow was able to weld on a new pole in only a few days.[108] Although the gondola was no problem, the balloon still had to be retrieved from Jean Piccard.

To E. H. Perkins fell the task of convincing Jean that it would be best to give over the balloon, no matter what his immediate hopes were or what he had promised the Chicago museum. Perkins urged Jean to let the exposition plan another flight quickly; if he did so, Dow might look kindly upon a third flight by Jean and provide the gondola. Willard Dow's plan was indeed that Dow would refurbish the gondola in short order to be reflown solo by Settle once again, and that, except for obligatory exhibitions of the gondola in Chicago and New York after the second flight, the gondola would be turned over to the Piccards for their own use. Perkins patronized Jean to control him and to keep him away from the press. By this time, however, Jean and Jeannette considered themselves a duo: They would be the Piccards who returned to the stratosphere.[109]

Jean took his time in submitting to Perkins's logic. Feeling betrayed and outnumbered, he lashed out at Settle and Compton in the privacy of his letters to Auguste, predicting that a second attempt by Settle would meet with the same fate. In Jean's opinion, Settle would again fail to prepare the balloon control systems properly; the abort was not an accident, but a result of incompetence, both in preparation and execution. Settle represented everything that was wrong with the flight. Jean believed that Settle controlled both Goodyear and Dow because both hoped for contracts from the Navy to replace the ill-fated Akron.[110] Jean's sense of betrayal extended also to Dow. He thought he had been promised a position with them after the flight. Now he learned that Dow had no interest in his services.

The key culprit in the aborted August 5 flight, the recalcitrant valve system, vanished in the confusion. When the Piccards obtained the balloon, it was minus the valve, and no one knew where it had gone. Through his lawyers, Jean demanded that it be found, but within two weeks the problem seemed moot to "A Century of Progress" people because Jean agreed to sell the balloon to the *Chicago Daily News* so that it could be refurbished by Goodyear. As L. B. Rock of the *News* put it to F. C. Boggs, Lenox Lohr's general assistant, Goodyear-Zeppelin would refurbish the balloon as agreed, and "a deal with Jean Piccard has been made which requires the title of the balloon, with his option to repurchase at a nominal price at the completion of the stratosphere flight, or not earlier than December 1." Rock added, "I understand from you also that the gondola has been made available for another flight without the opportunity of interference by Piccard in any way."[111]

But the sticky valve was still a sticky issue. During the rest of the month, Piccard's lawyers haggled with "A Century of Progress" lawyers about who was responsible for the device and whether it was included in the sale of the balloon to the *Chicago Daily News*. Piccard's point was well taken: If the valve was part of the balloon that he sold, he should expect to receive a valve when he repurchased the balloon back from the *News*. To further complicate the story, Jeannette Piccard took over the task of negotiating for the balloon and valve in late August, and wrote their lawyer that she recalled seeing the valve at the railroad site where the balloon landed, on August 5. It soon disappeared, and stories ran back and forth about its removal to Akron by either Settle or an agent of Goodyear. Threats of actions, counterthreats, and innuendo continued to fly between the lawyers, inflamed by Jeannette, until Lawrence Cole, representing the Piccards, got a straight answer out of Goodyear: They had both the balloon and valve all the while, and claimed that they told this to Jeannette in the first instance. By September 8, Goodyear announced that it would refurbish the balloon, and return both items to the *Chicago Daily News* and that both would eventually revert once again to the Piccards. Although the matter of the valve turned out to be a tempest-in-a-teapot, the episode is symptomatic of the state of emotions that then existed between all parties concerned with the flight.[112] With the fate of the valve at rest, the balloon was sold to the *News* and refurbished by Goodyear at no cost, and was returned to Chicago complete with a new valve by the end of September.[113]

Jean's dreams, running already on a shoestring, had to be deflated in stages. Another shock came on September 14 when Jean received a "surprising request" from E. H. Perkins of Dow that the word "solo" be stricken from the agreement and that another Navy officer be invited to fly with Settle. Out of "deference" for his friend Willard Dow, Jean reluctantly agreed that the second flight would not be a solo, but wished to establish anew that after the flight, Dow would restore the gondola completely, so that he and his wife Jeannette might have their chance.[114] Dow thanked Jean for his understanding, and tried to make him understand the Navy's situation:

I can appreciate that due to the ramifications of the United States Navy Department, it is quite embarrassing for them to propose to put on a so-called "solo flight" and I can say very frankly to you that we appreciate your co-operation in permitting the Navy to take up two persons rather than one. Also the fact that it has become an all Navy flight, rather than a flight sponsored entirely by science, puts an entirely different aspect on the whole venture.[115]

Unknown to Jean Piccard, Willard Dow was among those wanting Settle to have company and promoting a military man. In mid-September Lenox Lohr heard rumors that Settle wanted an assistant, although Dow's agreement with Piccard specified a solo flight. In reply, E. H. Perkins assured Lohr, "We have signed agreement with Piccard imposing no conditions on Settle or yourselves other than any aide must be active Navy or Marine officer. Gondola is available whether [solo] or not with Settle to make free choice."[116]

The altitude record could not be obtained if the vehicle could not be controlled. Ward T. Van Orman and other expert balloonists knew only too well that a solo flight was a foolish gamble. Even Settle had to admit this now, and recognized that he needed someone he could work with and trust completely. So E. H. Perkins and Willard Dow would have to cool Jean down, and even convince him to give up this foolishness of stratospheric flying:

After all is said and done, why should you and Mrs. Piccard risk your life for a venture of this type? I should like to urge you to give up the thought of such a proposed trip. There are many others, such as the Navy Department, who make it their business as well as their pleasure to make these "dare devil" hops and why not leave it to them to explore these regions; on the other hand, offering you an opportunity to take the results they obtain and correlate them and further laying plans for future experimentation.[117]

Why, thought Willard Dow, didn't Jean act like a proper scientist, like a Compton or a Millikan, who were, after all, interested only in the results? To Dow, Jean's plan to fly in the refurbished gondola in the summer of 1934 with Jeannette as copilot did not make sense. But to Jean, it made very good sense. He was not a Compton or Millikan, and, without somewhere to hang his hat, he was without a portfolio in science. Agreeing that an all-Navy flight "is quite a different proposition from a scientific one," Piccard added:

Such an attempt [a nonscientific all-Navy flight] would be an unnecessary risk for a civilian to run. Scientific investigation, however, has a commanding appeal. Mrs. Piccard and I cannot see that our lives are so very valuable as they stand now. Without a job, without a laboratory on the ground we are not in a position to render any service to humanity.[118]

Whereas Jean spoke to Willard Dow of service to humanity through science, to his brother he talked politics. Feeling outsmarted and betrayed once again, he admitted:

I loaned the cabin for a "solo" flight this autumn. A few days ago I was asked to agree to letting Settle take a Naval officer with him. It's the dream of the Navy. Had it been an order I should simply have refused, but if I refused [the request] he would have made the ascent alone, and if there were an accident the responsibility would have been all mine. Also, had I refused I would have cut myself off from any further collaboration with Dow. On the other hand, Dow promised, if I accepted, to give full collaboration in the event of a new ascent. I accepted.[119]

Jean had the satisfaction of believing that the aborted first flight gave the Piccards "a big moral advantage" and Soldier Field gate receipts "brought us $3000 net plus the repair of the balloon and the cabin." All the same he felt his turn would come. In fact, Jean was confused about where the money had come from. He had been reimbursed for the balloon only by the *Chicago Daily News*. Gate receipts from the August flight had been very disappointing. Only some 15,000 tickets had been sold, with a revenue of $6628. With expenses running over $17,000, this meant a loss of more than $10,000 which by agreement had to be made up by the *News* and NBC.[120]

The officer chosen to accompany Settle was Maj. Fordney, who had shown great initiative in saving the balloon. Fordney was originally asked to assist Settle in technical flight preparations, as he had both "scientific and practical mechanical knowledge." Thus he became the observer on the second flight of what was soon to be called *A Century of Progress*.[121] The Navy as well as Dow insisted that exposition officials nominate an "officer on duty" to make the flight; their first choice was Capt. Orvil Anderson, but Fordney was available immediately and was Navy.[122]

The flight had to take place before the fair closed in December. Once the initial inflation tests of the restored balloon in September indicated that it was air-worthy, the gondola was quickly repaired in Midland. But delays came from two unexpected quarters. First, Settle was extremely busy with other exploits, including the Gordon Bennett air races held in September as part of "A Century of Progress." He was also responsible for shepherding the visiting *Graf Zeppelin* from Akron to Chicago in November. The air races alone took all available manpower in Chicago in early September. As a result, F. C. Boggs asked E. H. Perkins to keep the gondola in Midland for the time being.[123] Another obstacle was that the scientific instruments were no longer available. Both Compton and Millikan had found quick use for their cosmic-ray devices after the first flight; some had been promised to Admiral Byrd for his Antarctic expeditions. Furthermore, Alvarez would not be able to supervise the use of the all-important direction-sensitive coincidence counters until the end of the month.[124]

When Settle was informed of this additional delay, he worried that the weather would be unfavorable if the delay went too far into October. Settle was right. Poor weather extended the delay into early November. What was needed was a large, stationary, and therefore predictable high-pressure cell, preferably over the entire northeastern part of the United States. Also needed was absolute calm during the long and critical inflation process; surface winds could not be in excess of 10 kilometers per hour. Caution had replaced bravado, but deadlines were deadlines.

Meanwhile, the Russians flew, and with great success. On September 30, 1933, a trio of Soviet balloonists—G. Prokofiev, E. Birnbaum, and K. Gudunov—achieved a new altitude record in excess of 18 kilometers under an enormous 24,000-cubic-meter balloon in a riveted aluminum alloy sphere made of 12 sections. Numerous scientific experiments were

FIGURE 3.7. Soviet gondola that flew from Moscow on September 30, 1933. Science Service collection, NASM.

said to have been performed on that flight, which was all the better for Soviet pride and Stalin's agenda.[125] Both L. B. Rock and Lenox Lohr now agreed that "The new Russian altitude record will be the mark to shoot at."[126]

As the November 12 fair closing drew nearer, Compton's patience grew thin, and he revealed his disappointment to Millikan. In late October Compton reported that all was ready, save for the dreadful weather:

[All is] in complete readiness for the flight at a moment's notice. They are only waiting for a proper turn of the weather. It would have been better to try this flight from California! In any case, our season is over on November 12 and if nothing happens by that time, I suppose we shall have to call the venture off. This would, however, be to us a big disappointment.[127]

Comic relief came from an overly imaginative press anxious for a story. Reporters creatively embellished upon an experiment designed to study the effect of cosmic rays on fruit flies by turning it into a study of sex changes in the crew due to cosmic rays. Before the flight, according to Luis Alvarez, ground crew members passed along a box of sanitary napkins to Settle and Fordney, just for laughs. But the highly absorbent material proved useful when serious condensation problems developed during the flight.[128]

The new flight could not be another fiasco. Soldier Field was abandoned in preference for the Goodyear dock just south of the Chicago fairgrounds. The crowds would not be so visible and pressing now, which if we are to believe Muskat's account, had provoked Settle to decide to fly on August 5.[129] Without the influence of the crowd, and with Goodyear in better control of the tedious launch procedure, most factors which had spelled disaster in August were gone. But they could not control the winds or the weather.

Bad weather persisted through October and into early November. The fair closing and the Piccard contract deadline both loomed large in Settle's mind. Chicago newspapers issued daily weather forecasts with hopeful predictions that Settle would fly tomorrow, but tomorrow never came.[130] Finally, it was time to make some hard decisions. Both Goodyear and the Navy preferred to launch from the restricted Goodyear-Zeppelin Airship Dock at the Akron, Ohio, Municipal Airport, while exposition officials and the *Chicago Daily News* naturally preferred the Goodyear dock in Chicago. But Akron had a balloon hangar 65 meters high, 100 meters wide, and over 300-meters long, more than adequate to inflate *A Century of Progress* in complete safety. The launch-inflation height of the balloon was 50 meters and there was plenty of room for all control roping.[131]

On November 13, exposition officials and backers met in Chicago to hear Settle's arguments. The fair had closed on the 12th, so with the link between the fair and the flight weakened and all contract agreements in limbo, Settle won approval to remove the whole operation from Chicago. With the support of the Navy, they prepared for a flight from Akron, where the facilities were far better, and clear weather was forecast. Two days after the fair closed, Fordney took his leave from "A Century of Progress," packed the gondola, balloon, and hydrogen cylinders, and rushed everything to the Akron, Ohio airport to take advantage of the predicted good weather. In Akron on November 16th, Fordney apologized to a bewildered F. C. Boggs, who claimed not to be aware of what had transpired. Nor, apparently, was the *Chicago Daily News*. Fordney admitted to Boggs, "It is unfortunate of course that you were not informed of the removal of the balloon, gondola and hydrogen." The Navy moved very fast when it wanted to.[132]

On November 17 inflation and rigging began inside the Goodyear Dock at the Akron airport. After several aborted attempts, the balloon was undocked at daylight on November 20 by a large naval reserve and civilian ground crew, and moved a short distance to the municipal airport. At 9:30 a.m. Settle and Fordney left the earth and set a new world's altitude record of over 18.6 kilometers (the National Aeronautic Association placed the height at 61,237 feet, only about 500 feet higher than the revised Russian altitude). They floated east during the day and descended that evening, landing in a marsh seven miles southwest of Bridgeton in southern New Jersey, only 40 miles from the Atlantic Ocean and certain oblivion.

Fordney handled the instrumentation during the eight-hour flight while Settle piloted the craft. Settle later reported that upon landing in the marsh under growing darkness, they both decided to stay with the craft for the

FIGURE 3.8. *A Century of Progress* lifts off from Akron near the Goodyear hanger, November 20, 1933. Science Service collection, NASM.

night and secure the critical barometers and other instruments as best they could. At dawn, they found a farm about three miles distant, and contacted the world. They handed over the precious record barographs to a member of the National Aeronautic Association, the scientific instruments were removed by R. J. Stephenson, and "the balloon and gondola were placed under the orders of their owners."[133]

Ironically, both Jeannette and Jean Piccard were among the first on hand at the landing site because it happened near their home in Delaware: "We went over there and Jean waded through several feet of water till he reached the gondola. Then he stripped and swam the bayou to get to the balloon itself. It was in beautiful condition . . . we went home delighted that our own prospects were so good."[134] According to Jeannette, the Navy salvage crew saved only the gondola and left the balloon to be retrieved by a local strawberry farmer, who demanded payment for services. It apparently suffered from water damage in the delay. Jean merely noted that the water was saline, and that no real harm was done.[135] Some scientific equipment was damaged by the moderately hard landing; the gondola was intact although heavily dented, but many of the spectroscopic

FIGURE 3.9. Chester Fordney and Tex Settle hold the altitude record mercury barometers at the landing site, Deep Neck, New Jersey, November 21, 1933. NASM.

plates that were taken were lost. Jean later criticized Settle for this, noting that he had neglected to include doors on the storage lockers, and failed to tie everything down properly.[136]

Exposition officials happily exploited the successful flight. The intrepid aeronauts and their gondola were paraded through many a main street and newspaper headline. To a United Press correspondent, the feat represented a win in a "race for supremacy in the stratosphere" that began with the first use of pressurized gondolas.[137] To the balloonists and especially to certain people in military circles, the successful flight demonstrated that ascents into the stratosphere were possible, survivable, and potentially valuable.[138] It also demonstrated for the first time how to coordinate the complex and costly details of flight preparation and execution. Portions of the scientific complement were underwritten by the Gaertner Company, the Bureau of Standards, and the Navy Bureau of Aeronautics, which supplied the spectrograph, navigation aids, barographs and thermographs, respectively. Kodak, the Weather Bureau, the Navy Bureau of Construction and Repair, and the Bausch and Lomb Company also provided technical support. A conflation of many drives and interests, some competing and complementary but most conflicting and divisive, *A Century of Progress* managed to reach the stratosphere and demonstrate that, indeed, science together with corporate America could do wonderful things.

Thus the exposition had finally done its job. Americans had entered the stratosphere to establish a world's record, in the name of science. At the end of November, ownership of the balloon and gondola reverted to the Piccards, and they immediately transferred title for the gondola to the Museum of Science and Industry, even though physically the gondola was soon to be returned to Dow once again for refurbishing.[139]

The Scientific Return

Settle proudly reported to his Navy patrons in late November that they had attained a new world's record. Although he felt the primary value of the flight to the Navy was the practical experience thereby gained "operating a craft at high altitudes, with the innumerable attendant detail problems," still he "hoped and expected that the scientific data and observations obtained will prove to be of technical value."[140] But this was the province of others.

R. J. Stephenson retrieved the scientific instruments from the landing site and sent them off to the various investigators for analysis. In December, Compton prepared a preliminary report that identified 11 separate studies: four on cosmic rays; one on the ultraviolet transmission of the atmosphere (the quartz spectrograph prepared by Henry Gale); and miscellaneous studies of the polarization and brightness distribution of the stratospheric sky, air samples for composition, temperature and pressure, radio propagation, and infrared photographic aerial reconnaissance.[141]

Compton reported that the self-recording cosmic-ray intensity experiments, mainly a series of shielded and unshielded ionization counters, worked reasonably well, but directional detectors, such as Alvarez's double Geiger–Müller coincidence circuit on board, did not give reliable results since the balloon rotated faster than expected, and was erratic in its motions. This unexpected motion also made it quite impossible to secure usable spectra of the sun with the quartz spectrograph, which was in a small adjustable mount. To compensate, Fordney had removed the spectrograph from its mount and pointed it to the Sun, but even this didn't work. The few plates that were exposed were broken on landing.

By April 1934, Compton had reduced his data and found evidence that cosmic rays were exponentially absorbed by his shielded counters. This suggested to him that alpha particles were a component of the cosmic-ray flux.[142] By this time, most physicists accepted the existence of the latitude effect. The question now was how reliable was the intensity/altitude curve in different longitudes, and, most important, what was its actual form? An unambiguous curve was required to distinguish between the charged particle or photon models for primary radiation, and Compton believed that his flight data helped to clarify one issue: Erich Regener's observation of a flattening of the altitude-versus-ionization curve at the very peak, which he thought indicated that his balloonsonde had entered a region of primary photons. Compton did not find the flattening in his flight data, which came from a higher geomagnetic latitude. To Compton, "it became clear that this was a geomagnetic phenomenon, and indicated that the corresponding portion of the radiation is electrical in nature."[143]

Between January and July 1934 Millikan's group examined their flight data alongside the results of their airplane cosmic-ray latitude survey of 1932. When they compared their unshielded counter data to Compton's shielded counter results, they found that virtually all of the ionization occurring "is due to secondaries produced within the atmosphere."[144] Millikan pointed out that because the ascent to 18.6 kilometers was so rapid, taking only 20 minutes, their self-charging electroscope, which had a cycle of 5 minutes, did not have enough time to take good readings at altitudes below the maximum. The balloon remained at maximum altitude for three and a half hours; therefore the readings there were "very reliable."[145]

Erich Regener was impressed with Compton's Fordney–Settle flight results, which he felt confirmed J. Clay's observations of and conclusions about the latitude effect. He told Millikan that Clay was now finding that the latitude effect was influenced by longitude, and that this was most likely due to the orientation of the magnetic axis of the earth.[146] But in Regener's view the new Compton results merely confirmed that the variations of the latitude effect with altitude and longitude were quite complex and needed elaboration in further flights. Regener thus argued along with Millikan that subsequent balloon ascents, manned or unmanned, should not be limited to the continental United States.

The Fordney–Settle flight demonstrated to Millikan the importance of having a flight from equatorial regions carry both shielded and unshielded counters to great altitude.[147] Millikan pointed this out to Albert Stevens, who was then well along in planning for his Explorer flights. Millikan admitted that the results from *A Century of Progress* were "pretty nice" and revealed a real latitude effect when compared to his unmanned ascents from Texas. But any new manned flight, Millikan argued, would be more valuable if conducted from a latitude far different from that of Settle's: Kelly Field near San Antonio, Texas, was Millikan's choice.[148]

Whereas both Compton and Millikan expressed satisfaction with the Fordney–Settle flight results, Jean Piccard naturally remained negative. With his own plans for 1934 under way, in direct competition with those of Stevens, Piccard was intent upon gaining scientific patronage and did not want either scientist to be satisfied now. Hardly a month after the flight, before any results were available from the raw data, Jean Piccard criticized the way Settle had executed the flight. Although he congratulated Settle publicly, privately he told Auguste that the flight yielded "very little scientific results." He outlined what he felt were blunders by both Settle and Fordney in handling the experiments, and said that the only experiment that worked was Millikan's automatic device. Jean took pleasure in telling Auguste this "serves Compton right!"

Even though Jean Piccard had maintained cordial correspondence with Settle about various options for air filtration and respiration before he was bumped from the flight, now all he could do was criticize Settle's carbon dioxide filter. However, Piccard had also come to appreciate the Navy's knowledge on the subject and its insight. Nonetheless, in bitter hindsight he felt that it was all wrong, and only reflected the stupidity of military minds: "The Navy and the Army are very stupid. However, that doesn't matter because all officers are part of an international rabble of parasites and . . . they are stupid everywhere."[149] Compton was even worse. He controlled the scientific reporting of the flight, whereas Jean deserved at least a coauthorship: "In [Compton's] role as scientist and Propaganda Christ he and I of course will appear in the title. In his role of a swindler he will of course, not do so. He is both." Compton was indeed the sole author of the notice summarizing preliminary results, but it was only an abstract report on the collective work of the participants, each of whom was free to report, as Millikan had done, on his findings. Both Auguste and Jean Piccard's names were prominently featured in Compton's abstract, but this was little consolation to someone who had been denied a flight into the stratosphere.[150]

Conclusions

The backers of *A Century of Progress* certainly were not motivated by the need to solve a scientific question; yet "A Century of Progress" officials wished to capture the attention of the public by performing spectacles in the name of science and thereby restore the nation's honor and faith in science and corporate America. Accordingly, they orchestrated a stratospheric showpiece to back up Frank Jewett and Michael Pupin's declaration that "American science and American industries welded to each other by scientific idealism are the most powerful arm of our national defense."[151]

A stratospheric ascent also fit the utopian image created by exposition planners and their scientific underwriters. One of the radio talks heralding the coming of the fair had proclaimed that "scientific research pays large dividends. There is no telling when a scientific Columbus is going to discover another America."[152] Auguste Piccard, a highly popular scientific explorer, reminded people of Columbus; in addition it was rumored that he was to be awarded a Nobel prize.[153] The flight of Piccard would provide hope to a nation in the depths of the Depression.

Public affairs officials felt there could be nothing better than the image of Piccard flying in the stratosphere from American soil. Capitalizing on an already proven attention getter was safer than betting on the visibility of a Nobel laureate, no matter how prestigious. Even without Auguste Piccard, the exposition had a powerful symbol in the gallant Navy flyer sent into the stratosphere by corporate America, along with the instruments for exploration designed by scientific America, an image favored by Irving Muskat in particular. This would be in keeping with the ultimate goal of the exposition, which was to restore faith in the idea that science and industry could be fused into an engine that would bring about national recovery.[154] If America could conquer the stratosphere, it could conquer the Depression.

The National Research Council, working hand in hand with the corporate entities backing the fair, also appreciated the symbolic value of conquering new territory. The nation would not be able to recover and science would not regain its position of respect unless someone could demonstrate the "power and universal applicability of the scientific method." Just as the Apollo Program offered a way of pulling America together during the cold war and restoring its "sense of mission," the flight of *A Century of Progress* was a means of signaling the expectation that better days were ahead, and that the path lay through the nurturing and exploitation of "science."[155]

If science itself could be linked with conquering the new realm of the stratosphere for the sake of social improvement, real or imagined (such as cheap long-range transportation), or if the flight could yield information about cosmic rays that might help humans harness limitless power from the atom, then people would believe that a new age of prosperity was on the horizon, "making possible the miracle of modern society."[156] Thus far, science had not lived up to its promise and had done little to stave off the Depression. To prove its value, science would have to create a vision of hope for the future, but to do so it needed a success, something with "display value." Just as the Soviet Union orchestrated aviation stunts to try to counterbalance its purges and demonstrate the efficacy of its ideology, corporate and scientific America hoped that it could reassure the man in the street by giving him something to be proud of, or something to look forward to. Among other feats, both nations used the conquering of the stratosphere for this purpose.

The stratospheric flight of *A Century of Progress* was a direct outgrowth of this strategy. Auguste Piccard demonstrated that such flights were possible, and that with proper support his first attempts could be bettered. He found this support through the FNRS while he courted Chicago interests on behalf of his brother. Eager not only for his proven expertise, but the instant recognition of his name, Chicago officials were quick to respond to, even anticipate, his interests. As eager as they were for Auguste Piccard's involvement, Chicago balked at his stand-in twin. Jean Piccard was a victim of exposition priorities and his own personality. Without Auguste, he was a nuisance. He had no real standing in the American scientific community, certainly not in cosmic-ray physics, despite having published numerous chemical papers and despite having some of his students go on to become organic chemists. He was still an outsider, and his actions certainly showed him to be an untrustworthy team player.[157]

Forest Ray Moulton, along with other exposition planners, could never be comfortable with Jean Piccard, let alone trust him with a visible role in their most visible and sensitive enterprise. As a student of the ardent moralist Thomas Crowther Chamberlain, Moulton was conditioned to believe that only "persons trained in the technically demanding, highly specialized work of a given field were qualified to have an opinion about it."[158] Jean's self-promotion and eagerness to be involved *ad hoc* in his brother's enterprise must have made Moulton and Willard Dow both suspicious. Dow must have wondered how a man, supposedly dedicated to the "supreme love of truth," could only consider conducting cosmic-ray experiments in the context of a manned balloon flight. At worst, Jean was perceived as a scientific pretender; at best, a pest with a certain usefulness who should be removed from the picture as soon as possible.

But what of Auguste Piccard? He too saw his contribution to cosmic-ray physics only in terms of manned exploration in balloons. Unlike Jean, Auguste had been active in physics, and boasted of several accomplishments. He continued by supporting Cosyns's flight,[159] as well as a later Polish endeavor to construct an enormous silk bag called *The Star of*

Poland which never got off the ground. Furthermore, he and his associates had supported South American plans for flying in the stratosphere.[160]

Auguste Piccard both created and won the race to the stratosphere, with the backing of the Belgian king; his early motives in developing his gondola and balloon centered on the need to solve problems in physics, which he had good cause to address and which he had addressed in traditional ways. He did not seek out the patronage of major scientists in Europe, although he accepted Kohlhörster's advice and aid after his first flight. But in preparing for and demonstrating stratospheric flight, Auguste Piccard the balloonist and physicist was wholly transformed into an explorer and entrepreneur. Cosmic rays could not hold his attention once he had tasted glory, and for the rest of his career, glory—albeit in the name of science and exploration—drove him to the depths of the ocean as well as to the heights of the atmosphere.

Piccard's challenge was met by Soviet scientists and its military, who were the next to cross the finish line, followed by corporate and scientific America, with its military as vanguard. Whereas Piccard's flights were fully civilian, the Soviet flights combined military fliers and civilian scientists, and the first American flights were fully military, as was the case in all the American manned spaceflights during the 1960s.[161]

Although the Compton–Millikan debate was exploited to legitimize manned flight, the flights themselves had virtually no impact on the progress of cosmic-ray physics, nor on the debate, which was largely moot by the time Settle and Fordney flew. This helps to explain why both Compton's and Millikan's involvement with *A Century of Progress* has not been treated by historians of science.[162] On the other hand, when one asks why and how scientists responded to the offer of a manned platform for their instruments, the public nature of the debate becomes clearer and one can better understand how scientists then approached matters of politics and patronage. The Carnegie's John Fleming needed a definitive test of both Millikan's and Compton's detectors; their differing results had to be reconciled to quiet criticism of Carnegie programs. Accordingly Compton orchestrated a dramatic demonstration that would reveal how the two detectors behaved under the same conditions. This would enable the Carnegie to determine which of the two it should adopt. In challenging Millikan to fly with him, Compton was, of course, taking a chance. In some ways, Millikan won the skirmish because the flight failed to provide an adequate test. As intractable as ever, Millikan publicly maintained that his new detector had been vindicated and that he had met Compton's challenge. Compton, on the other hand, probably felt quite frustrated by the whole "fiasco." But Compton won the race in the long run: his model for cosmic rays dominated, and his high-pressure detector, the "model C" electroscope developed in collaboration with R. D. Bennett, became the Carnegie standard for many years.[163]

Is it possible to give the "A Century of Progress" effort high marks as a voyage dedicated to scientific gain? The Soldier Field launch site was an exposition priority chosen without regard for safety or scientific value. The number and nature of the crew was at first determined by a desire to gain maximum altitude. Exposition officials hoped that the flight could be executed during scientific meetings held in Chicago, but were unwilling to accept, or were unaware of the need, to promote any special conferences or symposia to review the results. There was also no centralized effort to provide barometric or other housekeeping data to the experimenters so that they could reduce their data, and no effort was made to provide major reports or catalogues or congresses in conjunction with the re-opened "A Century of Progress" activities in Chicago, other than a small exhibit on the flight mounted next to Auguste Piccard's *FNRS* in the Hall of Science.[164]

The importance of the flight of *A Century of Progress* lies not in its scientific return, which was meager considering the effort and costs involved. Rather it gave American scientists their first chance to send aloft heavy instrumentation under a balloon, and thus provided a handful of them with a foretaste of the types of trade-offs to be expected in any large-scale engineering venture based on motives outside the realm of science. The scientists involved knew that the promoters of the flight were not driven to solve a scientific problem. Yet they were willing to let themselves be used, or found themselves trapped into participating. At first, Compton may well have sincerely believed that the manned flight might prove something. But Millikan certainly did not. Both knew that the *A Century of Progress* was meant to demonstrate that science could be done during manned balloon flight. Whether science needed it or not, a manned laboratory could be transported into the stratosphere, and could be made to work after proper planning and preparation. Both men were also aware of the political implications of participating in a visible and dramatic venture that could restore public faith in science. At the very least, by flying together they could demonstrate that the scientific spirit of inquiry was alive and well.

Settle's second flight with Fordney was a great improvement over his first solo flight. Publicly it was a success, but it did not convince scientists that science could be done in the stratosphere in a shirt-sleeve environment. After the accomplishments of Auguste Piccard and the Russians, these flights had only shown that man could fly in the stratosphere and survive. In the end, the flight of *A Century of Progress* merely taxed scientific expertise and manpower. The very existence of the gondola demanded the presence of scientific experiments as a means of legitimizing the effort: The frame had been provided, now it had to be filled.[165]

Compton, of course, had been wedded to the fair and therefore to the flight, and had to see it through. He enthusiastically embraced it at first, and convinced both his subordinates at Chicago as well as his rival Millikan to come along. Millikan, far more cautious about the effect of man upon his instruments, made them automatic and insisted that they be isolated: man was to be avoided at all costs. Reluctant to admit that he was trapped by circumstances, at one point he rationalized that his participation gave him a chance to send a shielded counter into the stratosphere, since he lacked either the initiative or sufficient funding to do it in a larger unmanned sonde. Millikan and Compton expressed some satisfaction with their results—Millikan, for example, regarded them useful but marginal—but their experience did not make them enthusiastic patrons of manned flight, certainly not in light of their attendant frustration. Neither scientist urged a return of *A Century of Progress* to the stratosphere, although Millikan argued that if manned flights did continue, they should be conducted in other latitudes. Compton began developing unmanned balloonsondes whereas Millikan continued with balloonsondes and further flights with Stevens and Piccard. Both led research teams that concentrated on traditional ground-based cosmic-ray and particle physics research. Millikan, however, still had to clari⌐ʹ what he thought was an unacceptable form for the altitude-intensity curve, which required him to send the same equipment back into the stratosphere when the opportunity arose. He later argued that he continued with manned flights in the Explorer series because the results he and Compton had obtained on *A Century of Progress* had to be either corrected or completed.[166]

At the outset, the prospect of a stratospheric manned balloon flight was closely tied to the image of Auguste Piccard. Jean Piccard tried in vain to assume the central role in the "flight of the century," which was to be a showdown between two American titans of physics. But it was the spectacle of the flight itself, as an American adventure, that drove exposition backers. The personal dreams of Jean and Auguste Piccard were secondary to the spectacle, but this did not deter either Piccard from pursuing their destiny.

Notes

1. Jean to Jeannette Piccard, 10 March 1933, black letterbox of Jeannette Piccard, PFP/LC.
2. Auguste Piccard to F. R. Moulton, 16 March 1932, PFP/LC. This letter was likely prepared with Jean's or Jeannette's assistance.
3. F. R. Moulton to Auguste Piccard, 18 March 1933, PFP/LC.
4. A. W. Stevens to R. A. Millikan, 21 July 1933, RAM/CIT, R23 F22.5.
5. De Maria and Russo (1987), p. 9.
6. Irving Muskat to Auguste Piccard, 16 March 1933, PFP/LC. The FNRS continued to provide generous support for both Piccard and Cosyns. See 3 February 1933 entries in the FNRS Steering Committee minutes files, in annex 1 attached to Paul Levaux to the author 10 August 1988.
7. Auguste Piccard to F. R. Moulton, 16 March 16 1932, PFP/LC; folder 1-1201, "Auguste Piccard," COP/UIC.

8. Jean to Jeannette Piccard, 8 April 1933, black letterbox of Jeannette Piccard, PFP/LC.
9. F. R. Moulton to Jean Piccard, 31 March 1933, PFP/LC.
10. Telegram, Jean Piccard to Moulton, 5 April 1933, Folder 1-1201, "Auguste Piccard," COP/UIC.
11. Later, in the face of Jean Piccard's demands, Dow made it clear that it would donate the gondola but retain full title. See Dow Company to Century of Progress, 2 May 1933, Folder 1-4762, "Dow," COP/UIC.
12. Jean Piccard to Mr. Pickard (du Pont de Nemours & Company), 8 April 1933, PFP/LC. In fact, to Jeannette, Jean admitted that he was ready to sell 2,800-square-meters of the balloon after the flight to Du Pont for $15,000 so that they could cut it up for advertising gimmicks. Du Pont did not think much of this idea either. Jean to Jeannette Piccard, 8 April 1933, black letterbook of Jeannette Piccard, PFP/LC.
13. Unknown in Chicago, however, was the fact that Cosyns was having his own problems getting his gondola ready, even with Auguste's aid. At first they thought of using the old FNRS gondola, but then with considerable FNRS assistance, Cosyns and L'Hoir advisers decided to try to fabricate a magnesium sphere and were just getting started in the new venture. See telegram, Auguste to Jean Piccard, 10 April 1933, PFP/LC; and Crouch (1983), p. 609. On the political context behind the Soviet initiative, see Bailes (1976), pp. 58–59. On continuing FNRS support and the attempted use of magnesium by L'Hoir, see 3 February 1933 entry from the FNRS Steering Committee minutes files, "Attribution d'un subside extroadinaire..." and 31 October 1933 "Examen d'une demande de subside emanant de Mr. Max Cosyns." in Annex 1 to Paul Levaux to the author, 10 August 1988.
14. On the value of the Piccard name, see, for instance, E. Ross Bartley (Director of Promotion) to Irving Muskat, 6 June 1933, COP/UIC.
15. Jean to Jeannette Piccard, 12 April 1933, black letterbook of Jeannette Piccard, PFP/LC. Original spellings.
16. See representative specification and requirement descriptions and research reports in "LTA: Balloon Fabrics," NASM Technical files.
17. See Jean to Auguste Piccard, 3 April 1933, PFP/LC.
18. Crouch (1983), p. 607; Jean Piccard (1933), pp. 30–31.
19. William H. Gross oral history interview, 23 September 1987, pp. 2–5, NASM.
20. Jean Piccard (1933), p. 31; Jean to Auguste Piccard, 3 April 1933, PFP/LC.
21. William H. Gross oral history interview, pp. 5–6, NASM.
22. Ibid., pp. 6–7; Titterton (1937), pp. 200–204. Titterton was more conservative about magnesium's properties: The heating requirement was something of a concern, and his data showed that weld strengths were only 60 percent the strength of the parent, and not the 90 percent or above claimed by Dow engineers or Piccard.
23. William H. Gross oral history interview, pp. 4–6; Titterton (1937), p. 200.
24. Morrill and Rich (1933), p. 4.
25. Titterton (1937), p. 207; William H. Gross oral history interview, p. 7.
26. William H. Gross oral history interview, pp. 4–9.
27. See, for instance Jean Piccard to T. G. W. Settle, 20 April 1933, PFP/LC.
28. See Jean Piccard to R. F. Moulton [sic], 29 April 1933, PFP/LC.

29. A. H. Compton, "Marco Polo 1932," quoted in De Maria and Russo (1987), p. 31.
30. A. H. Compton to R. A. Millikan, 9 May 1933, RAM/CIT, R23 F22.4.
31. A. H. Compton to R. A. Millikan, 14 February 1932, RAM/CIT, R23 22.3. For a description of Compton's ionization chamber, see Compton (1933), pp. 387–403, especially pp. 389–390. Earlier, John C. Merriam of the Carnegie Corporation wrote Millikan, "It would be interesting to have your work and that of Compton fitted together so as to aid and supplement each other." See J. C. Merriam to R. A. Millikan, 18 January 1932, RAM/CIT, R20.17, quoted in De Maria and Russo (1987), p. 42, n. 81.
32. De Maria and Russo (1987), p. 43.
33. R. A. Millikan to John A. Fleming, 6 May 1933, RAM/CIT, R23 F22.4.
34. De Maria and Russo (1987), p. 12.
35. Neher (1985), p. 91; Bowen and Millikan (1933), p. 695.
36. Johnson (1932), p. 665.
37. On Johnson's proposals to verify his suspected east-west effect in 1932, see Merle A. Tuve to John Fleming "Memorandum for the Acting Director," November 15, 1932; and "Memorandum re T. H. Johnson's Cosmic-Ray Telescope Work," December 13, 1932, in Department of Terrestrial Magnetism file #108 "MAT" DTM.
38. See Thomas H. Johnson to John A. Fleming, 6 June 1933, RAM/CIT, F22.4 R23.
39. De Maria and Russo (1987), pp. 11–14, nn. 17; 20.
40. "Cosmic Ray Romancing," New York Times, 18 September 1932, 2:1, reported in Science 76, (1932): 276, and quoted in De Maria and Russo (1987), pp. 47–48.
41. Kargon (1982), p. 156.
42. De Maria and Russo (1987), p. 76.
43. Kargon (1982), pp. 156–157.
44. The name of the craft itself went through a number of transformations, from Piccard–Compton to Fordney–Settle to A Century of Progress.
45. R. A. Millikan to John A. Fleming, 6 May 1933, RAM/CIT, R23 F22.4.
46. A. H. Compton to R. A. Millikan, 9 May 1933, RAM/CIT, R23 F22.4.
47. Numerous historians have studied the Millikan–Compton controversy, notably Kargon (1981, 1982); Galison (1987); and most recently, De Maria and Russo (1987), p. 40, 63ff.
48. R. A. Millikan to John C. Merriam, 16 April 1933, RAM/CIT, R23.
49. See correspondence, RAM/CIT Roll 23, 1933 through 1935; see folder 22.7 on his collaboration with Serge Korff and others.
50. R. A. Millikan to Mark E. Ridge, 20 December 1932, RAM/CIT, R23 F22.3.
51. Millikan to Fleming, 6 May 1933, RAM/CIT, R23 F22.4.
52. Kevles (1974), p. 399.
53. R. A. Millikan to A. H. Compton, 15 May 1933, RAM/CIT, R23 F22.4.
54. R. A. Millikan to A. H. Compton, 24 May 1933, RAM/CIT, R23 F22.4.
55. Millikan to Compton, 3 July 1933, RAM/CIT, R23 F22.4.
56. Charles Sloan to Jean Piccard, 10 June 1933; Jean Piccard to Waldemar Kaempffert, 27 May 1933, PFP/LC.
57. Dempster MacMurphy, "Piccard Flight May End Compton–Millikan Debate on Cosmic Ray Properties," Chicago Daily News, n.d., clipped to Compton to Millikan, 27 May 1933, RAM/CIT, R23 F22.4.

58. Draft speech for Lenox Lohr, 4 August 1933, p. 3, COP/UIC.
59. Irving Muskat to Willard Dow, 27 April 1933, "Dow" folder, COP/UIC.
60. Telegram, Muskat to Jean Piccard, 25 May 1933, folder 1-1201, "Auguste Piccard," COP/UIC.
61. Muskat to L. B. Rock, 31 May 1933, folder 1-3103, "Chicago Daily News, 1933," COP/UIC.
62. E. Ross Bartley (Director of Promotion) to Irving Muskat, 6 June 1933, COP/UIC.
63. J. Piccard to L. B. Rock, 24 May 1933, PFP/LC.
64. See Muskat to Perkins, 19 June 1933; to L. B. Rock, 22 June 1933, folder 1-3103, "Chicago Daily News, 1933," COP/UIC.
65. Jean to Auguste Piccard, 3 April 1933, PFP/LC.
66. A. H. Compton to Jean Piccard, 7 July 1933, RAM/CIT, R23 F22.4.
67. A. H. Compton to E. H. Perkins, 3 June 1933, copy to Piccard; Jean Piccard to Settle, 11 June 1933, PFP/LC.
68. See, for instance, Tex Settle to Jean Piccard, 16 June 1933, PFP/LC.
69. Jean Piccard to O. B. Hanson (NBC), 13 May 1933, PFP/LC.
70. See, for instance, telegram, Auguste to Jean Piccard, 10 April 1933, PFP/LC.
71. See "Programme d'une Nouvelle Ascension," 30 November 1932, in Annex 2, and entries from the FNRS Steering Committee minutes files for 3 February 1933 in Annex 1 to Levaux to the author, 10 August 1988.
72. Jean Piccard to T. G. W. Settle, 22 April 1933. The name was later changed. See Jean Piccard to Auguste Piccard, 17 June 1933, PFP/LC.
73. Telegram, Auguste to Jean Piccard, 26 May 1933, PFP/LC.
74. Muskat to Willard Dow, 27 April 1933, COP/UIC.
75. Muskat to L. B. Rock, 31 May 1933, folder 1-3103, "Chicago Daily News 1933," COP/UIC.
76. Press release, 21 May 1933, "PICCARD," Piccard folder, COP/UIC.
77. Irving E. Muskat to Jean Piccard, 2 June 1933, 13 June 1933; Jean Piccard to Irving Muskat, 8 June 1933, PFP/LC. In fact, Jean had heard as early as May 18 that if Auguste did not sign the contract, the Chicago planners would fly Settle and Alvarez. Jean was bewildered that they would "cut off their noses to spite their faces and do it with Settle and Alvarez without mentioning us." See Jean Piccard to Jeannette Piccard, 18 May 1933, PFP/LC.
78. Jean Piccard and Irving Muskat to Auguste Piccard, "On Board S.S. Champlain due in New York June 21, 1933," (draft), 17 June 1933, PFP/LC.
79. Jean Piccard to T. Settle, 12 July 1933, PFP/LC. See Goetzmann (1986), pp. 150–151.
80. T. G. W. Settle to J. Piccard, 27 April 1933, PFP/LC.
81. Settle to H. B. Henricksen, n.d., 1933. General Correspondence, 1933 INA Folder, RG 167 NBS/NARA.
82. Owens (1985), pp. 186–187.
83. Weiner (1970), p. 33.
84. Jean to Auguste Piccard, 19 September 1933, PFP/LC. Translation by Benjamin Pearce.
85. Alvarez (1987), p. 29. Jean apparently did not get along with anyone. See Stehling and Beller (1962), pp. 222–223.

86. Alvarez (1987), p. 30. Jean's letters to Jeannette regarding Alvarez's role and negotiations with Compton during June are too vitriolic to be useful. According to Jean, Alvarez at one point accepted that he would not fly, while Compton continued to haggle over gate receipts in return for flying his cosmic-ray devices. See Jean to Jeannette Piccard, 22 June 1933, black letterbox of Jeannette Piccard, PFP/LC.
87. Jean Piccard to "A Century of Progress," 21 July 1933, folder 10-10079, COP/UI.
88. T. G. W. Settle to Lenox Lohr, 22 July 1933, folder 10-10079, "Jean Piccard Balloon Ascension," COP/UI.
89. Van Orman (1978), pp. 170–171.
90. Excerpt from Compton telegram, in F. C. Boggs to Rock, 24 July 1933, folder 1-3103, "Chicago Daily News," COP/UI.
91. B. Williams to J. N. Stewart, 12 July 1933, 28 July 1933, folder 1-6463, "Goodyear Tire and Rubber," COP/UI.
92. T. Settle to Jean Piccard, 22 July 1933, PFP/LC.
93. Telegram, Jean Piccard to E. H. Perkins (Dow Chemical), 3 August 1933, PFP/LC.
94. William Dever to the File, 4 August 1933, folder 1-1201, "Auguste Piccard," COP/UI.
95. Draft speech for Lenox Lohr, 4 August 1933, p. 3, "Auguste Piccard" folder, COP/UI.
96. Maravelas (1980), p. 16.
97. Waldemar Kaempffert, "The Week in Science," fragment, in "Balloons—Science and Technology file," NASM Technical Files.
98. Rydell (1985), p. 542; Smith (1983); Maravelas (1980), p. 16.
99. Stehling and Beller (1962), p. 225. See also Don Piccard, "A Chilly Day in 1933," (1933), RR/MSI; Van Orman (1978), pp. 170–178; "Dares Defeat in Flight For Stratosphere," newspaper fragment, n.d., PFP/LC.
100. Willard Dow to "A Century of Progress," 2 May 1933, "Dow" folder, COP/UIC; Willard Dow to Jean Piccard, 29 April 1933, PFP/LC.
101. Don Piccard, "A Chilly Day in 1933" (1933), RR/MSI.
102. Jean Piccard to O. T. Kreusser, 11 August 1933, RR/MSI.
103. J. C. Boggs to Jean Piccard, 8 August 1933, folder 1-1201, "Auguste Piccard," COP/UI.
104. Irving Muskat to A. H. Compton, 11 August 1933, folder 1-1201, "Auguste Piccard," COP/UIC.
105. Compton to Millikan, 12 August 1933; Millikan to Compton, 7 August 1933, RAM/CIT, R23 F22.5.
106. F. C. Boggs to Jean Piccard, 7 August 1933, PFP/LC; folder 1-1201 "Auguste Piccard," COP/UIC.
107. Jean Piccard to Willard Dow, 11 August 1933; Jean to Auguste Piccard, 16 August 1933. PFP/LC.
108. Stehling and Beller (1962), pp. 225–226; E. Perkins to J. Piccard, 6 September 1933, PFP/LC.
109. E. H. Perkins to Jean Piccard, 8 August 1933. Conditions accepted by Piccard noted on copy. See also Jean Piccard to Dr. Harkins, 18 August, 24 August 1933, PFP/LC.

110. Jean to Auguste Piccard, 24 August 1933, PFP/LC. Translation by Renata Rutledge.
111. See L. B. Rock to F. C. Boggs, 8 August 1933, folder 1-3102 "Chicago Daily News 1933," COP/UIC.
112. Representative correspondence concerning the valve may be found in the Poppenhusen, Johnston, Thompson and Cole file maintained by Donald Piccard, which he kindly made available. See for instance Lawrence Cole to Jean Piccard, 25 August 1933; W. Dever (COP) to Cole, 2 September 1933; Cole to Jeannette Piccard, 5 September 1933; Jeannette Piccard to Cole, 6 September 1933; L. W. Baker (Goodyear) to Cole, 6 September 1933; Cole to Jeannette Piccard, 8 September 1933.
113. William Dever to L. B. Rock (Chicago Daily News), 24 August 1933, folder 10-10079, "Jean Piccard Balloon Ascension," L. B. Rock to Goodyear (F.M. Harpham), 29 September 1933, folder 1-6461, "Goodyear Tire and Rubber," COP/UIC.
114. Jean Piccard to Willard Dow, 14 September 1933, PFP/LC.
115. Willard Dow to Jean Piccard, 23 September 1933, PFP/LC.
116. E. H. Perkins to Lenox Lohr, 16 September 1933; telegram, Lenox Lohr to E. H. Perkins, 16 September 1933, folder 1-4762 "Dow," COP/UIC.
117. Dow to Piccard, 23 September 1933, PFP/LC.
118. Piccard to Dow, 26 September 1933, PFP/LC.
119. Jean to Auguste Piccard, 19 September 1933, PFP/LC.
120. Memorandum, n.d., circa 4 August 1933, folder 1-1201, "Auguste Piccard," COP/UIC. The 15,000 tickets agrees reasonably well with popular accounts that stated that 20,000 people viewed the launch and a few moments of flight.
121. Boggs to Roberts, 21 August 1933, folder 1-1201, "Auguste Piccard," COP/ UIC.
122. See Orvil Anderson Collection, MAFB 168.7006-10, microfilm ed., R1933–1935, F1151.
123. F. C. Boggs to E. H. Perkins, 5 September 1933, folder 1-4762, "Dow," COP/UI.
124. A. H. Compton to Irving Muskat, n.d., circa 20 August 1933, folder 1-3732, "Compton," COP/UIC.
125. For reviews of the Soviet flight of 30 September, see "The Soviet Stratostat," *Flight* (December 21, 1933), pp. 1287–1288; Philp (1937), pp. 69–72; Crouch (1983), pp. 609–610. On Stalin's agenda, see Bailes (1976).
126. L. B. Rock to Lenox Lohr, 2 October 1933, folder 1-3102 "Chicago Daily News 1933," COP/UI.
127. Compton to Millikan, 26 October 1933, RAM/CIT, R23 F22.5, p. 2.
128. Alvarez (1987), p. 30.
129. F. C. Boggs to E. H. Perkins, 22 September 1933, COP/UIC.
130. Lawrence Cole to Jean Piccard, 17 November 1933, Don Piccard Papers.
131. See "Settle Packs Up His Balloon; May Fly From Akron," *Chicago Tribune* (13 November 1933), clipped to Lawrence Cole to Jean Piccard, 13 November 1933, Donald Piccard Papers.
132. F. C. Boggs to Settle, 15 November 1933; Fordney to Boggs, 16 November 1933; J. N. Stewart to Boggs, 15 November 1933; Boggs to Goodyear-Zeppelin, 14 November 1933, folder 1-6460, "Goodyear Tire and Rubber," COP/UIC.

133. Lt. Comdr. T. G. W. Settle to Chief of the Bureau of Aeronautics, 27 November 1933, "Report on Stratosphere Balloon Flight," Orvil Anderson Collection, MAFB, copy in Don Piccard files. See also Stehling and Beller (1962), p. 228; Philp (1937), p. 73.
134. Mrs. Jean Piccard to William Rosenfield, 28 November 1933, PFP/LC.
135. Jean to Auguste Piccard, 1 December 1933, PFP/LC.
136. Jean Piccard to E. H. Perkins, 9 December 1933, PFP/LC.
137. W. G. Quisenberry, "News Report," quoted in Van Orman (1978), p. 167.
138. "Into the Stratosphere," *Aviation 32*, (December 1933), p. 383.
139. "Temporary Receipt, #1283," 1 December 1933, RR/MSI.
140. Lt. Comdr. T. G. W. Settle to Chief of the Bureau of Aeronautics, 27 November 1933, "Report on Stratosphere Balloon Flight," Orvil Anderson Collection, MAFB, copy in Don Piccard files.
141. Compton (1934b), pp. 79–81.
142. Compton and Stephenson (1934), p. 564.
143. Compton (1936), p. 1130.
144. Bowen, Millikan, and Neher (1934), p. 646.
145. Ibid., p. 645.
146. Erich Regener to R. A. Millikan, 10 November 1934, RAM/CIT, R45 F42.11; Rossi (1985), pp. 68–70.
147. Millikan to Serge Korff, 24 January 1934, RAM/CIT, R23 F22.6.
148. Millikan to Albert W. Stevens, 26 March 1934; to Thomas W. McKnew, 22 May 1934, RAM/CIT, R23 F22.6; F22.7.
149. Jean to Auguste Piccard, 12 December 1933, PFP/LC. Translation by Renata Rutledge.
150. Compton (1934b), pp. 79–81.
151. Pupin's remarks to exposition backers, quoted in Rydell, (1985), p. 530.
152. Floyd Karker Richtmyer, quoted in Rydell (1985), p. 530.
153. Jean to Auguste Piccard, 1 November 1933, PFP/LC. Jean had read in the *New York Times* that Auguste was to receive the prize for 1933. He was nominated for physics in both 1932 and 1933 by Léon Counson and Franz Cumont, respectively. Werner Heisenberg, Erwin Schrödinger, and P. A. M. Dirac won in those years. See Crawford, Heilbron and Ullrich (1987).
154. Rydell (1985), pp. 525–526.
155. Smith (1983), pp. 179, 192.
156. Kevles (1978), p. 237.
157. See, for instance Jean Piccard to Willard Dow, 11 August 1933, PFP/LC.
158. Hollinger (1984), chap. 5, p. 145. See also quotations from Chamberlain, pp. 148, 153.
159. After considerable delay, which included a real setback when the L'Hoir magnesium gondola exploded and burned during laboratory tests in August 1933, Cosyns finally flew with Vander Elst in a new aluminum sphere on 18 August 1934 from Belgium to an altitude of just over 16,000 meters. After this, the FNRS decided to close out its support of stratospheric flight. The old Piccard–Cosyns balloon that had flown three times into the stratosphere burned in a fire in 1937. See 31 October 1933 entry from the FNRS Steering Committee minutes files "Examen d'une demande de subside emanant de Mr. Max Cosyns," in Annex 1, and "Les Ascensions Stra-

tospheriques et le Professeur Auguste Piccard," n.d., in Annex 2 to Paul Levaux to the author, 10 August 1988.

160. The *Star of Poland* was to have a 113,000-cubic-meter balloon constructed by the Military Balloon Plant at Legjonowo, Poland. It was damaged by a fire that erupted during its first inflation on 14 October 1938, and was completely destroyed one year later when Warsaw itself was bombed by the Nazis. See Auguste Piccard (1950), p. 107; Albert Gilmor, Military Attaché, U.S. Embassy, Warsaw, Poland to Chief, M.I.D., WD, Washington, D.C., 9 January 1935, Piccard Folder, NASM Technical Files; and Philp (1937), pp. 200–201. Argentine scientists also were influenced by Auguste Piccard. Working with Vander Elst, a pupil of Auguste's, who had flown with Cosyns in 1934 and had performed liaison with the Polish effort in 1938, scientists at Argentina's Cosmic Physics Laboratory at San Miguel in collaboration with the National University in Buenos Aires and 11 other scientific and military groups in Argentina, Peru, and Chile, planned to construct a 2.5-meter aluminum gondola in three sections, an exact copy of Piccard's *FNRS*. They were hoping to perform cosmic-ray experiments, as well as sky brightness and ozone studies, but never got the project off the ground. See *Primera Ascensión Estratosférica* (ca. August 1939), "Balloons-Science and Technology," file A2003600, NASM Technical Files. My thanks to Jose Villela for helping with this document.

161. According to news accounts previously noted, the first Soviet flights were conducted by a combined military crew and civilian scientists. At one point, however, Auguste Piccard and FNRS colleagues claimed that the first Soviet flight was all military and did no science. See 31 October 1933 entry from the FNRS Steering Committee minutes files "Examen d'une demande de subside emanant de Mr. Max Cosyns," in Annex 1 to Paul Levaux to the author, 10 August 1988.

162. Those secondary sources scanned include Galison (1987); Kargon (1981, 1982); Kevles (1974; 1978); De Maria and Russo (1987); and Seidel (1978). Ziegler (1986) has discussed Piccard's flight, and scientists who provide reviews sometimes note the flights. See Pfotzer (1974, 1985); Sekido and Elliot (1985).

163. See archival folder, "Model C Electroscope," in Department of Terrestrial Magnetism Files, courtesy Louis Brown; Seidel (1978), p. 284.

164. See Knight (1986); and *Handbook of the Century of Progress 1934*, pp. 28–31, folder 16-219, COP/UI.

165. See Smith (1983), p. 183.

166. Millikan (1935), p. 24.

CHAPTER 4

Continued Dreams of Jean Piccard

FIGURE 4.1. Jeannette and Jean Piccard peeking out through gondola manhole for reporters. Courtesy of the Museum of Science and Industry, Chicago.

W. F. G. Swann and the Bartol Connection

In October 1933, divorced from the proceedings in Chicago and wondering about the future from his home in Marshallton, Delaware, Jean Piccard attended a lecture by the physicist W. F. G. Swann, hoping to meet him. Swann was an active cosmic-ray physicist, and, as director of the Bartol Research Foundation of the Franklin Institute in Philadelphia, was a potentially valuable scientific patron. Bartol was close to Jean and Jeannette Piccard's home and just possibly, Jean might find a professional home there.[1] After Swann's public demonstration of the wonders of the cosmic ray, Jean sought him out and asked if he could visit Bartol. By December, Jean had invited himself into Swann's laboratory and was delighted to find himself accepted by the congenial physicist, who seemed willing to provide research space and a collegial atmosphere, if not a salary. A grateful Jean Piccard assured Swann that he would organize his time in Marshallton "so that I can do some real work."[2] Swann endorsed Jean Piccard to his board as a temporary unpaid research associate of the Bartol Research Foundation, arguing that Piccard was collaborating in the scientific agenda of the institution. This position provided the status Jean so much desired and enabled him to use Bartol facilities to prepare devices for his balloon flight, such as the life-support systems.[3]

Jean was most fortunate to find William Swann. A sympathetic, well-established teacher of such physicists as E. O. Lawrence and J. Beams during his tenure at Minnesota, Chicago, and Yale, and now leader of a growing experimental group (and an ardent popularizer, as his demonstrations of cosmic-ray penetration inside a New York Bank vault for a CBS Radio audience attest), Swann was a ready supporter of Jean's dream of a flight into the stratosphere.[4] In fact, Swann was quite familiar with balloon technology and approved of its scientific use. He had studied problems of fire prevention and static electricity in balloons and dirigibles for Millikan's Signal Corps group in 1918, and equipped balloons with simple electroscopes to determine the rate of electrical discharge of the balloon system itself.[5] After World War I, Swann continued to be interested in manned flight and the problems of static electricity in dirigibles, and was quick to see their value in cosmic-ray research.[6] Swann thus became a major scientific underwriter of Jean Piccard's balloon ventures. He endorsed the flight in order to secure the support of commercial interests in Detroit, and Jean found him most helpful in convincing Millikan to cooperate.

Thus as both *A Century of Progress* and *Explorer* flew in late 1933 and 1934, Jean and Jeannette Piccard worked from their base at Bartol and from their Delaware home nearby to gain sufficient patronage to refly the exposition's balloon and gondola. Their entrepreneurship and subsequent success stands in marked contrast to the institutional efforts of the Army, Navy, and National Geographic. That the Piccards succeeded at all was due to their enormous persistence, and as we shall see, considerable confidence, pluck, and luck.

The Piccard Rhetoric

Jean Piccard never gave up hope that he would fly in the stratosphere. As already mentioned, after Settle and Fordney flew from Akron in November 1933, the Piccards gained title to the balloon and gondola. All the while, Jean and his wife Jeannette were planning for a flight in the summer of 1934. By the end of 1933, they were deep into plans for refurbishing the balloon and gondola, and were searching for patronage. Auguste encouraged their effort, wishing his brother and sister-in-law well in their quest for glory: "Hopefully you will make your flight ahead of other competitors. It would be nice, if the name of Piccard through Jeannette, would once more be placed on the record list of the F.A.I."[7]

However, many obstacles had to be overcome before they could fly. Not only did they have to refurbish the balloon and gondola, but Jeannette had to obtain a proper pilot's license. Jeannette would need training and many hours of practice flying before she could legally take the balloon up. Energetic and forceful, she seemed to have a better chance of obtaining a pilot's license than Jean, who was preoccupied with restoring the gondola and balloon and convincing scientists to provide instruments to fly. They certainly did not trust anyone else. Auguste's stated desire to see the Piccard name restored in the record books hints at the true motivation of all three Piccards.[8] But Jean and Jeannette were also driven by bitterness, directed largely at everyone associated with the Chicago flights. Jean gloated over Compton's difficulties before the successful November 1933 flight, and even after the flight Jeannette was quick to claim that it had been a failure, and had harmed their own chances: "It has lowered, we feel, the general interest while it has accomplished nothing for science that was worth the effort."[9]

Subsistence and employment problems also caused anxiety. By the end of 1933 Jean had lost his position with the Hercules Powder Company, which put an unhappy end to an ever-worsening relationship. This only made him all the more determined to obtain subsistence through patronage for their stratospheric flight. As a result, both Piccards tried to be all things to all people. When Millikan argued that "one single flight in [an equatorial] region would give more important data than could be obtained from a hundred flights in this latitude," Jean was the only one to say he would fly anywhere Millikan wanted, if funding could be found. He would "do anything which I can reasonably do for a scientific experiment on cosmic rays." This included flying from Brazil or even Mexico City "if a closer study would show that a landing would be safe."[10] Both Piccards agreed to fly from Detroit without hesitation when funding appeared there. Their contact in Detroit was a balloon enthusiast named Ed Hill who had been a Gordon Bennett race winner in 1927 and ended up working in a truck factory in Detroit.[11] Through Hill, Jean and Jeannette were introduced to Detroit, and eventually obtained backing there, and Jeannette ultimately obtained her pilot's license there, under Hill's guidance.

Although they had to struggle to secure support for refurbishing the old *A Century of Progress* gondola and balloon, and flying it, they did manage to obtain enough pledges from the Detroit Aero Club and actual funding from the People's Outfitting Company of Detroit and the Grunow Radio Company to begin working toward a summer 1934 flight. But the initial funding from these local business interests fell far short of what they would ultimately require. Jeannette therefore redoubled her efforts at entrepreneurship. She designed and had prepared commemorative stamps, souvenir programs, and folders for sale, and set up a lucrative news release agreement with the North American Newspaper Alliance.[12] Jeannette knew well the rhetoric of patronage, and was an effective spokesperson for their ventures:

When we make our flight in the Summer we will not attempt a new altitude record. We will try chiefly for scientific results and will try only to reach a height that will enable us to get such results. Altitude records have their place in the development of aviation. They are splendid achievements. To us, however, altitude is only a means to an end. The end itself is pure science, the search for Truth, from which practical science derives.[13]

Jeannette regaled her potential patrons with the many problems that had to be overcome, all of which required funding and aid which was not forthcoming from big firms like Goodyear. But this only strengthened her resolve, as she pointed out in ending her January 1934 letter to Mrs. Ulysses Grant McQueen: "The greater the difficulties, the more interesting the conquest!" And when asked by her father in June 1934 why she persisted, what her "*real* reason for going on" with ballooning was, she admitted that:

There are many reasons, some of them so deep seated emotionally as to be very difficult of expression. Possibly the simplest explanation is that we got started along this road and because I am I, I cannot stop until I have won."[14]

In contrast to Jeannette's rhetoric and passion, Jean admitted to Auguste privately:

I do not want to become a specialist in cosmic rays through the use of Geiger tubes. That would require a much deeper knowledge of the matter. It is useless to work in a field where one is handicapped from the start. On the other hand, photography would not require anything new from me.[15]

The "difficulties" that fascinated the Piccards were not those associated with doing the best possible science. Geiger counters were showing great promise, but were less tractable than ionization chambers. Yet Jean was unusually resistive to anything that might distract him from his primary interest: that of building the vehicle and making the voyage. This stands in stark contrast to the rhetoric that Jean prepared the previous September for Willard Dow's consumption, which suggested that his only wish was to be of service to humanity.[16] Jean wanted to leave all matters of science to his scientific collaborators, such as Swann and Millikan. As far as the science was concerned, Jean avoided making any additional effort needed to understand the new techniques of scientific observation.

Throughout the winter, Jean prepared special flight instruments for the old *A Century of Progress*, including devices for detecting the acceleration of the balloon during flight, aspect indicators, and improved valves. Swann helped Jean develop his flight plan to ensure a reasonable scientific return, and even went so far as to intercede eventually with Dow on Jean's behalf to try to persuade the firm to provide the additional services Jean thought were due him.[17] However, there were serious differences of opinion.

Problems from the Start

Difficulties arose immediately. As we saw in Chapter 3, they were largely the result of misunderstandings about support for rebuilding the gondola and balloon. Jean claimed that the "A Century of Progress" and its backers had agreed to bear all costs of preservation, shipping, handling, and repair; the *Chicago Daily News* and Goodyear thought otherwise. Jean even quarreled with the Dow Chemical Company. Dow had agreed only to restore and test the gondola, which turned out to be a simple job, but when Jean demanded alterations and additions to integrate the experiments of Swann and Millikan, Dow balked.[18] Jean and Jeannette persisted with contentious replies, and finally in exasperation the Dow staff agreed to some of their demands in August 1934, but only after the Piccards signed a document stating that they would make no more demands after this.[19] Predictably, Piccard felt cheated by Dow. He believed with all sincerity that Dow had benefited through their relation, complaining to Swann: "I really had believed that the introduction of Dowmetal into the physical laboratories of several colleges had been a good thing for them." The Dow staff indeed worried about the image of Dowmetal: but they feared that the inexperience of both Piccards could spell disaster, and nobody wished that type of publicity.[20]

When the National Geographic Society and the Army Air Corps announced *Explorer* in January 1934, Jean knew that he had to keep Willard Dow aware that he was still in the race. In order to keep Dow happy, Jean wrote to congratulate him upon receiving a contract for the Explorer gondola, and predicted that his firm "will [soon] have many orders for commercial stratosphere planes," adding that "there is not the slightest reason that the stratosphere, once opened to human beings, should remain only the field of free balloons."[21] To Jean, Dow had the lead in its experience with using Dowmetal for stratospheric flight, now that they had the contract for *Explorer* as well. But since his own gondola was on exhibit in Chicago through the winter and spring by agreement, Jean also nagged Dow to restore it as soon as possible. He became especially anxious when he learned that his gondola would have to wait until the Explorer gondola was completed in the Dow shops, which would be not before June 1934.[22] This delay aggravated Jean, but was quickly overshadowed by larger problems: insufficient funding, the license Jeannette needed, and Goodyear's growing unwillingness to restore the balloon.

Determining who was responsible for what concerning the balloon restoration was not simple. As with the valve episode, haggling continued between Lawrence Cole, Jean's lawyer in Chicago, and officials from the *Chicago Daily News* and from Goodyear through 1933 and into 1934 over every step of the restoration, from its retrieval from the New Jersey marshes to its evaluation for flightworthiness. Jean insisted that the balloon envelope be shipped to Detroit so Ed Hill could evaluate the damage and determine what Goodyear had to do to make it right. Hill found that it was in good shape generally, but that many rips had to be repaired. But as legal fees, fire and liability insurance premiums, and a host of other incidental fees continued to pour in, both Jean and Ed Hill reversed their opinion of the state of the balloon, and by July, 1934, decided that it had been badly handled at the New Jersey landing site, and therefore the *Chicago Daily News* was liable for repair costs up to $3,000. Denials, threats, and innuendo followed, but finally, by the end of July, the balloon was shipped to Goodyear for repair, partly underwritten by the *Daily News* and partly by Piccard. The restored balloon reached the Ford Airport in Dearborn in late September.[23]

In spite of their problems, both Jean and Jeannette appeared upbeat in public and seemed confident that they would overcome all these difficulties. True entrepreneurs, they invested both their time and what little savings they had into their dream, encouraging others to think of what their flight would mean for long-distance commerce and for science. In glowing prose, Jean predicted that long-distance commercial flight would take place in the stratosphere. The mysterious stratosphere would be converted into a "superhighway" once forays with manned balloons taught aeronauts how to survive there, and how to make machines that could fly there safely and cheaply.[24]

The Collaboration and Sponsorship of Swann and Millikan

One person they did not have to convince was W. F. G. Swann. His longtime interests in atmospheric electricity and in relativity came together in his attempts to explain the behavior of cosmic rays, which he first regarded as photons but then as charged particles, most likely electrons after Compton's work. Walking a thin line between giants, Swann incorporated elements of both Compton's and Millikan's theories and observations to argue that sufficiently high-energy electrons moving at close to the speed of light would appear to behave like photons and pass through the earth's atmosphere to depths greater than would be normally expected from charged particles moving with lower energies. In May 1933, Swann pointed out, "if we accept the arguments of Millikan and Compton together, they practically demand the assumption of charged particles of a non-ionizing character . . ."[25]

Although an experimenter, Swann was prone to both mathematical and physical theorizing. In developing his non-ionizing primary model in 1933, he proposed that electrons, accelerated by the intense magnetic fields such as might be found on the surfaces of stars more active than the sun, were a possible source of cosmic rays. As star spots or stellar flares developed, their rapidly changing magnetic fields could accelerate charged particles into space.[26] A critical part of Swann's theory was the assumption that primary electrons had to possess a certain level of kinetic energy to become non-ionizing. His theory was based on observations of secondary particle energies by his Bartol staff during mountain expeditions, which led him to conclude that the threshold was above 10^{10} electron volts.

Although Swann's theories found little support in the physics literature of the time and Swann quietly dropped his idea of non-ionizing particles when the muon was discovered, his views constituted a response to what seemed to be a crisis in quantum mechanics and did influence the design of his instruments for the Piccard and Explorer flights. The problem was that the rules of quantum mechanics could not explain cosmic-ray behavior, especially how deeply they could penetrate matter, if they consisted of the known fundamental particles.[27] To complicate the matter, as Swann and others began to realize, much of the hard (or high-energy) component of the cosmic-ray spectrum that they were observing did not consist of the primaries themselves, but of secondaries produced by collisions of primaries with particles in the earth's upper atmosphere. If physicists could determine how secondaries were produced, this would reveal the character of the primaries. In addition, Swann felt strongly that his non-ionizing particles had to be incorporated into a general scheme of atmospheric absorption: At some point in their passage through the atmosphere the non-ionizing relativistic electrons had to become ionizing.[28] Swann's idea was both abstruse and unique; most physicists at the time speculated that two particles might be involved: a new one producing cosmic-ray "showers," and high-energy electrons that penetrated deeper than the laws of physics allowed. Swann thought that he could describe what was happening in terms of one particle.

Swann came to this concept in stages, after analyzing the observations of his Bartol associates. Thomas Johnson, working with J. C. Street, had used large Geiger counter arrays on a mountain top to show that the total contribution to the charged particle cosmic-ray flux may come from large bursts known as Hoffmann Stösse.[29] These Stösse looked like explosions when caught in photographic emulsions: One high-energy particle somehow exploded, or burst, into tens or even hundreds of lower-energy particles. As Peter Galison has pointed out, quantum theory could not account for such behavior, but most physicists at the time thought these bursts could explain what had been observed on the ground as "showers" of cosmic rays, with dozens arriving at the same time.[30]

Swann wanted to know how much energy was bound up in the largest Stösse, which he assumed were to be found at the highest levels of the atmosphere. But the Stösse detector that Swann developed was far too bulky to be carried on a balloonsonde. The collecting chamber had to be quite large to capture the entire event, and so consisted of a large pressurized spherical nitrogen gas tank, two electroscopes connected to electrodes within the tank but separated from it by containers holding 20 kilograms or more of lead shot, and photographic cameras for recording the movement of the electroscope fibers.[31] Unlike most experimenters at the time, who turned to cloud chambers and plate stacks to analyze the structure of the multiple-particle beams, Swann designed the Stösse chamber to determine the frequency and energy range of the events themselves. In particular he hoped to discover the maximum energy for a Stösse event, which would indicate the energy level at which non-ionizing particles became ionizing and thus produce what he and others thought were showers.

Swann originally intended to have his Stösse chamber fly on *Explorer* in the summer of 1934, but it was removed from the flight at the last moment because it was too large. He then asked two other Bartol physicists, C. G. and D. D. Montgomery, to take his chamber to Pike's Peak for a season of observing. By the end of the summer they had concluded that Stösse production depended on altitude, but they also found that the nature of the dependency was extremely difficult to reconcile with the corpuscular theory if the primary particles had the mass of an electron or a positron. Burst frequency increased with height faster than did the cosmic-ray intensity itself.

The Montgomerys' burst observations drew Swann's attention to Stösse. He reported that their work had "a most profound bearing upon the nature of cosmic-ray phenomena, and has led me to formulate a new theory of [the] cosmic-ray mechanism" which in this case was a new set of special assumptions and conditions created to reconcile his original theory with the new observations.[32] He had originally argued that primaries lost energy in two ways: (1) the usual losses due to ionization once their energy was reduced by passage through the high atmosphere

FIGURE 4.2. Swann's Stösse chamber, intended for the Piccard and Explorer II flights. Boxes marked A are cameras. NGS.

and (2) an additional loss, "which was proportional to the energy of the primary ray itself and which was responsible for the production of secondary rays among which we may include groups of rays or what are now called showers."[33] The Montgomerys' observations indicated that the largest Stösse events should reside at stratospheric levels in the atmosphere, and predicted what their energy range should be. Swann altered his proportionality assumptions to fit, but needed confirming observations at high altitudes; if he could measure the energy of the very largest Stösse events, he would know at what energy levels his primaries became ionizing.

Much of Swann's efforts on Piccard's behalf centered around determining how best to plan Piccard's flight profile to gather the type of data Swann required. He knew that his cosmic-ray instruments had to remain at specific altitudes to measure the soft component of the radiation spectrum, and he also hoped to extend the Montgomerys' burst rate curves. In addition, Swann wished to obtain accurate high-altitude data on the east-west effect, a special interest of Thomas Johnson. He also hoped to refine the information on the variation of cosmic-ray intensity with height in the vertical direction, compared to its horizontal intensity. He and his Bartol staff therefore prepared a wide array of detectors in addition to his Stösse chamber. They built an array of Geiger counters set in a series of coincidence circuits which were virtually identical to those he prepared for *Explorer*. Swann, as the director of Bartol, was probably one of the few people who could afford to produce multiple sets of instruments for these manned flights while supporting the usual laboratory, field, and mountain observations that continued to be their mainstay.

One of the most telling characteristics of any stratospheric adventure, in order to evaluate its usefulness for cosmic-ray studies, was its flight profile. Just as ship captains and their scientist/passengers often could not agree on an itinerary, the former loath to linger at places long enough to collect comprehensive data on local flora and fauna, the flight profile for piloted craft depended on the objective of the flight. Along with his contemporaries in Chicago, the Navy, and the Army Air Corps, Jean Piccard knew that the combination of balloon filling ratio and ballast complement required for maximum altitude produced a quick ascent, which was not what the scientists wanted. Swann, Millikan, and the other scientists needed to gather sufficient amounts of accurate data at all levels in the atmosphere, especially if they wished to obtain reliable intensity curves. With instruments that had to be recharged time after time and that could be affected by secondary production from the metallic skin of the gondola, there was good reason to calculate the amount of time the instruments would have to tarry at each altitude in order to obtain reliable readings. Millikan had not been satisfied with the rapid ascent of the *A Century of Progress*, and felt that only the data returned from the maximum altitude were reliable. As a result Swann assumed the responsibility for calculating the look-time requirements for the Piccard flight. He first calculated the expected secondary production of Dowmetal and determined, to his dismay, that accurate look-times were going to be on the order of two days "before the accuracy sacrificed by the statistics of the problem would be less than the absorption effect of 2 millimeters of Dowmetal."[34] For this reason, Swann argued that Piccard should not worry about a system of fans to rotate the balloon and gondola: "It would be best to plan arrangements for keeping the balloon as far as possible in one assigned direction." Swann also knew that the Explorer gondola was going to be rotated for other experiments, so that this type of maneuver did not have to be attempted on Piccard's flight as well. Piccard, however, wanted to perform rotation maneuvers because his brother did it. Jean remained insistent on this point, and went through trials in an open basket flight with Jeannette in mid-May using small fans to rotate the baskets.[35]

As spring turned into summer, and as *Explorer* established a new world's record and demonstrated the dangers of flight when its huge hydrogen balloon exploded, the Piccard flight plans moved into the fall. Swann tried to cheer up Jean and Jeannette in June noting that one advantage of the delay was that "the way is tremendously smoothed for the preparations for your flight, because we know what we want to do and how to do it."[36]

The lessons Swann learned while preparing for the Explorer flight would serve him well, and his own enthusiasm for such activity carried him without hesitation through the Explorer crash (see Chapter 5). Tom Johnson, now assistant director at Bartol, admired Swann's devotion to manned flight. Hearing of the crash of *Explorer*, Johnson wrote to Swann asking for news of the progress of his work with Piccard, adding: "I certainly admire your spirit in going on with the work after the discouraging circumstances of the Stevens flight."[37] Swann's enthusiasm and support for manned flight carried through the crash as well as any momentary disappointment from the cancellation of his Stösse device. While awaiting the flight of *Explorer*, after he knew that his Stösse experiment was canceled, Swann wrote to Jean Piccard from the South Dakota launch site saying that he still fully supported Piccard and that he would like to fly the chamber in Piccard's flight.[38]

Swann's enthusiasm stood in marked contrast to Millikan's stiff demands on Piccard. Millikan remained diffident when, while waiting for the Fordney–Settle flight to take place, he was courted by both Albert Stevens and by Jean Piccard. Curiously enough, Millikan, despite all his bluster and his many other research commitments around the world, still agreed to supply both with instrumentation. A number of things may account for this. First, as we shall see in Chapter 5, although he had little personal use for manned scientific ballooning, he wished to promote Caltech initiatives in military weather reconnaissance as well as new campus programs in meteorology and aerodynamics that would serve the aircraft industry, from which he hoped to attract both students and patronage from the military.[39] Millikan thus cooperated with Stevens and the Army Air Corps for political reasons, but the case for why he supported Piccard, who had no military backing and was at odds with the military, is quite another matter. Millikan was a voluble showman receptive to private initiative. He relished the attention and notoriety given him by such exploits, which might bring added support for his own programs. Funding for science was still very hard to come by, even at Caltech, which was comparatively untouched by the ravages of the Depression, and these manned flights brought in useful publicity.[40] Publicly, Millikan planned to use Piccard's flight first to check on Settle's flight, and then to confirm whatever data they obtained on *Explorer*. But most of all, Millikan wished to be in control, something he did not enjoy when he flew with Compton.

In trying to be all things to all people, Piccard from the outset planned an ambitious agenda, although it never approached that of Stevens's *Explorer*. In addition to the instruments that Millikan would provide, he agreed immediately to include Swann's heavy Stösse chamber for detecting bursts, which required the same amount of lead as Millikan's shielded apparatus, some 270 kilograms worth, as well as ranks of Geiger counters for directional information.[41] This strategy backfired with Millikan, who charged that Piccard was not aware of the complexities of trying to cram a great amount of equipment into the gondola. Millikan demanded that Piccard reduce his instrument payload, or he would pull out, and that in any event, Swann's heavy Stösse chamber was an unnecessary duplication: "All the data on 'bursts' which can be obtained at all will in my judgment be obtained from our recording instruments." Millikan added that his chambers were more appropriate for balloon work. What Swann had in mind, Millikan correctly pointed out, was best left to longer-term observations from mountain tops where "one can observe for hours or days at a time."[42] At one point, Millikan insinuated that if Swann was going to go ahead with his Stösse chamber and counters, he should be given the whole pie and Millikan could happily find other things to do. He wrote to Swann that he had little faith that Piccard "could gain proficiency with recording electroscopes" quickly enough, and that "if we both try to do it all we shall load your balloon so that we will not any of us get anything."[43]

In a response carefully crafted by Swann, Jean Piccard was able to win Millikan over. He argued that the planned experiments were unique and did not duplicate each other. More important, the lead shot required for one experiment could be used for all of those that required it if the integration design was planned correctly; the Stösse experiment in particular required a dense medium that would produce secondary showers, and the lead itself would later be used as ballast. Finally, the total amount of lead ballast required for the shielded detectors was still less than what was required for control of the balloon, so that no altitude would be sacrificed.[44] The most Piccard got out of Millikan in January was that if his chambers were available when Piccard was ready to fly, they would be given over.

Three days before the fateful Explorer flight in late July 1934, Millikan wrote Piccard asking if his plans were still alive. He had not heard from Jean since early in the year. Millikan felt that, in any case, his three instruments, now in Rapid City under the care of Victor Neher—since "I was very anxious to have no slip-up of any kind on the technical preparations"—would probably be available after they flew on *Explorer* if they did not have to be modified or repaired.[45]

Jean still had not secured all the funding he hoped for when *Explorer* crashed, and complained to Millikan that, while Stevens was able to gather in over $50,000, "We have received up to now only $3,500." Jean added: "Would it be possible for you to find any one sufficiently interested in cosmic rays to make a subscription to our flight? We are still about $4,000 to $5,000 short."[46] Millikan was now unable to provide his electroscopes, which had perished in the crash, and, although willing to provide one unshielded instrument that was sitting around, became quite incensed by Piccard's nosing around for funds when everyone else was concerned with the implications of the crash itself:

As to the financing, as in the case of the Stevens and Settle flights, we are putting in all we can in the supplying of the instruments. If those who are interested in the flight from the sports and the publicity angles cannot do the financing for this mode of approach to the stratosphere problem, we can get the most important elements of the scientific problems involved so far as cosmic rays are concerned from the method which has already given us good results, namely, the sending up of light automatic recording instruments by sounding balloons, and shall have to turn to this method for our further data.[47]

Millikan would only go so far with Piccard, or any of the others advocating manned flight for science. It is clear that he did not share Swann's fascination with the enterprise, nor his general sympathy for Piccard. Millikan's rebuke, however, meant little if funding was not found. Through the fall, while funding problems loomed, Jean and Jeannette Piccard continued to follow every available Detroit lead while they prepared for flight.

Readying for the Flight: The Wrinkles Deepen

In desperation, Jean wondered if the Franklin Institute might fund the flight if he gave the gondola to Philadelphia instead of Chicago. Lawsuits and loyalty prevented this, so he asked his brother if the old *FNRS* might be displayed at the Franklin, which might shake loose some cash support. Auguste would have none of this either.[48] So Jean continued his search, and meanwhile tried to be as frugal as possible by scrounging everything. One of his best sources was Albert W. Stevens, the driving force behind the Explorer flights. As we shall see in Chapter 5, Stevens and Piccard freely exchanged designs for life-support devices and valving systems, and on several occasions Stevens agreed to pressure-test devices Piccard sent him, and provide miscellaneous hardware.

During August, hoping that funds would materialize and doing all they could to ensure that they would, Jean and Jeannette prepared logistical charts looking for places where they might economize or employ volunteer labor. Everything was to be housed at or near the Dearborn Inn close to the Ford Airport southwest of downtown Detroit, and there had to be personnel available for "mob control."[49] Jeannette contacted possible patrons, including members of the Du Pont family, and fought with vendors when memorabilia she had contracted for did not meet her standards or schedule. Jean tended to last-minute design problems, mainly the valve for the old Chicago gas bag, and getting the bag itself into flyable condition.

The newly refurbished and repainted gondola now showed no traces of its "A Century of Progress" history. Sitting at the Dow plant in Midland, Michigan, it was now all white and proudly displayed the Piccard name with blank spaces for patron identification. In mid-August, E. H. Perkins still did not know when the Piccards wanted it, and confided to Earnest Teberg of the Chicago Museum of Science and Industry that the repainting had reduced the gondola's historical value.[50] The Dow Chemical Company was still reluctant to furnish new load rings, drift rings, cupboards, air locks, drag ropes, and seats to Piccard's specifications, as they felt that the refurbishment, painting, and testing were all that had been agreed to.

More wrinkles came when Millikan decided that he had obtained a satisfactory record from the first Explorer flight. Millikan cabled Piccard that further data were "unprofitable," and that he saw no reason to continue, but since he had agreed to fly, he would do so if Piccard insisted. Piccard insisted.[51] But the edge was off Millikan's interest, and Piccard knew it was only a matter of time before he might change his mind. Piccard also knew that his brother's design for the connecting ring and rope system between the gondola and balloon was considered dangerous by Arnstein and his colleagues at Goodyear-Zeppelin in Akron. Jean felt secure enough on this latter point to tell Swann that the Akron people were being very stubborn and would soon see his way.[52]

Matters brightened when A. W. Winston of Dow's metallurgical laboratory fabricated the ancillary parts Piccard required, at no cost.[53] Although the gondola was a reality, money was not, so Jean wrote articles for the North American Newspaper Alliance which helped to fill the gap.[54]

The Flight in October 1934

The gondola and all instruments were at the Ford Airport by September 15. Swann, with an assistant, busily integrated their cosmic-ray devices into the gondola while Ed Hill made the final decisions about ground operations. And while Jean did what he could to keep track of the balloon, funding for its restoration and for preparing the Ford Airport launch site still remained uncertain. Even with last-minute donations from several individuals, and a considerable amount of volunteer manpower from local enthusiasts, income was still considerably below expenses. The airport itself provided free housing and some services, the Edison Company donated and installed the ground anchoring system, and 700 cylinders of hydrogen were supplied at low cost by the Burdett Oxygen Company. Finally, clouds lifted when Grunow Radio advanced the difference (of several thousand dollars) against the revenue expected from public exhibitions after the flight.[55]

A first flight attempt early in October was canceled owing to poor weather. After another aborted attempt on October 22 when the weather still proved uncooperative, the Piccards finally left the earth before daybreak on October 23rd but not before some maneuvering of the balloon on the field.[56] The launch itself was far from straightforward. Albert Stevens, between his duties preparing for *Explorer II*, took time to observe and help in Detroit. He watched the preparation and inflation of the bag and felt uneasy about what he thought was a poorly orchestrated process. By Auguste and Jean's design, a large ring at the base of the balloon skin (its appendix) kept it open to allow for some mixing of air, which the Piccards argued would increase the stability of the bag.[57] Stevens and the Goodyear balloon designers believed this to be a dangerous procedure, especially in the wake of the explosion and crash of the first Explorer. Also, another innovation by Jean Piccard, from his Hercules days, was to release the main ground control lines with small TNT charges. This combination of pyrotechnics, magnesium sphere and 700 cylinders of explosive hydrogen left many people uneasy. At launch, not all charges released the balloon simultaneously, and so it began to drift down and swing on the one remaining ground line. As Don Piccard recalls, after some drift, the ground line was released and the craft moved up into the clouds with his mother visible through the still-open escape hatch.[58]

After launch, Stevens followed in an airplane to track the balloon. Watching it rise from a gloomy field and disappear into low clouds, he felt that the Piccards had nerve flying at all. Airport control was nonexistent, and the sky turned out to be heavily overcast although the forecast had been promising. Stevens wondered why in the world the Piccards wanted to do this. As he reported to Chester Fordney:

> The Piccards deserve a lot of credit [for conducting the physics experiments]. Furthermore, not very many people would care to go up through 2000 feet of clouds, and float most all day, knowing that they had to come down through the clouds. Fortunately, they did not drift much and they landed, as you know, in southeastern Ohio.[59]

Stevens was impressed that the Piccards insisted upon carrying the heavy scientific experiments because the devices certainly limited their maximum altitude. He knew, of course, that the lead shot was still needed as control ballast during descent, but his preoccupation with the altitude attained—which turned out to be 17,800 meters during some eight hours of flight—belies what he knew were the Piccards' own priorities.

Jean Piccard later claimed that he was misled about weather conditions. Their "weather experts" told them that the required conditions of no ground wind and clear skies to the east had been met on October 22, so they launched. But when they traveled through the heavy cloud cover and after daybreak saw nothing but clouds, "we knew that the danger of being blown over the open ocean was quite real," as Jean told Victor Neher, Millikan's associate.[60] An ocean landing under cloud cover would have ended in almost certain disaster. Piccard thus shortened their float time and limited their altitude in the blind hope that they would not travel as far as the Atlantic coast. By altering the flight profile he compromised the scientific work, since they still made a push for maximum altitude.

The Piccards were very lucky indeed to land intact, although the gondola itself suffered some damage and the balloon was heavily ripped by trees. Upon landing in a wooded area on a farm approximately 5 kilometers southwest of Cadiz, Ohio, the lower portion was crushed up to the working floor level, but little else in the cabin was affected. Jean apparently was shaken up somewhat; x-rays taken once they were back in Dearborn showed a minor fracture of one of his ribs, and small fractures in the left foot and ankle.[61] The gondola, balloon shards, and instruments

FIGURE 4.3. Jeannette Piccard talks with reporters at the landing site near Cadiz, Ohio, October 23, 1934. NASM.

were hauled back to Dearborn for disbursement once the proper trailers were driven to the landing site in Ohio. Back at the Dearborn Inn, Jean and Jeannette planned for their immediate future, which included a round of banquets and congratulatory ceremonies. Jean reported to Earnest Teberg of the Museum of Science and Industry, "Now our gondola has been twice in the stratosphere, it is rather tired and will soon be back in your museum for a last and long rest," but not before it was paraded around the country by its sponsors, the Peoples Outfitting Company and Grunow Radio.[62] The museum was not concerned about the tour or the fanfare, but it did worry about commercializing what was still their artifact. Piccard had painted the names of Grunow Radio and Peoples Outfitting onto the gondola along with his own name, and had removed all former names, including that of Dow Chemical.[63]

Soon after the return to Dearborn, all the scientific equipment was shipped off to Millikan and the Bartol by Swann's men. Bartol's G. L. Locher began to reduce the orientation and flight data, as well as the cosmic-ray data, while the Piccards enjoyed a heady but brief period of rejoicing and recognition. Stevens reported to Swann that in all the speeches made at one banquet given for the Piccards in Dearborn, no one mentioned the fact that the gondola carried Swann's and Millikan's instruments, although the Piccards, he added, were probably too modest to talk about the science:

They discussed the popular things of interest about the flight . . . but no one told the important facts at this banquet that the Piccards had successfully carried your Geiger Counter apparatus and the Stösse Chamber to a very high altitude and had used it over a considerable length of time . . . I had all I could do to keep from getting up and stating these things.[64]

Swann sent a copy of this letter to Jean Piccard, possibly to remind him of his stated reasons for flying, but also because Stevens ended his report by stating that the science no doubt would soon be reported: "Personally, I believe that the flight of the Piccards' ranks with any stratosphere flights that have been made. Very possibly, it may be the most valuable one made so far from the standpoint of scientific results." Whatever Stevens really thought of the enterprise, his words to Swann were phrased to indicate his appreciation of the science that was done. To his military colleague Fordney, on the other hand, Stevens confided that the scientific apparatus compromised the altitude they could reach.

Scientific Results of the 1934 Piccard Flight

From the beginning, the scientific data suffered because of Jean's alteration of the flight profile, Jeannette's unplanned and impulsive maneuvers during the flight, and the lack of complete records of the actions taken during the flight. For example, when Gordon Locher examined the altitude and orientation data, he found that during ascent and descent, Jean had let out ballast unevenly with the result that the gondola rocked badly. The amplitude was not great, but it affected the data on the direction-dependent cosmic-ray flux Swann desired. Nevertheless, Swann hoped that Piccard could reconcile the time-dependent data with his own flight log, since the logs provided for Locher had only a record of exposure times and compass readings.[65]

Swann prepared a preliminary report on their observations that Piccard could use in a planned article for *Science*, and also suggested that Piccard prepare a report for the Franklin Institute, since he was already planning to send a note to the *Journal of Industrial Engineering*. Swann's preliminary results from his 16 triple-coincidence cosmic-ray telescopes (which contained some 178 tubes) were confined to measurements of the vertical intensity of radiation at 16,400 meters and the ratio of horizontal intensity to vertical intensity at that altitude. He also obtained an integrated intensity curve from launch to 16,400 meters, but no mention was made of the results from the Stösse device, nor was there any comment on the azimuthal variations they were looking for. These calculations could not be made until the gondola orientation had been figured out.[66] None of Swann's preliminary results contained anything new, but he hoped that further analysis might be fruitful. There were no observations beyond the 16,400-meter level, according to Locher's reduction books. The time records used by Locher indicated that the Piccards had remained at that level for some 2 to 3 hours before they ceased scientific observations to try for an altitude record around 9:30 a.m.

Notwithstanding Stevens' admiration for what the Piccards had accomplished in taking along such a battery of scientific instruments, they most certainly still hoped to snag a new world's altitude record. Although their "souvenir booklet" proclaimed that the reason for the flight was to study "the mysterious 'Cosmic Rays' [which might provide knowledge of] new sources of energy . . . of immense value to humanity," in fact as they prepared for the flight, the Piccards agreed to give the North American Newspaper Alliance exclusive rights to their stories. Their compensation was to be determined by how high they went, with a $1000 bonus if they broke the altitude record.[67] As Jeannette later wrote in a handwritten account of their journey: After they made observations at some stable height, she "shot our last sandbags [to] try to make an altitude record."[68] In addition to the sandbags, they also started to empty the lead shot ballast from the counter shielding to lighten the load as much as possible. The ballast release, however, was poorly designed for accurately metered flow. Jean had to scoop the material out with his arm, pushing it toward the sealed outlet in the gondola every time he wanted to release some of the lead. He therefore had no accurate record of how much lead was released as a function of time. Moreover, he never emptied the ballast chamber completely, so there was no opportunity for Millikan's un-shielded experiment to take clear readings.

Millikan and Victor Neher put little faith in the data returned from the Piccard flight. Neher wrote Jean in November saying that, although the record looked clear, they could not reduce it without proper time calibration (the clock had to be started manually after launch), fiducial time calibrations during flight, times of releasing the lead shot, and the corrected barographic record. Neher congratulated Jean and Jeannette on their success, and hoped the data would be forthcoming.[69] Jean could not supply the data requested by Neher because, apparently, he never recorded it. As already mentioned, the Piccard flight plan had to be modified "en route" when they discovered, with daybreak, that they were still faced with heavy cloud conditions and began to fear their lives might be in danger. They had no idea where they were; Stevens had lost them in the clouds, they were out of radio contact, and, as Jean claimed to Neher: "We had therefore the alternative of making good cosmic ray records and putting them down in danger of being drowned in the ocean or smashed on the ground—or of going up less high and coming down earlier."[70] Obviously flustered and worried about their fate, the Piccards failed to take proper barometric or even time readings during their altered course, even during the critical times when they decided they could stop in their ascent to gather in cosmic-ray data. During their first stop, which they cut short for fear of drifting too far east, they took no barometric readings that could be correlated properly with the cosmic-ray instruments, but hoped that the variations they did record in other experiments might be used to work backward to recalibrate the barometric record. Jean also failed to turn on Millikan's apparatus in time to obtain a complete altitude record, and this, along with his failure to dispose of the lead properly and take reliable barograph readings, left much for Millikan to criticize. Millikan also knew that the barograph readings he had seen showed serious inconsistencies. So when Jean asked for his report, Millikan gave him nothing that he could quote without embarrassment.[71]

Piccard tried in vain to excuse himself, claiming that they had to retain some ballast to ensure that they could control the gondola during descent, since Jeannette had released all the external ballast bags for their altitude attempt. But he had already admitted that his own design of the ballast outlet box for the shielded cosmic-ray apparatus made it difficult to get all the ballast out quickly, and that they had no time to clean the lead shot out before descending. All these factors conspired against him, Jean felt; he even rationalized that "We had believed that you [Millikan] had sufficient data concerning the unshielded instrument at hand, therefore we made no effort to get new data in that respect."[72] Millikan, with barely concealed contempt, had the final word: "I do not think, therefore, that we can get data of importance out of our record."[73]

Swann thought about encouraging Millikan to cooperate later in March but then decided against it when he saw the magnitude of the ineptness before him.[74] Even though he once mentioned that his own data for the ratio of vertical to horizontal intensity as a function of altitude from all the manned flights were in "extraordinary agreement," Swann saw many discrepancies between the results obtained during the Piccard flight and those of Settle and Fordney, the first Explorer, and Regener's unmanned observations. The two other manned ascents yielded moderately consistent data for Millikan, but Regener's differed from all of the others, especially in the rate of increase of ionization with altitude. Swann pointed this out to Jean later in 1935. Piccard, now bitter about Dow, thought the differences might be due to the Dowmetal skin of the gondola, which he thought created a different secondary spectrum than the one Regener might have observed. Swann considered the situation more complicated, and attributed the differences to secondary effects from the total amount of material in the gondola. Experiments that his staff had recently undertaken led Swann to suspect that Jean's design for the casing that held the counters and the shield material was at fault.

Piccard's planned review of the scientific results never appeared in *Science* or in the *Journal of the Franklin Institute*. A brief one-sentence statement in the *Science-Supplement* merely mentioned that their flight had reached 17,850 meters and returned data for Swann. Millikan was not mentioned at all.[75] Although Jean had hoped for complete success, he admitted soon after the flight that the scientific return was not going to be great. He said as much in a public lecture during Christmas 1934 to the American Institute of Aeronautics, but there is no evidence that this dampened the Piccards' sense of accomplishment. Although they feared for their survival during flight, they still pushed for maximum altitude taking actions they knew would compromise the science. As a consequence, they failed to establish a new record, and soured their relations with Millikan. Nevertheless, both Jean and Jeannette had at last tasted stratospheric flight.[76]

After the flight, Jean and Jeannette's attention returned to their family, finances, and professional futures. They hoped that their flight would bring with it a flood of offers from institutions worthy of their talents and name. What ensued was as much a result of the continuing intellectual famine during the Depression as it was the Piccards' inability to appreciate how faculty are appointed to posts at American universities.

Notes

1. Jean to Auguste Piccard, 23 October 1933, 1 November 1933, PFP/LC. Translation by Benjamin Pearce.
2. Jean Piccard to W. F. G. Swann, 12 December 1933, Swann Papers, APS.
3. W. F. G. Swann to Jean Piccard, 12 January 1934. See also letters in folder 1934–1947, PFP/LC.

4. Swann (1935a), p. 259. See also comments by M. Stanley Livingston, in Stuewer (1979), pp. 133, 135, 315.
5. W. F. G. Swann to R. A. Millikan, 8 September 1918, RAM/CIT.
6. See William Swann to R. A. Millikan, 1 March 1923, RAM/CIT, R33 F30.2. Swann's study was titled "Causes and Prevention of Fires in Balloons." See also Swann to Millikan, 28 July 1917, RAM/CIT, R5 F5.2; and "NRC Division of Physical Sciences Report 1918," p. 11, RAM/CIT, R5 F5.11.
7. Auguste Piccard to Jean Piccard, 7 June 1934, PFP/LC. Translation by Renata Rutledge. "F.A.I." refers to the Fédération Aéronautique Internationale, the international body that validated aviation records.
8. Maravelas (1980), p. 17.
9. Jeannette Piccard to "Dear Bill" [William Rosenfield], 8 December 1933. See also Jean Piccard to Auguste Piccard, 1 November 1933, PFP/LC. Translation courtesy Benjamin Pearce.
10. R. A. Millikan to Jean Piccard, 12 December 1933; Jean Piccard to R. A. Millikan, 18 December 1933, and 19 December 1933, PFP/LC. Both Mexico City and Brazil were mentioned in Stuart A. Rice to Isaiah Bowman, 18 June 1933 and in Jean Piccard to R. A. Millikan, 18 December 1933, PFP/LC.
11. Jeannette Piccard, "He Taught Me How to Fly," 12 April 1958, pp. 1–2, Speech and Writing file, PFP/LC. In this account as well as in memorabilia from the Detroit flight, Jeannette indicated that Hill won the race in 1927. Earlier records say it was 1926.
12. See "Stratospheric Flight 1934" folder, PFP/LC; Merritt Bond to Jean Piccard, 10 April 1934, and Jean Piccard to Bond, 28 April 1934, PFP/LC. See also Crouch (1983), pp. 626–631.
13. Jeannette Piccard to Mrs. Ulysses Grant McQueen, 4 January 1934, PFP/LC.
14. Jeannette Piccard to Dr. John Ridlon ("Dearest Daddy"), 19 June 1934, PFP/LC.
15. Jean to Auguste Piccard, 23 October 1933, PFP/LC.
16. Jean Piccard to Willard Dow, 26 September 1933, PFP/LC.
17. W. F. G. Swann to Dow Chemical, 14 February 1934, PFP/LC.
18. See, for instance, Jean Piccard to E. H. Perkins, 9 December 1933; L. B. Grant to Jean Piccard, 7 March 1934; E. H. Perkins to Jean Piccard, 11 June 1934; Jeannette Piccard to E. H. Perkins, 13 June 1934; E. H. Perkins to Jean Piccard, 15 June 1934. PFP/LC.
19. Jean Piccard to E. H. Perkins, 16 June 1934, and L. B. Grant to Jean Piccard, 21 August 1934, PFP/LC; Piccard to Swann, 8 September 1934, Swann Papers, APS.
20. E. H. Perkins to Earnest J. Teberg, 17 August 1934, RR/MSI.
21. Jean Piccard to Willard Dow, 30 January 1934, PFP/LC.
22. Jean Piccard to E. J. Teberg, 28 May 1934, RR/MSI.
23. Jean Piccard to Lawrence Cole, 3 July 1934, Cole to Piccard, 7 July 1934, Jeannette Piccard to D. T. Ganun, 1 August 1934, 31 August 1934, "Piccard Stratospheric Flight 1934," statement, 15 September 1934, Donald Piccard Collection.
24. Piccard (February 1934), n.p., copy in PFP/LC.
25. Swann (1933b), p. 946.

26. Swann (1933a), p. 217; See also Swann (1955), p. 36. Bruno Rossi in the 1950s felt that Swann's hypothesis was still tenable, but Hans Alfvén and M. S. Vallarta in hindsight argued that it was unrealistic. See Rossi (1953), p. 67; Alfvén (1985), p. 428.

27. When W. Heitler and H. Bethe were unable to explain how cosmic rays could penetrate to the depths observed if they were electrons, they questioned whether the quantum rules might break down for particles of very high energy or if a new type of particle existed. As the historian Peter Galison recently described the problem, "At the time there were *two* real alternatives: quantum mechanics was correct and the particles were protons, or quantum mechanics was incorrect and the particles were electrons." See Peter Galison (1987), pp. 103–107, especially p. 106. See also Jánossy (1948), pp. 16–17. But protons as primaries created serious problems as well. The east–west effect indicated to Swann's associate Thomas Johnson that the particles had to be positively charged, but their penetration properties could not be explained if they were protons (Galison (1987), pp. 106–112). If the particles were positively charged, and behaved as if they were more massive than electrons, then the proton remained a likely but problematic candidate, or something existed in between. Galison has reviewed how the in-between particle, the "heavy electron" now known as the muon, was finally identified—the ultimate answer to the problem. This state of affairs in the mid-1930s led Swann to suggest that primaries were non-ionizing relativistic electrons, and this played a strong role in the design of his experiments on both Piccard's flight and on *Explorer*. For the detailed development of Swann's ideas, see, for instance Swann (1932), pp. 540–541; Swann (1934), pp. 828–829; and for general background, see Galison (1987), sections 3.6–3.11.

28. Swann (1934), pp. 828–829.

29. Street and Johnson (1932), pp. 142–144.

30. Galison (1987), p. 110.

31. Swann, Montgomery, and Montgomery (1936), p. 26, fig. 15.

32. Swann (1936b), p. 39.

33. Swann (1933c), p. 1026. See also Swann (1935b), pp. 575–577, especially p. 576; Swann (1937), p. 434.

34. W. F. G. Swann to Jean Piccard, 22 December 1933, PFP/LC.

35. Albert Stevens to W. F. G. Swann and to J. Piccard, 8 April 1934, PFP/LC; Jean Piccard to W. F. G. Swann, 21 May 1934, PFP/LC.

36. Swann to Piccard, 13 June 1934, PFP/LC.

37. T. H. Johnson to W. F. G. Swann, 4 October 1934, Swann Papers, APS.

38. Swann to Piccard, 5 July 1934, PFP/LC.

39. See R. A. Millikan to William R. Blair, 22 July 1933; Millikan to Gen. Benjamin D. Foulois, Chief of the Air Corps, 3 August 1933; J. H. Newton, U. S. N., to Millikan, 20 July 1933; Millikan to Cmdr. F. W. Reichelderfer, 5 September 1933, RAM/CIT, R33 F30.5. See also Bates and Fuller (1986), pp. 34–35.

40. See Kargon (1982), chaps. 4, 5; Kevles (1978), chap. 12. The Caltech Archives maintains a clipping file from the 1930s that reveals that these flights, and Auguste Piccard, received much local publicity.

41. Jean Piccard to Millikan, 4 January 1934, PFP/LC.

42. Millikan to Swann, 17 January 1934, PFP/LC.

43. Ibid.
44. Jean Piccard to Millikan, 22 January 1934, PFP/LC.
45. Millikan to Jean Piccard, 25 July 1934, PFP/LC.
46. Jean Piccard to Millikan, 11 August 1934, RAM/CIT, R23 F22.8.
47. Millikan to Jean Piccard, 16 August 1934, PFP/LC.
48. Jean to Auguste Piccard, 13 August 1934, folder 1934–1947, PFP/LC .
49. "Memorandum for Piccard Flight," 15 August 1934, folder 1933-1934, PFP/LC .
50. E. H. Perkins to Earnest J. Teberg, 17 August 1934, RR/MSI.
51. Cable, Millikan to J. Piccard, 24 August 1934, PFP/LC.
52. Jean Piccard to Swann, 18 August 1934, Swann Papers, APS.
53. Winston was Dow's chief engineer in its metallurgical laboratory and was the designer of both the Piccard gondola for Chicago and the Explorer gondolas. He was most involved in Dow's participation, and kept detailed logs and extensive scrapbooks on the manned effort through these years. See Winston Collection, Dow Archives.
54. Jean Piccard to Merritt Bond, 25 September 1934, PFP/LC.
55. See "Piccard Stratospheric Flight, 1934," 15 September status report, Don Piccard Collection, and Jean Piccard to Earnest Teberg, 15 December 1934, RR/MSI.
56. Jean Piccard once said they flew on October 27. The October 23 date is confirmed by the National Aeronautic Association's official report. See "N.A.A. Directing Official's Report on Piccard Stratosphere Flight," 30 October 1934, Piccard folder, LTA files, NASM Technical Files; and Crouch (1983), p. 629.
57. "The Piccard Stratosphere Flight," souvenir booklet, p. 5, discusses this "startling innovation." PFP/LC.
58. Don Piccard to the author, 18 July 1988.
59. Albert Stevens to Chester L. Fordney, 19 November 1934, PFP/LC.
60. Jean Piccard to H. V. Neher, 13 October 1934 [sic, notation indicates 13 November], RAM/CIT, R23 F22.9.
61. R. D. McClure (Henry Ford Hospital) to Dr. John Ridlon (Newport, RI), 3 November 1934, PFP/LC.
62. Jean Piccard to E. Teberg, rec'd 13 November 1934, RR/MSI.
63. E. Teberg to Jean Piccard, 16 November 1934; JRWP to Files, 1 June 1936. "Memorandum of Piccard visit to the Museum of Science and Industry," RR/MSI.
64. Stevens to Swann, 16 November 1934, PFP/LC.
65. See "Stratosphere Flight 1934," PFP/LC; also in Piccard folder, Swann Papers, APS. See also W. F. G. Swann to J. Piccard, 7 December 1934, Swann Papers, APS. The record of the flight profile examined in both collections does not agree in detail with one kept by Jeannette Piccard, copy kindly provided by Donald Piccard. The main point of discrepancy is that the logs used by Locher indicate clearly that the Piccards stayed at about 16,000 meters for two hours, and then made a try for maximum altitude. Jeannette's rather sketchy log does not support this. A handwritten account of the flight and landing by Jeannette, also provided by Donald Piccard, does confirm that they made observations above 12,000 meters, and then pushed for an altitude record. See "Copy of Original Log of Dr. Jeannette Piccard, BALLOON FLIGHT, October 23, 1934," p. 46.

66. W. F. G. Swann and Gordon Locher, "Measurements of the Angular Distribution of Cosmic Ray Intensities in the Stratosphere with Geiger–Mueller Counters," MS attached to Swann to Piccard, 7 December 1934, PFP/LC.

67. Quoted from "The Piccard Stratosphere Flight," p. 3. If the Piccards failed to reach 7.4 miles they would not receive anything. If they rose higher than 8.6 miles (45,400 feet or 14,000 meters) they would receive $600 for signed stories. And for an altitude between 7.4 and 8.6 miles, they would receive approximately $1 "for each 0.002 miles." See "Agreement Between N.A.N.A., Inc. and Dr. Jean Piccard," 7 May 1934, Donald Piccard collection.

68. See Jeannette Piccard, "Copy of Original Log of Dr. Jeannette Piccard, BALLOON FLIGHT, October 23, 1934," p. 46, Donald Piccard Collection.

69. H. V. Neher to Jean Piccard, 8 November 1934, RAM/CIT, R23 F22.9.

70. Jean Piccard to H. V. Neher, 13 October 1934 [sic, notation indicates November 13], RAM/CIT, R23 F22.9.

71. R. A. Millikan to Jean Piccard, 21 December 1934, PFP/LC.

72. Piccard to Millikan, 17 February 1935, RAM/CIT, R23 F22.10.

73. Millikan to Piccard, 9 March 1935, RAM/CIT, R23 F22.11.

74. See draft, Swann to Millikan, 22 March 1935, Swann Papers, APS.

75. See listing under "aeronautics" in "Some Advances in the Sciences During 1934," *Science-Supplement 81*, (4 January 1935), p. 11. For Swann's testimony, see "Cosmic Rays," n.d., p. 3, Briggs Folder, Swann Papers, APS. For reviews by Piccard, see Piccard Speech and Writing File, "Piccard Stratosphere Flight 1934," (press release), "Results of our Stratosphere Flight," (1934); and "Our Stratosphere Flight," PFP/LC.

76. Jean Piccard [Talk to American Institute, Christmas 1934 n.d.], Speech and Writing File, PFP/LC. For her own part, Jeannette regretted not being able to steer clear of the woods on the Ohio farm where they landed. She later admitted that this was her one failing, as it destroyed the balloon and seriously damaged the gondola. See Jeannette Piccard, "Copy of Original Log of Dr. Jeannette Piccard, BALLOON FLIGHT, October 23, 1934," pp. 44–45, Donald Piccard Collection.

CHAPTER 5

The National Geographic Society—
Army Air Corps Explorer Flights

FIGURE 5.1. Albert W. Stevens demonstrating operation of K-6 aerial camera. AAC/NASM.

Although the National Geographic Society was the chief supporter of the Explorer flights, it did not initiate them. *Explorer* was the brainchild of an Army Air Corps photogrammetrist. In February 1932, Capt. Albert W. Stevens, chief of the photography laboratory of the Army Air Corps Materiel Division at Wright Field, Dayton, Ohio, and a leading aerial photographer, used Auguste Piccard's success to argue that the Army should revitalize its program in balloon research, which had been discontinued after Hawthorne Gray's death.[1] Gen. B. D. Foulois agreed to let Stevens conduct initial feasibility studies and cost estimates, which were completed by May.[2] The Army viewed the Navy's prominence in the flights of *A Century of Progress* with some envy; William E. Kepner, who was to become part of the first complement riding *Explorer* in 1934, later recalled that the Fordney–Settle flight gave:

ample precedent for the Army Air Corps to enter the balloon race to the stratosphere. If beating the Navy record was not enough, there was the national prestige and attention, and more importantly, the considerable serious scientific study already done at Wright Field, Dayton, Ohio, in high altitude research.[3]

Stevens was a highly motivated aerial reconnaissance specialist and ardent flyer. Since 1928 he had dared high-altitude flights in aircraft to set records both for altitude and for photographing distant objects. Since reconnaissance aircraft were vulnerable to ground fire and attacks from fighter aircraft, it made sense to develop long-range systems, and ultra-high-altitude balloon reconnaissance was an irresistible challenge.[4] In the tradition of George W. Goddard, he and Wright Field colleagues constantly experimented with high-altitude, long-distance photography, employing near-infrared films and filter combinations to cut through haze to bring out distant horizon detail. He also tried other plate–filter combinations to study cloud formations at all accessible heights in the atmosphere. During the summer of 1929, Stevens and Lt. John D. Corkille were authorized by the War Department to take a 22,000 kilometer aerial photographic tour of the Pacific Northwest. During the tour Stevens managed to photograph Mt. Ranier from a distance of 360 kilometers using an Army K-6 camera, a new world's record for long-distance photography. An Air Corps spokesman described Stevens' work and added:

The great value of high altitude or long-distance photography in time of war is obvious. It would mean the ability to secure layouts of enemy territory far beyond the reach of enemy anti-aircraft guns. But the value extends to peace time work also . . . increased penetration means the possibility of mapping far greater areas in a single operation with the use of the four or five-lens Army mapping cameras. This would be infinitely valuable to surveyors.[5]

The Army approved the study at end of the year, but limited its involvement to manpower, logistics, and the use of present facilities; it would not cover the cost of the balloon and gondola, and it would not give formal approval until Stevens had secured additional support, in both funding and expertise.[6] This could not appear to be a military adventure; somehow, Stevens would have to find private support for what would be seen as a patriotic initiative that the Army would help make possible. While preparations were being made for *A Century of Progress* in the summer of 1933, Stevens searched for support. He knew Tex Settle through mutual contacts at the National Geographic Society and had provided some aerial cameras for Settle's flight. Most important, Stevens was a friend of the society's Gilbert Grosvenor, often providing him with aerial photographs for the magazine as well as for his own personal use, and he knew well Grosvenor's interests in aviation. As Stevens later told his friend, the physicist Lewis M. Mott-Smith, the Geographic Society had, in the past six years, devoted some 2,000 pages to aviation-related topics.[7] The National Geographic Society, well known for its support of voyages of exploration on land and sea, also cooperated with the Smithsonian Institution and the Navy to observe solar eclipses on land and from aircraft and a wide range of geophysical phenomena in remote corners of the globe. The Explorer series was the society's entry into manned ballooning.[8]

Initial Planning and Invitations

Stevens, as entrepreneur, wrote of his hopes to Millikan in July 1933, emphasizing the scientific advantages of such a flight, but when he presented his friend Grosvenor with a formal proposal for National Geographic support, he argued that the project was "for the purpose of attaining what may be regarded as perhaps the maximum altitude that it is ever physically possible to attain by man."[9] Stevens hoped that his planned balloon would fill out to some 85,000 cubic meters, five times larger than the *A Century of Progress* balloon. This vastly larger gas bag, "regarded as the maximum that it is feasible to construct," would be able to send aloft a 2.8-meter Piccard-type gondola (which was 0.6 meters larger than Chicago's gondola), to 23,000 meters, "even though [it was] carrying a gondola of sufficient size to enclose the many instruments that are necessary for a thorough study of the atmosphere."[10]

Before submitting his proposals to the National Geographic and to scientists such as Millikan, Stevens had given considerable thought to the experiments one would want to include. First, of course, he developed an elaborate plan to obtain accurate altitude estimates by photographic means, and he felt that this program set apart his proposal from all other ballooning programs then known to him. Stevens pointed out to his potential supporters that all previous balloon ascents, including the openbasket flights that he had made with Professor Mott-Smith of the Rice Institute in Houston, concentrated mainly upon cosmic-ray measurements. His flight would have a broader scientific agenda:

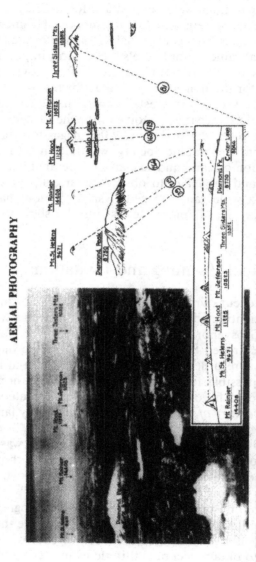

AERIAL PHOTOGRAPHY

LONG DISTANCE PHOTOGRAPH OF MT. RAINIER, WASHINGTON, FROM VICINITY OF CRATER LAKE, OREGON
A Distance of 266 miles, far Beyond the Range of Human Sight
This photograph was taken from the airplane at an altitude of
20,000 feet, using a slightly modified Air Corps camera
fitted with special film and infra-red filter.

FIGURE 5.2. Promotional demonstrations circa 1930 of Stevens's aerial reconnaissance capabilities. Appended materials discuss in-flight photographic processing and sending images by wire to facilitate rapid examination and analysis. AAC/NASM.

QUICK-WORK PHOTOGRAPH OF PRESIDENT HOOVER'S HOME AT PALO ALTO

This photograph was taken, finished in the air, and transmitted by wire to The President within one hour and thirty minutes.

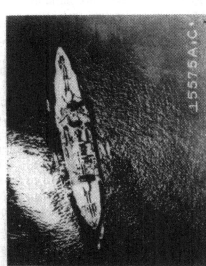

NIGHT FLASHLIGHT PHOTOGRAPH OF BATTLESHIP "WYOMING" AT NEWPORT NEWS

HIGH ALTITUDE PHOTOGRAPH OF WRIGHT FIELD FROM 32,400 FEET

FIGURE 5.2. (continued)

Two balloon flights have been made by [Auguste] Piccard; a flight is about to be made by Settle in Chicago; a flight is being planned by Belgian and Dutch engineers; even Soviet scientists have announced plans for a flight with a balloon carrying a cylindrical gondola. It should be noted that all of these flights have as their announced purpose measurement of cosmic rays, with little or no mention of plans to get other information.[11]

Public spectacle had to be avoided. He would not be subject to the onerous constraint placed on the Piccard–Compton flight to launch from a place like Soldier Field in Chicago—under "conditions of possible strong public pressure." He admitted that the Chicago flight would be of "distinct scientific value" but thought that little attention was being given to general studies of atmospheric conditions and properties.[12] Although Stevens was entirely correct in pointing out that all publicity surrounding the Chicago flight centered on the cosmic-ray measurements, we have seen that at the last moment Compton added other experiments, including some employing a spectrograph. Stevens was quite sensitive, however, to the complex patronage problems and unreasonable flight demands that the Piccard–Compton flight faced. Thus to Millikan he added that he was looking for more thoughtful financial backers, and hoped to find one large appropriation rather than many small ones:

We hope to keep this a purely scientific project, and do not wish to get mixed up in any advertising schemes or undesirable publicity. Furthermore, we plan to make the flight from a rather remote place . . . where the public can be kept at a safe distance . . . With such a large balloon a flight can only be made under the most perfect conditions and with no possibility of interference by the public, newspaper men, motion picture operators or others. In this respect the coming flight from Chicago will be difficult.[13]

The National Geographic Society as Sponsor

Stevens hoped that the National Geographic Society would agree to the above conditions. It supported exploration, had its own vehicle for the dissemination of reports on its activities, and was in the business of documenting exotic and especially photogenic expeditions worldwide. Founded in January 1888 by Gardiner Greene Hubbard, who drew together government scientists, explorers, and influential laymen in the salons of Washington's Cosmos Club, the society from the beginning wished to both define and promote geography and geographical expeditions.[14]

In its first years, the society remained a quiet organization of scholars. When Hubbard died, his son-in-law Alexander Graham Bell took over and expanded the scope of both the society and its magazine; he also increased the magazine's appeal to the common man. Bell brought in Gilbert H. Grosvenor as assistant editor, and in short order Grosvenor became editor and architect of the plan for broadening the base of the *Geographic* magazine. Grosvenor hoped to promote geography and exploration by bringing images of faraway peoples and places to *Geographic* readers. The society supported many of the activities described in its magazine, through revenues from membership subscriptions. By 1918 its membership had reached 650,000, and by the time of the Explorer flights in 1934–1935, it was in excess of 1,000,000. In effect, the society's success created a constant need for new initiatives in exploration, because other popular magazines, although not devoted exclusively to exploration, gave more attention to it in their pages.[15]

The society wished to feed well-crafted and visually rich accounts of voyages of exploration and discovery to its vast and hungry market. But they had to be well-planned and exquisite success stories that accented positive and noncontroversial themes. Above all, as Philip Pauly has pointed out, these expeditions and the stories following them had to portray the "romantic exploration of the unknown" and show that "the adventure was at least as important as the information gathered."[16]

Support for a manned balloon flight fit Grosvenor's policy. Beyond exploration, such a flight complemented the society's keen interest in aerial photography. Since World War I, Grosvenor had been advocating stories for the *National Geographic* that exploited the airplane. Time and again, Stevens's expertise was called upon for visual spectaculars, first in black and white, then in color. However, early color films required long exposures, so they could not be used on fast-moving powered aircraft. Stevens saw that a manned balloon might solve this problem, and although the society had employed dirigibles since 1930 for color photography,[17] Stevens's proposal promised a perspective never before captured on color film.

Although the society sought public attention and significant visual firsts, it also held to the dignified vestiges of its origins as a patron of thoughtful geographical research and exploration, for only in doing so could it maintain its credibility. In all respects, the society would prefer anything but the circus Stevens correctly predicted would take place at Soldier Field a few months hence. Thus it was in a position to support what was being promoted as a return to the stratosphere in the name of science.

Securing Scientific, Military and Financial Patronage

Stevens predicted that his balloon and gondola would cost $50,000, or roughly twice what the Chicago balloon and gondola had cost. He wanted the National Geographic Society to underwrite most, if not all, of that amount, and the Army Air Corps to promise logistical support.[18] While he searched for financial support, he courted scientific participation, starting with Mott-Smith, whom he knew would agree, and Millikan, whom he needed for visibility.

Stevens wanted Millikan to stimulate others to participate and thus help him to round out his ambitious agenda. In addition to continuing cosmic-ray experiments, Stevens proposed that the main scientific experiments include measurements of the variation of pressure with altitude compared to photographic determinations of altitude; photography of the solar spectrum with quartz spectrographs; determinations of the variation of temperature with height and the atmospheric electrical gradient using a recording electroscope; bolometric measurements of solar radiation; radio propagation studies; air sampling and the containment of samples using highly evacuated chambers; humidity measurements; the study of the survivability of plant spores in the high atmosphere; and an extensive evaluation of wind patterns using automatic cameras to monitor the drift of the balloon, to test "the general belief . . . that there is a general drift easterly at high altitude over the United States."[19]

With $1,000 donations from Eastman Kodak, Sherman Fairchild, the Fairchild Aerial Camera Corporation, and Cornelius V. Whitney, President, Pan American Airways, as well as $1,000 out of his own pocket, Stevens hoped that the National Geographic would provide at least $25,000. He also considered asking the Franklin Institute of Philadelphia for support, but to Millikan he worried about potential rivalries and the " 'division of credit': We will try to promote the idea of a flight in the interests of science, with no wrangling over the commercial or selfish angles of the project."[20] Stevens therefore waited for the society to make up its mind.

To sweeten the pie for Millikan, Stevens added that chances were good that the balloon would still be useful for unmanned flights after the manned ascent. The enormous bag could easily carry very large instruments to heights exceeding those attainable by normal sounding balloons. In effect, Stevens invited Millikan to buy into the manned program with the hope of securing lifting power far in excess of anything he could hope to obtain otherwise. In response, just after Settle's Chicago embarrassment, Millikan tried to educate Stevens, as he did Jean Piccard, about the value of a flight of properly shielded counters in the equatorial zone to determine the cosmic-ray energy spectrum and its variation with height and latitude. In fact, because of the fate of the Settle flight, Millikan thought that he was in a strong position to force Stevens into an equatorial flight, and that if he insisted, possibly his demand would be met. Millikan retrieved his instruments from Chicago and held them hostage while he reviewed for Stevens the latitude and east–west effect problems. He warned Stevens that all previous balloon attempts had yielded ambiguous results since they had not carried enough shielding material to sufficient heights. He needed some 270 kilograms of shielding, which could also be used as ballast. Millikan added that the sounding balloon attempts by himself and Regener could not carry sufficient shielding to adequately analyze the high-altitude spectrum of cosmic rays.[21]

Millikan had a number of reasons for letting both Stevens and Piccard pursue him, as we saw in Chapter 4. Before Settle's second flight with Fordney, Millikan knew that he still had to meet Compton's challenge. His airplane latitude survey, conducted by H. V. Neher and other associates, had obtained reliable data only to 6800 meters; and his higher altitude flights of small electroscopes, supported by the Weather Bureau, had not yet returned totally satisfactory data, even though they agreed well with Regener's low-altitude results.[22] But his lecture-by-correspondence with Stevens reveals as well his keen desire to obtain the high-energy spectrum of cosmic rays, available only by carrying heavily shielded instruments to extreme heights.

Also as mentioned in Chapter 4, politics was another motive. Millikan chaired a committee of the National Academy of Sciences Science Advisory Board (SAB) that was charged with making recommendations for strengthening the Weather Bureau. The SAB endorsed closer contacts between the Weather Bureau and other government agencies, and supported Millikan's proposal to start collecting three-dimensional aerological data for the assessment of air mass information. It also endorsed Millikan's planned educational program for meteorologists at his California Institute of Technology, which boasted a new center where the air mass theory was being put into practice. Millikan made sure that Gen. B. D. Foulois read a copy of the SAB recommendations, and continually sought the favors of Foulois and his military colleagues for flight services and support for his research and policy recommendations. Since Foulois endorsed Stevens's manned ballooning, there was good reason for Millikan to endorse it as well by agreeing to participate.[23] Millikan had long promoted better relations between the Weather Bureau, the Signal Corps, and the Air Corps and often utilized their services in concert for his cosmic-ray work and relied upon their support for his growing aeronautical and meteorological programs at the California Institute of Technology. Accepting Stevens's invitation was, in effect, a compliment to his military patrons' support for his own scientific research; it was in their own interest to have Millikan's tacit endorsement through his cooperation. Beyond the data that might result from the flight, Millikan's political debt obliged him to go along for the ride.[24]

Soon after contacting Millikan, Stevens sought the advice and endorsement of Lyman Briggs, the newly designated director of the National Bureau of Standards as well as confidant and adviser to Gilbert Grosvenor and the society. Stevens also approached other members of Briggs's bureau whom he knew would be sympathetic to his proposal. A key bureau staff member was W. G. Brombacher, chief of the Aeronautics Instruments Section. Brombacher and his colleagues had been constant suppliers of information for the Departments of Commerce and State, the National Advisory Committee for Aeronautics (NACA) and other aviation interests. One of their specialties was instrumentation for lighter-than-air craft. Thus Brombacher had already aided Tex Settle and Wiley Post by providing calibrated barographs and was also an influential member of the National Aeronautical Association Panel which was charged with revising Fédération Aéronautique Internationale (FAI) regulations for barometric altitude standards.[25] The bureau was, therefore, well known to balloonists and aviators, and was a likely place for Stevens to look for support as well as for technical assistance and advice. Brombacher could not supply funds, but was more than willing to supply both technical information and actual instruments if the flight became a reality.[26] Stevens knew that Brombacher would be consulted by Briggs before the director made any decision. Properly informed of the particulars, Brombacher encouraged Briggs to endorse the proposal to the society. Briggs duly informed Grosvenor that Stevens's proposed flight was "essentially a scientific expedition, to be carried out in accordance with plans carefully thought out, and as such I believe it is worthy of the hearty support of the National Geographic Society."[27]

Knowing that his staff was eager to participate, Briggs was motivated to both endorse and provide scientific and technical patronage for Stevens's plan; beyond his personal scientific interest—in the 1920s Briggs had contributed to the improvement of the earth indicator compass for air navigation—he headed an organization that had devoted much energy and resources to aviation and aeronautics. His National Bureau of Standards staff was ready to design, build, test and calibrate the many meteorological devices sought out by Stevens to round out the scientific complement.

But Briggs had another reason for getting his bureau involved. In 1933 the National Bureau of Standards was experiencing the worst of the Depression; that summer half his staff had been furloughed, and appropriations were down by 40 percent.[28] At a time when every federal scientist was beginning to fear for his job and program, when there was talk of a merger between the bureau and the NACA, and when some claimed that there was duplication of effort in aeronautics between the two institutions, getting involved in a visible way with the Army and the National Geographic was an excellent way to demonstrate the unique talents and abilities of the bureau to serve national needs.[29] The 60-year-old Lyman Briggs, a former student of H. A. Rowland at The Johns Hopkins University and a civil servant since the turn of the century, proceeded with quiet determination to restore the health of his organization. One of the first obstacles he had to overcome was the criticism leveled by President Roosevelt's Science Advisory Board, which called upon the bureau to reduce its commercial regulatory activities, in the face of growing industry objections, and concentrate upon "the development of fundamental standards for science, medicine and industry."[30] Briggs had to maintain his work force in the face of budget reductions at the nadir of the Depression, as well as answer to a growing series of congressional and special committee investigations of bureau activities. Not an especially rugged or ruthless manager, Briggs resorted to creative solutions that played to the strengths of his staff. He had two personal enthusiasms, one of which was scientific exploration, which he shared with some of his staff.[31]

Succeeding George K. Burgess on the Board of Trustees of the National Geographic Society, Briggs saw Stevens's proposal as a creative solution to the bureau's needs. What better way to preserve his best staff than to take a central role in a stratospheric flight? Brombacher, L. B. Tuckerman, and since 1919 Hugh Dryden, who had headed the aviation physics section, were only the most prominent among a large staff fascinated with flight. And there were other specialists in some jeopardy who could contribute their expertise in optics, radio, and thermal physics and thereby could be kept visibly occupied in the name of science.

With Briggs's strong endorsement in December 1933, the National Geographic Society agreed to support Stevens to $25,000, subject to the society retaining "absolute charge of distribution of all news releases, including photographs and the first magazine publication of results."[32] The society also wished to form a scientific and technical advisory committee that would help Stevens decide "on scientific equipment and what to endeavor to record on flight." The original committee roster of eight included Lyman Briggs as chairman, C. E. Kenneth Mees (senior vice-president and director of research at Eastman Kodak), Millikan, and finally Compton, who soon dropped out. Grosvenor and members of his staff completed the membership. The Scientific Advisory Committee ensured the society a governing role in the project and maintained its place in the interlocking network of boards and panels that made up so much of scientific Washington's power structure. Briggs and Grosvenor were soon joined on the National Geographic Society Board of Trustees by Brig. Gen. Oscar Westover, once the Army formally approved *Explorer*.

The sum provided by the National Geographic Society to Stevens, and thereby to the Air Corps, was by far the largest single contribution from an outside source. For the affluent society, however, this was not an unusual show of support. During the same period, it provided four times the amount for Adm. Richard E. Byrd's expeditions, twice as much for expeditions to study Inca remains, and comparable amounts to preserve the giant Sequoia.[33] With the society's support, Stevens collected what had already been pledged, and with some additional donors brought the total to $39,000, with several thousand more still possible. To make up the difference, Stevens approached broadcasters, asking them to buy into the program with the prospect of receiving and rebroadcasting transmissions from the balloonists during flight.[34]

With funding pledged by the society, Stevens made his formal proposal to the office of the Chief of the Air Corps. It had taken him 11 months to secure private support and develop a sufficiently well-defined plan, as well as obtain the endorsement of scientists willing to embark on the venture. All the meteorological instruments were available from Air Corps and bureau stock, and the scientific instruments were to come from the participating institutions, at no cost to the Army Air Corps. Test flights of the instruments in aircraft were necessary and would require the support of Air Corps manpower and equipment, but only a few flights would be needed from Wright Field. Stevens listed some 16 areas of scientific research he hoped to pursue, but admitted that it might not be possible to do everything in one flight. He pointed out that since many of the instruments "are largely automatic, much of the . . . information may be obtained in one flight."[35]

Stevens's personal goal was to refine the relationship between barometric pressure and altitude, critical knowledge for the Air Corps aerial reconnaissance program at Wright Field. He pushed this in his proposal for logistical support, predicting that it may well "compel the revision of part of the present pressure–altitude tables." He then followed with 15 single-spaced pages of description of each experimental area he hoped to attack.

By mid-January, 1934, Westover, who by then was acting chief of the Air Corps, approved Stevens's December proposal, which included the services of Maj. William Ellsworth Kepner and Lt. Orvil A. Anderson as flight personnel.[36] Westover was now made a member of the Advisory Committee, along with W. F. G. Swann, director of the Bartol Research Foundation of the Franklin Institute. Swann, however, accepted Grosvenor's invitation in a letter to Briggs stipulating that he wished to work closely to coordinate efforts aboard *Explorer* with those he had already committed to Jean Piccard. This was the first Briggs, Stevens and the National Geographic apparently knew of Jean Piccard's plans.[37]

With everything in hand, Stevens and his sponsors announced the "National Geographic Society—U.S. Army Air Corps Stratospheric Flight," in mid-January 1934. Kepner—a recognized Army balloonist, past chief of the Materiel Division Lighter-Than-Air Branch at Wright Field, former meteorologist and teacher at the Balloon School at Omaha, and at the time Wright Field's purchase branch chief—was designated pilot, with Stevens as scientific observer. Among Kepner's many exploits, which made him well qualified as a resourceful balloonist, was his successful flight through a strong cold front in a damaged Army blimp in October 1928.[38] Stevens had assumed in July that the Air Corps would choose Kepner, or Orvil Anderson if Kepner was not available, since they were "considered the Army's best balloonists."[39] There was never any option other than a proven military flier on active duty.

Searching for Scientists

Stevens identified his scientific agenda in his formal proposal to the Army Air Corps. Counselled by Briggs and his associates, Stevens emphasized those areas in which they would improve on the performance of the Akron flight of Settle and Fordney, and highlighted activities of direct interest to the Air Corps. With National Bureau of Standards advice and expertise, he could improve upon both temperature recording and air sampling techniques. Settle's recording thermometers did not function properly, so much of the data depending upon them were of questionable value; also, his air samples were later found to be contaminated by the cabin atmosphere. Stevens added that cosmic-ray experiments with Millikan's new shielded electroscopes would be more efficient than the directional observations Compton had attempted with Geiger counters, since the Compton counters were not fully functional during the flight—there simply was not enough time to allow them to work correctly.[40] Finally, on Briggs's advice, Stevens urged that the balloon crew make spectroscopic observations of the vertical distribution of ozone in the atmosphere. Such an experiment had failed on *A Century of Progress*, but Stevens had the optical expertise of the bureau behind him. Even though Harvard University's Theodore Lyman subsequently tried to discourage them, arguing that the balloon could not rise high enough to penetrate the ozone layer to examine the ultraviolet solar spectrum itself, Stevens and Briggs felt that an ultraviolet spectrograph would be a useful addition to the scientific arsenal because one only had to travel part way through the ozone layer to observe its distribution.[41] These were really two independent goals: in the latter, the solar spectrum served only as background to study the absorption characteristics of atmospheric ozone.

Stevens negotiated for cosmic-ray experiments with Millikan, Swann, and Mott-Smith, while Briggs quietly rounded out the scientific payload by exploiting the talent within his bureau. Beyond barographic measurements and chemical studies of the upper atmosphere, he found an advocate for spectroscopic studies of the ultraviolet solar spectrum and the earth's atmospheric ozone layer in the bureau's Fred L. Mohler. In early February Mohler began seeking advice and assistance from senior spectroscopists such as R. Ladenburg of Princeton University, R. W. Wood of The Johns Hopkins University, and C. L. Pekeris of the Massachusetts Institute of Technology (MIT). The first lent advice on both practical and theoretical spectroscopic technique, the second agreed to provide diffraction gratings, and the third information on light scattering in the upper atmosphere. Both Briggs and Mohler contacted the Gaertner Scientific Instrument Company in February asking if the instrument used on the Settle flight might be available, or if they were interested in providing an automatic version.[42]

The spectrographic complement was originally the responsibility of Mohler, but one of his advisers, Brian O'Brien, director of the University of Rochester's Institute of Optics, assumed primary responsibility for the spectroscopic experiments by the end of March. This transition came after Mohler and Stevens met in Rochester to discuss designs and coordinate responsibilities.[43] O'Brien emerged as the primary player along with his Institute of Optics patrons Bausch and Lomb and Kodak. In all, six commercial firms, including Bausch and Lomb, Folmer and Schwing, the Taylor Instrument Companies, and Eastman Kodak, provided counsel, manpower, and optical and photographic instrumentation. With the added patronage of the society, O'Brien and his Rochester staff, including H. S. Stewart, were in a unique position to create highly sophisticated and powerful spectroscopic devices. These are discussed in Chapter 6.

O'Brien, a friend of Briggs, was delighted to be a part of the flight.[44] He had no previous experience with ballooning, but was a specialist in physiological optics and therefore was keenly interested in ozone. His research included studies of the ultraviolet energy distribution in the lunar spectrum and in the spectrum of terrestrial skylight, seasonal changes in the photochemical effects of sunlight, and biological effects of high-energy radiation. He was also an excellent experimental physicist with experience in optical systems and testing, and so was an ideal candidate for scientific collaborator.[45] To be sure, there were some differences of opinion—Mohler preferred narrow-band filters for reducing scattered light rather than complex double-dispersing spectrographs—but O'Brien, with Bausch and Lomb backing was now in charge, and was anxious to send full-scale laboratory instruments into the stratosphere.[46]

In hindsight, O'Brien had no difficulties with participating in a manned scientific flight. He felt that having a man fly with the instrument along would help if something went wrong, and his presence also meant that one could fly "real" instruments and not little "toys." O'Brien recalled that at the time, photoelectric sensors were still too crude and one needed larger laboratory type dispersing systems to capture the solar and sky spectrum properly.[47] O'Brien's background and institutional and corporate affiliations naturally led to the use of highly sophisticated optical systems and recording devices. Little would be demonstrated by sending up simple narrow-band filter detectors that could only determine how much energy was transmitted through the atmosphere in different ultraviolet spectral bands as a function of altitude. Laboratory-scale high-dispersion systems represented the state of the art in spectroscopy, and that was the playing field that O'Brien and the Institute of Optics relished.

The players on the cosmic-ray front were not so enthusiastic. Millikan continued to complain, and Compton would not budge. Stevens, however, got along well with Professor Lewis M. Mott-Smith of the Rice Instutute, Houston, Texas (now Rice University), who was planning to send up a set of electroscopes, and, of course, William Swann was happy to participate. But Rice itself was not forthcoming as a supporter, and even though Stevens still hoped Mott-Smith could participate, he also wanted to have Millikan, Compton, and Swann on the roster.[48] Earlier in the summer of 1933, Compton, echoing Millikan, refused to participate "unless the flight be made in other latitude[s] than this."[49] Compton took a harder line than Millikan on this critical point, especially after the problems he encountered with the Chicago and Akron flights. Compton turned to balloonsondes in a highly public manner later in the year. A popular account treated it as something new to science, and as the proper way to do science:

Professor A. H. Compton plans to study the cosmic rays at great heights in a new way. Instead of sending up huge spherical gondolas loaded with a battery of instruments and a crew of observers, he will rely on small unmanned balloons which will carry a cosmic-ray meter together with a small radio transmitter. The readings of the instrument will be automatically sent back to the ground to physicists comfortably seated at desks in laboratories.[50]

Compton clearly did not have a satisfactory experience in the Fordney–Settle flight, and with the close of the "A Century of Progress" Exposition, conducted his research in what was to him a more natural mode, on his own terms. *Science* magazine quoted him in August 1934 as saying that he had no intrinsic distaste for manned flights, which he admitted could carry larger instruments than his radiosondes. But even though radiosondes would not replace manned flights of heavier equipment, they would provide critical information in latitudes where manned flights were not possible, such as the polar regions.[51] Compton would probably have been more encouraged to work with Stevens had it been possible to make the flight from an equatorial region, but he was not anxious in any event to relive the problems encountered before and during the flight of *A Century of Progress*.[52] Typically, when Compton turned to unmanned balloon-sonde flights on his own initiative, he found that he had to pay for the flight equipment out of his own pocket.[53]

Stevens, in response to Millikan and Compton, felt that it was not possible to "make our first flight from anywhere else than in this country; it would be foolhardy to attempt it anywhere else until we know more how things are going to work out."[54] He added that there was much to study in the same latitudes other than cosmic rays and held out hope that Compton would come around. Even if he did not, Stevens felt that Mott-Smith's detectors, and possibly one or two identical ones from Millikan, would provide a full complement. Stevens did not appreciate at first what Millikan had in mind.

Millikan criticized Stevens, as he had Piccard, for trying to pile in too many cosmic-ray experiments. He argued, "if you will let me say so I think you are making a mistake in shopping around to get us all who are interested in cosmic rays to send up essentially the same kind of experiments."[55] Millikan required 270 kilograms of lead shielding alone, and was not at all happy with the idea of sharing the lead shielding, or having the shielding used as ballast, which meant that it would have to be fine-grained. He argued that this would affect the experiment. He tried to clarify what Stevens really needed, and what would be good for cosmic-ray research:

There are two cosmic ray jobs to do. One is to get the directional effects handled in some way, and Mott-Smith is as good as anybody for that; the other is to get the ionization experiments done. These have to do mainly with getting the hardness as a function of altitude, and that can easily be done with our self-recording instruments. It does no good whatever for us to send up one instrument. We have done that already in sounding balloons, we did it in the flight last summer, and we can get nothing more out of it. What I would be glad to do in the whole ionization matter would be to send up three of our self-recording instruments, one without lead, one with a lead shield 5 cm. thick, and one with a lead shield 10 cm. thick. These are the instruments that we have got already largely built.

Millikan argued that each experiment should have a clear and concise purpose, and that the best idea would be to measure the cosmic-ray intensity spectrum with shielded coincidence circuits. If Mott-Smith or Swann wanted to do it all, it was fine. Conversely, he would take the whole project on, but only on his terms. As we have noted, Millikan was pursued simultaneously by Piccard, who asked for anything Millikan could release for his Detroit flight, and quite logically was growing impatient with both his pursuers. Piccard already had the endorsement of Swann, and so confusions again arose concerning who would do which cosmic-ray experiments.

But before Millikan's criticism reached Stevens, Briggs called an advisory committee meeting in Washington at the Cosmos Club, to which were invited Swann, as well as John Fleming from Carnegie's Department of Terrestrial Magnetism, Millikan's old watchdog, and members of the National Geographic staff. A plan identical to Millikan's was developed there, largely by Swann, whose interests paralleled Millikan's. Stevens was therefore able to respond to Millikan with great enthusiasm, and as a result, Swann became a coordinating influence with Briggs's backing. Swann, as the director of Bartol, had far more backing than did Mott-Smith at Rice, and invited Mott-Smith to collaborate in the directional experiments, leaving the three self-recording instruments to Millikan. This suited Millikan and everyone else.[56]

The Launch Site

By late January the Scientific Advisory Committee for *Explorer* had expanded, as did the number of experiments planned for the flight, scheduled now for the early summer 1934. A major question—where to launch—was still not settled; Kepner and the U.S. Weather Bureau were asked to find the best possible site, and so Kepner led a nationwide tour of some 40 sites, mostly in the northern Great Plains, that were anxious to play host. The experience of Settle's first flight convinced the society, Stevens, and Kepner that they should not be "tied up with any publicity scheme" and that they had to have the freedom to take off both when and where they thought conditions were best to ensure success. At the first full meeting of the advisory committee on February 21, 1934, all agreed that one "must return [alive] in order to have the mission a success."[57] There would be no compromise with safety.

Grosvenor and his associates began to worry about preserving the society's image, to say nothing of the lives of the fliers, when news arrived that a January 30, 1934, Russian stratospheric flight, which had reached some 22,000 meters, ended in tragedy when the gondola fell from 12,000 meters, killing all three Russian balloonists.[58] Accordingly, the society insisted that it was not in charge of the American effort and would not be responsible for the safety of Stevens and Kepner, but was only supporting the interests of the Army Air Corps.[59] The launch site would be free from any society influence, and the Scientific Advisory Committee would be completely free to make the final choice.

Many of the 40 cities and communities that vied for the honor of becoming the base of operations did try to apply pressure on the society and Stevens. In Denver, the publisher and editor of the *Denver Post* wrote to Grosvenor offering the use of the airport, hotel facilities, and literally the keys to the city. Then the mayor of Denver wrote the society and George Dern, secretary of war, trying to apply political pressure. Eventually the governor of Colorado and Chamber of Commerce spokesmen urged that Denver be named the site. A mile-high airport, great publicity, and local support did not sway Kepner or the society, who consistently replied that they had to consider, above all else, the reports of the Weather Bureau and the maximum safety and scientific return of the mission. Denver was one of the finalists, but it came behind two quite obscure locations: In second place was Lander, Wyoming and in first was "the wonderful natural bowl" near Rapid City, South Dakota.[60] The society played no significant part in the choice of the South Dakota site. It was true to its word.

Rapid City not only had a deep natural depression nearby which was large enough to allow the balloon to be prepared and deployed while protected from winds by a stable inversion layer of cold air, but it also had a large Army post (Fort Meade) about 60 kilometers away. The post provided efficient housing and deployment of the 100 or so Army personnel and considerable materiel that would be involved in the flight.

After Kepner toured all the sites with Orvil Anderson, who was later designated "alternate pilot," he testified to the advisory committee that after being shown two locations in the Rapid City area that were not acceptable and after touring possible sites around Denver, Wyoming, and Utah,

By this time I was getting pretty discouraged and finally one of the [Rapid City] party said, "Why not take him to the spot we had in mind?" They were just putting on a show, getting me discouraged and then taking me to the spot they really thought would suit us. They use plenty of psychology out there. This spot was a spectacular sight and when I first saw it from the top of the surrounding cliffs I could hardly keep from jumping up and down I was so pleased.[61]

Kepner did not show his enthusiasm at the time, knowing that extensive windage and air current data still had to be gathered. But after sending up small hydrogen balloons, burning old tires and watching the smoke streams, and conferring with weather specialists detailed from Fort Meade, Kepner was convinced that this old gold mining area in the Black Hills called "Bonanza Bar" should be their launch site, and the committee agreed.[62]

The Rapid City site did not excite Millikan, who wrote both to Stevens and to McKnew, of the National Geographic Society, when he received advance copies of the news releases on the choice. Rapid City was at almost the same geomagnetic latitude as Akron, the site of the Fordney–Settle flight. He thought that at the very least, they should fly from somewhere in the southern part of the country. Realizing that his admonition would probably not make them change their plans, he felt it was his duty "to inform the whole group connected with this flight that so far as cosmic-ray data is concerned, in my judgment a flight from the lower border of the United States, for example from Kelly Field, San Antonio, Texas, is bound to yield cosmic-ray information of greater significance than can a flight made from the northern border of the United States."[63]

Millikan's opinions were, of course, already well known to everyone. He said the same things to Stevens in March after Neher had analyzed the Fordney–Settle flight data, and had received a polite response from Stevens as well as from Grosvenor, which intimated that the society could only afford one series of flights, and that they had to take place under the strictest control to ensure a safe return. Only the appropriately dubbed "Stratobowl" outside Rapid City satisfied them.

The Explorer Gondola and Balloon

Both the gondola and balloon for the Explorer flights were larger and refined versions of *A Century of Progress*. In his original proposal, Stevens was confident that the gondola could be built in 90 days and the huge balloon in 120 days; about 150 days would be needed for testing. By late February 1934, total donations amounted to $45,000, five thousand below Stevens's goal; accordingly, Dow charged some $2,000 for the gondola, and Goodyear-Zeppelin just under $25,000 for the balloon and all connecting systems, both far below cost. Goodyear-Zeppelin and Dow were happy to assume the loss, "because of the advertising value . . . and because it considered this in the nature of a contribution to the project."[64] See Fig. 5.7a.

Dow was confident that little needed to be changed this second time around, but Goodyear-Zeppelin had a larger scale-up task. Stevens's altitude target for *Explorer* was 24,500 meters with a payload of at least one ton. In order to meet this goal, the Goodyear-Zeppelin design and research engineers, headed by Karl Arnstein, vice-president for engineering, determined that *Explorer* would require a balloon not smaller than its planned 85,000-cubic-meter capacity. Goodyear-Zeppelin's largest balloon to date had been the 17,000-cubic-meter *A Century of Progress* bag. Now they were to build one five times larger in volume requiring almost three times the fabric surface area. As before, they chose rubberized cotton as the fabric but would use it in three weights for different parts of the balloon. The upper four-fifths of the balloon was constructed of medium-weight material, and the remainder of the lightest material. The heaviest material would be concentrated in load-bearing sections carrying the two catenary bands that connected ground ropes as well as the 160 bridles of 1/4-inch manila yacht rope that held the gondola.[65]

The nearly spherical bag was made from 40 separate gores, and each gore consisted of 88 panels of cloth with a total surface area of 9,670 square meters, not including the material for the girdles. Over 13 kilometers of seams were glued with rubber cement, with strengthening tape cemented on top and sometimes on both sides. Appendages to the bag included three small inflation ports, two exhaust values near the top of the bag, a rope-operated rip panel at the top of the bag, and the girdles. The fabrics and the seam designs were subjected to batteries of environmental stress tests. Arnstein's staff knew that the rubber coatings would deteriorate when exposed to cold, and found that the underlying cloth tended to crack in early tests.

On the ground, at an elevation of some 1,200 meters in South Dakota, the balloon would be filled to 7 percent capacity with hydrogen. As a result, at launch and for some time afterward, the balloon would be carried aloft by a relatively small bubble of expanding hydrogen. This put additional stress on the uppermost part of the bag and left the lower parts folded. Accordingly, Goodyear-Zeppelin technicians examined how stress on the balloon changed as its "gas head" expanded, and paid close attention to how the folds would open under low-temperature conditions. The design staff conducted numerical simulations and also constructed mechanical analogs to predict how the shape of the balloon would change and how meridional stresses would change as a result. Using what the simulations taught them, they then folded, creased, and cooled test samples of cloth, and drew them apart abruptly to look for tears and weakening. The samples showed no significant weakening at first.[66] The balloon passed all its initial tests. Only later, during its first flight, did the balloon fail. When folded pieces were hard-frozen first and then pulled apart in later, more severe tests, they were found to tear and crack frequently. This problem continued to plague the project to the end.

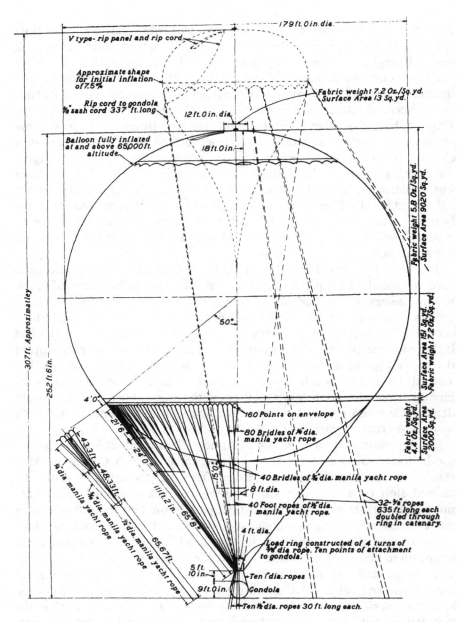

FIGURE 5.3. The first Explorer balloon, configurations at launch and at pressure height. The balloon's pressure height was expected to be some 20,000 meters. At this altitude, the expanding hydrogen was expected to fill the balloon, and ballast would have to be thrown overboard to allow the balloon to rise to its maximum theoretical height of 24,500 meters. NGS.

FIGURE 5.4. Cementing reinforcing seams onto the balloon gores. Science Service collection, NASM.

The Gondola

In terms of scale and construction, the gondola differed far less from its predecessor than did *Explorer's* enormous balloon. Dowmetal was again used for the skin and for all the fittings inside and outside the skin. Because Stephens was calling for an ambitious array of scientific instruments and wanted the gondola to carry at least two and possibly three passengers, the sphere was enlarged to 2.54 meters, although Stevens originally had hoped for a 2.75-meter sphere. At one point he heard that Fordney and Settle had been "cramped" in *A Century of Progress.*

To provide rigidity comparable to that of *A Century of Progress,* the larger sphere required 35 percent thicker walls. Dow engineers decided that the eight orange peel gores and two polar caps should be formed out of 0.5-centimeter plates. As before, the gores would be welded with an oxyacetylene torch and the seams ground smooth.

Kepner and Stevens collaborated on gondola specifications. After a visit by Kepner to Midland in the early Spring to discuss specifications with Dow's A. W. Winston, Stevens followed up with a long description of their requirements for manholes (which had to open quickly), glass port-

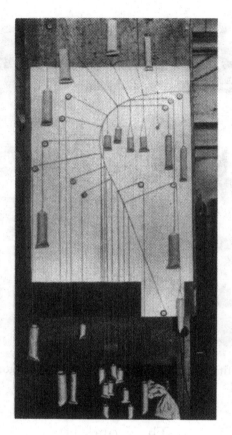

FIGURE 5.5. Goodyear simulations of the balloon-skin shape and stress analysis.
NGS.

holes, shot ballast discharge systems (which had to be greater than Settle's
to ensure control), flooring, shelves, camera openings, load rings, valve
control cord openings, and the accommodation of a new pneumatic balloon venting valve control system that he designed for the gondola. He
also listed general requirements for instrument apertures and electrical
connections, and integration needs for scientific instruments and paint.[67]
While the gondola superficially may have looked the same, as far as Dow
was concerned, Stevens's ambitious agenda and critical attention to every
detail distinguished *Explorer* from its predecessors.

Stevens helped the Dow engineers and the participating scientists integrate their instruments. He asked Millikan, Swann, Johnson, Mott-
Smith, O'Brien, and others for rough dimensions and weights, and then
blocked out proper space within the gondola looking for possible mechanical and electrical interference. He also had dummy instrument con-

FIGURE 5.6. Fabrication of the first Explorer gondola at the Dow plant, Midland, Michigan. Note the elliptical manholes. Dow.

tainers fabricated at Wright Field, and then sent them to each scientist for testing and evaluation. By providing integration services, Stevens knew exactly what each experimenter was up to. He was kept fully informed about exact design specifications as well as environmental demands, handling requirements, and the operational impact of each instrument on available hands during the flight. The interior design of the gondola had to be organized such that each instrument would be usable to its fullest advantage. Every device had to be completely understood, and Stevens constantly asked his scientific collaborators to be on the lookout for ways to make the inflight operation of their instruments as simple as possible; specifically he asked Millikan's people for detailed information on inflight recharging of the electroscopes and whether an automatic method suitable for the gondola environment would be possible.[68]

For emergencies, both Kepner and Stevens wanted two large elliptical manhole openings in the gondola. Each was to be fitted with cast Dowmetal double doors that could create a reliable seal but still could be opened quickly. Everything that went into the gondola, from people to instruments and ballast, had to fit through the 46-by-51-centimeter doors set above the equator of the gondola. Stevens wanted an unobstructed view of the landscape and sky, and asked for 10 Pyrex viewing ports around the gondola, 5 above and 5 below the gondola equator. One would face the zenith so that the crew could see how the balloon was behaving, and another looked straight down through the 60-inch false floor for the cameras. Above and parallel to the floor were three shelves for equipment, shaped to accommodate each of the experiments. Two of the instruments, the skylight and horizon light spectrographs constructed by Brian O'Brien, required special ultraviolet-transmitting quartz windows.

The completed 318-kilogram gondola was stress-tested in much the same way as its predecessor. The water pressure test revealed only a few tiny leaks at first. After these were repaired, the gondola passed its two-day air pressure test. In late April Stevens announced that the Dow Chemical Company had finished the gondola and that it would be filled with some 20 experiments, the heaviest being the electroscopes and counters suggested and prepared by Millikan and Swann, which required the lead shielding.[69] After the gondola was removed to Rapid City in June, Dow engineers conducted more pressure tests to ensure that the gondola was still sound following shipment and instrument casing integration; they proudly reported that it held a safety factor of 18.7.[70] Once at the Stratobowl, the gondola was fitted with the scientific instruments themselves.

The lessons learned by Auguste Piccard and his cohorts in their European flights, which were confirmed by the behavior of the *A Century of Progress* gondola, led the engineers to paint *Explorer* white on the top and black on the bottom as the best combination to stabilize its temperature during flight. The black bottom of the *Explorer* would keep it warm by absorbing scattered sunlight, whereas the white top would resist the direct impact of incoming solar radiation.[71] The life support system of the first *Explorer* was based on aircraft respiration devices. Like the *A Century of Progress* system, the *Explorer* system evaporated a combination of liquid oxygen and liquid air through a carefully controlled set of evaporation coils. This provided a 45/55 oxygen/nitrogen mix that would not render the cabin atmosphere explosive.[72]

After the respiration system was installed in *Explorer* and it was shipped to the Rapid City site, repeated tests of the sealed gondola with its passengers inside revealed that even though Stevens's external venting system helped to maintain constant pressure and allowed toxic gases to escape, the humidity of the environment increased to unacceptable levels, which could not be controlled by passive venting. Stevens knew that high humidity had been a problem on *A Century of Progress*. So while they all waited for the right weather conditions to arrive and all systems in Rapid City to be completed, Briggs and Swann helped Stevens design and con-

Their traveling home, high above the Earth..

DOWMETAL SELECTED
FOR BOTH LIGHTNESS
AND STRENGTH

DOWMETAL, because of its strength and extreme lightness, was chosen for the gondola to be used in the stratosphere flight sponsored by the National Geographic Society and the U. S. Army Air Corps. The story of this remarkable flight is told in this issue of the National Geographic. Strength was required to provide a gondola which would carry all heavy instruments and equipment, and which could withstand heavy air pressure from within. This strength had to be secured in the lightest possible form, since every hundred pounds saved meant 1500 feet higher possible flight with the same balloon.

Both strength and lightness were obtainable by the use of Dowmetal. In addition, properties needed for the fabrication of the metal into the form desired were inherent in Dowmetal. *Rolled sheet* was formed into "orange peel" sections. These sections were *welded* together into a one-piece metal ball. *Castings* were *welded, riveted,* or *bolted* into the shell

for manholes, windows, and other requirements. Bars and tubes were *extruded* for use as supports for racks and shelving. Nearly every standard form of metal fabrication and machining was used in this Dowmetal traveling air-home, just as it is used in the manufacture of portable machinery, fast moving machine parts, motorized transportation units, and in a great many other places where strength is combined with lightness.

Our booklet describing the uses and properties of DOWMETAL, the world's lightest structural metal, may be had upon request.

THE DOW CHEMICAL COMPANY
Dowmetal Division · MIDLAND, MICHIGAN

★   ★   ★   ★   ★   ★   ★   ★

FIGURE 5.7a. Dow Chemical Company advertisement prepared for the *National Geographic*, 1934. Dow.

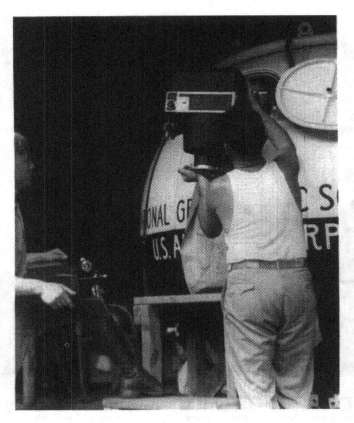

FIGURE 5.7b. Integrating the instruments into the gondola at the Stratobowl. Stevens's aerial cameras are being loaded, while one of O'Brien's spectrographs awaits, held by Gustave Fassim. Photograph by Richard Stewart, NGS.

struct a backup humidity control system using fans and sodium hydroxide as a desiccant and air purifier. The following year, their *ad hoc* design was refined as a total environmental air-conditioning system for *Explorer II*. The sodium hydroxide filter system handled both respiration and perspiration products. To provide better humidity control, they added a condensation coil to the second system, between the fan and the desiccant. Finally, Stevens and an associate, Oscar Steiner, refined the automatic pressure release valve inside the gondola so that it could be preset to any desired pressure.[73] The air-conditioning system also contained air samplers and humidity gauges that provided a continuous record of air quality in the gondola during the flight to assess flight performance. Small self-recording meteorographs used by the U.S. Weather Bureau on unmanned balloonsondes were placed within the gondola to record internal pressure,

temperature, and humidity during flight, and a huge thermometer dial
was designed by Stevens to indicate balloon temperature.

During the design and construction of the life-support system and associated balloon and gondola hardware, Stevens shared his knowledge
and experience with Jean Piccard, who at that time was readying the old
A Century of Progress. Jean both asked for support and offered his services
to the National Geographic in January 1934 when he learned of the
planned flight. Although no official relationship developed, they did share
ideas, and Stevens provided Piccard with services.[74] Stevens and his associates had designed a new balloon valve system utilizing pneumatic
pressure that was better than Auguste's mechanical rope system. Jean
acknowledged Stevens's offer to make a valve for him, but preferred his
own modification of his brother's design, using the usual ropes drawn
through a tightly sealed greased rubber gasket.[75] Jean in turn tried to offer
information on desiccants and air filters and chemicals that would absorb

FIGURE 5.8. Balloon thermometer dial designed by Stevens to be read visually
through a vertical window in the gondola and an opening in the bottom of the
balloon. NGS.

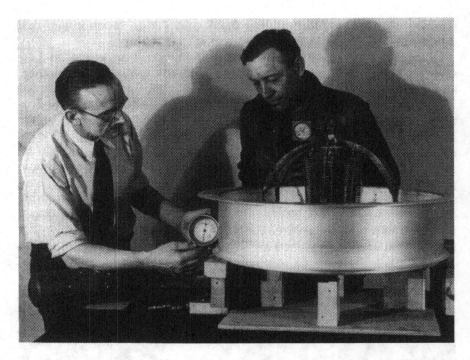

FIGURE 5.9. Pneumatic valve to release hydrogen from the balloon. This valve sat at the top of the balloon, and was operated from the gondola by means of a pressurized gas line. NGS/NASM.

carbon dioxide. Stevens appreciated the tips, but decided to stay with standard sodium hydroxide pellets, on the recommendation of Navy chemists. However, the two did trade useful information on the types of ballast screens that could prevent clogging, and ballast release systems, and maintained a spirit of cooperativeness throughout 1934 and 1935.[76] Jean found Stevens especially helpful when he was scrounging for miscellaneous but critical equipment, such as surplus hydrogen cylinders. Stevens and Piccard never discussed the scientific instruments; their exchanges were limited to general encouragement and to details of the housekeeping systems.

Public Rationale for the Flight

Gilbert Grosvenor, introducing a series of society radio talks in April 1934 about the flight, cited the society's traditional support for "exploration and research begun when it was chartered 46 years ago for the 'increase and diffusion of geographic knowledge.' "[77] George H. Dern, former governor of Utah and now secretary of war, reviewed the history of ballooning and argued that although small motorized balloons and airships seemed destined to continue their part in military aviation tactics, "The object of all stratospheric flights is scientific in nature." But Dern added, "whether for routine use in Army air maneuvers, or for possible later use in actual national defense, lessons learned on this stratosphere flight may be of great value to the Army."[78] In May, 1934, Oscar Westover, now assistant chief of the Air Corps, expanded on Dern's rationale by predicting that stratospheric flights would be of great use in aerial reconnaissance, which he saw as one of the most important scientific goals of the mission, "much more than to achieve merely a new altitude record."[79] To Westover,

The Army's interest naturally will concentrate upon the adequacy and efficiency of the balloon equipment for the purpose, the physical fitness, the efficiency and ability of the crew of the balloon to carry out the purpose of the flight; the instruments to be used and the methods to be applied for collecting scientific data of permanent value; the thoroughness of preparation for the successful launching of the balloon on this flight and its prompt recovery on landing; and finally, the deep human interest all people will have in the achievement of such a spectacular altitude while at the same time accomplishing every technical and scientific achievement desired from the flight.[80]

Both Westover and Dern knew that achieving an altitude record was not enough, but that it would be a useful demonstration of capability. They also saw, as did the National Geographic, that loading up the gondola with so much large equipment for the experiments, such as the lead shot for the cosmic-ray telescopes (which had to be carried to maximum altitude after much of the dedicated ballast, some 2500 kilograms of it, was expended) would limit altitude. But these experiments were necessary, not for their scientific return, but as a test of the military fliers' ability to perform complex tasks during stratospheric flight. Although Westover noted the technical difficulties of constructing the huge balloon that would lift the fully loaded gondola, he emphasized the logistical lessons that the flight would provide. To Westover and Dern, the flight represented more than a push for a world's record or a commitment to scientific exploration. It was a study in logistics, a test of Army Air Corps planning and developmental capabilities.

In June 1934, with the launch planned for sometime in July (the exact date depending upon readiness and proper weather conditions), dramatic news releases and radio programs were so well received that Grosvenor told John Oliver LaGorce, his immediate subordinate, "The scientific value of the Stratosphere Flight together with the huge amount of publicity accruing to The Society make it advisable to proceed with a second flight later this summer from the same base, if the first expedition is successful."[81] There was little doubt in Grosvenor's mind that the stratosphere flight would be the major popular initiative of the one-million-member society in these times, and so did not hesitate to advance $10,000 for a second flight of the same balloon and gondola.[82]

Scientific Priorities and Mission Plan

When the Scientific Advisory Committee met in May 1934 to choose a launch site, various members expressed concern that the scientific agenda was too heavy. In addition to flying the balloon, the two balloonists would have to perform the experiments and were going to be overworked. Lyman Briggs echoed this concern: "If we have got too much on board or too much planned we will leave part of it out.[83] Briggs, as the scientific conscience of the mission, said that if the Army wished to add a third man to the crew, "that is another matter." Otherwise, the committee was ready to reduce the scientific agenda of the flight. General Westover argued that "every effort should be made to secure as much data as possible," and approved a third crew member to assist Captain Stevens. Kepner felt, however, that the decision on the crew complement could wait until "they had a chance to see exactly how these instruments function in operation."

Even though most of the instruments were designed to be self-recording, flight tests of the instruments in aircraft and trial partial inflations of the balloon in the newly designated Stratobowl revealed how complex the flight was going to be. Thus the Army and the Scientific Advisory Committee soon decided that a minimum of two people would be needed to fly the huge balloon, and a third crew member was needed to perform the experiments. Capt. Orvil A. Anderson was added as "alternate pilot" in late June.[84]

That Briggs was willing to reduce the scientific agenda, while the Army, represented by Westover, wished to maximize the effort (at the expense of attaining higher altitude) by adding a third crew member, is a clear indication of their priorities. Briggs wished to ensure the overall success of the mission, at the expense of low priority experiments, and diplomatically offered the Army a way out, if it wished. The Army, on the other hand, was not interested in scientific priorities but knew that a third crew member would offer a better test of an operational system. Only the society was interested in an altitude record, but did not say so at the time.

Between June 20 and July 10, all of the instruments were integrated into the gondola and the balloon system was tested. A 1,000-cubic-meter test balloon carried the tethered gondola aloft on July 7 for launch trials, and gave the ground crew experience with inflation techniques. Tests in the sealed gondola on the ground began on the same day; the three crew members operated all of the equipment over a period of six and a half hours. By July 9, all tests and integration had been completed, and the gondola was christened *Explorer*. All was in readiness for the actual flight.[85]

FIGURE 5.10a. Interior of the first Explorer gondola. 1, 2, 3, 4, 22: flasks to collect stratospheric air samples; 41: vertical window to view balloon; 43: outside pressure altimeter; 47: inside pressure altimeter; 6, 30, 44: observation windows; 34: instrument to measure vertical motion; 40: manhole; 48, 49, 50, 51: cosmic-ray counters; 12: barometer; 38: horizon spectrograph; 39: cosmic-ray electroscope shielded with 270 kilograms of lead; 36, 37, 46: batteries; 45: photographic camera to record output of cosmic-ray counters; 52: switcher for cosmic-ray counters; 31: vertical observation port; 32: ballast air lock. From Stevens (1934), p. 407, NGS/NASM.

Not everything went as planned. Some problems arose when Swann and Briggs helped to modify the humidity control system in the days just before launch. Swann wrote to Piccard from the balloon camp mentioning that they worked day and night on these problems. For one thing, the cosmic-ray experiments originally planned were all ready, but, as dis-

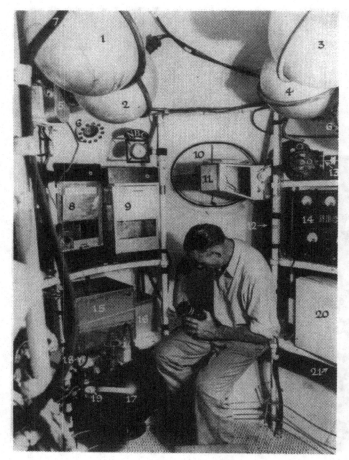

FIGURE 5.10b. Interior of the first Explorer gondola showing Stevens. 7: balloon gas valve pressure hose; 5: portion of Millikan's unshielded electroscope; 10: manhole; 11: recording camera for barometer; 13: motion picture camera; 8, 9: automatic cameras to photograph instrument dials; 14: NBC radio transmitter; 15, 16, 20: batteries; 18: pressure dial; 19: gas cylinder pressure dial for balloon valve; 17: one of the liquid oxygen flasks; 21: observation port. From Stevens (1934), p. 407, NGS/NASM.

cussed in Chapter 4, his bulky Stösse chamber was too large to fit and had to be removed at the last moment.[86] Although these were definite annoyances, the general mood remained upbeat and positive.

While *Explorer* waited for good weather conditions, which took two weeks to come, radio broadcasts from the headquarters cabin in the Stratobowl beamed a message of excitement and expectation far and wide. The National Broadcasting Company, which had bought into the expe-

FIGURE 5.11. Lyman Briggs and W. F. G. Swann at the Stratobowl. Swann is holding a tiny electroscope chamber. Swann (1935a), p. 261. NASM.

dition and now supported the pre-launch programming, hoped to maintain direct contact with the balloonists during the flight. During one broadcast, the NBC announcer introduced Kepner and Briggs, who then asked Stevens, Swann, O'Brien, and Neher to talk about the scientific experiments that were ready to fly. Briggs noted that "these men have spent months in building special apparatus for this flight and came here personally to install it in the gondola." They also worked during the integration and testing phases, and now were waiting for the flight with the rest of the contingent.[87]

The first experimenter to speak was Swann, who, with Gordon L. Locher and T. H. Johnson had prepared devices to detect the directionality and intensity of cosmic rays. Swann believed that charged particles in the secondary cosmic-ray spectrum would be anisotropically bent into streams by the earth's magnetic field and that these streams could be detected by his 144 Geiger–Müller tubes, which were grouped into 16 directionally sensitive "telescope" arrays. Swann summed up his philosophy:

In cosmic rays, as in many other things, circumstantial evidence is all that is available to tell the history of new figures on the horizon of science. The rays carry no insignia of office other then their deeds; and, of them it may be most truly said; "by their fruits alone shall ye know them."[88]

Brian O'Brien spoke next about his set of three quartz spectrographs built and modified for the flight by staff from Bausch and Lomb in conjunction with H. S. Stewart of Rochester and F. L. Mohler of the National Bureau of Standards. O'Brien felt that he and his colleagues had been given an "unusual opportunity" to send these instruments 24 kilometers above the earth's surface to examine the ozone distribution in the stratosphere. He explained how the total amount of ozone could be calculated from the ground, but emphasized that its distribution could only be measured by traveling through it: "thus we hope to add another piece to our knowledge of this all important blanket of air about us."[89] In a recent interview, O'Brien recalled the excitement of the preparations at the Stratobowl. He was convinced that the indirect measurements of the distribution of ozone with altitude made by Götz and his colleagues in Germany during the 1930s needed to be confirmed, and that his precision spectrophotometry provided the best means of obtaining the true curve.[90] O'Brien also remembers his delight when a cousin wrote to tell him how excited she was when she heard his voice on the radio. She had not been expecting anything of the kind, and was not even aware that he was involved. Not many scientists were allowed such national exposure, and to some, this was quite a thrill.

Finally, Victor Neher, Robert Millikan's colleague at the California Institute of Technology, described their shielded cosmic-ray experiments, which were intended to measure "the penetrating powers, the intensity and atom-destroying powers of cosmic radiation."[91] Remaining as neutral as he could, in deference to Millikan's position in the debate over the nature of cosmic rays, Neher identified the two known components of cosmic rays—the "hard" and "soft" components—noting that only the former penetrated to earth while the latter could only be detected at the very top of the absorbing atmosphere. Their three sets of electroscopes had 15-centimeter, 10-centimeter, and no shielding, in order to determine the energy ranges of the incident "soft" cosmic radiation. "The great size of this balloon permits for the first time at this altitude the study of the radiations with accurate instruments surrounded by thick shields," which Neher hoped would help to "distinguish between two or more theories" concerning their origin.[92]

The overall purpose of these radio talks was to convey the wonder and mystery surrounding the abstruse questions scientists studied and to indicate how this mission into the stratosphere, an equally mysterious realm, would help to answer them. The scientists spoke in appropriately reverent tones, and for their part, the National Geographic Society and military speakers reminded the audience of the great pride one and all would share in the enterprise. The society also emphasized that *Explorer* was a carefully planned program, and that after its flight, the society would support

FIGURE 5.12. Small highly-efficient motors and solenoids were incorporated into automatic systems to trip camera shutters at specific time intervals, wind photographic film, and perform other repetitive tasks. NGS.

the processing of the data from the stratosphere. Most of the recording electroscopes and Geiger counters, as well as the spectrographs, used photographic film, driven by tiny automata, to record their observations. The general plan of the flight program was to have the Scientific Advisory Committee coordinate the reduction and dissemination of the results. As an earlier news release put it:

When the gondola returns to earth with its hundreds of feet of photographic film recording observations in the stratosphere, it is this committee that will correlate the data, study it, and put it into shape for use by the scientists of the world. Of the thousands of facts gleaned from the stratosphere by busy men and busy instruments, a few, undoubtedly, will help to penetrate still farther the veil of blue mystery that has hung over the earth as long as there has been an atmosphere.[93]

While everyone waited for the flight to begin, scientists and crew knew only too well that they would need a good measure of luck if the venture was to succeed. Positive prose and detailed contingency planning notwithstanding, they all knew that the three crew members, as well as the instruments, were riding in a thin magnesium sphere under a huge amount of explosive hydrogen gas.

Launch of *Explorer* and Its Fate

Explorer was finally launched from the Stratobowl in the Black Hills at 5:45 a.m. on July 28, 1934, and carried its host of scientific equipment over 18 kilometers into the stratosphere.[94] The launch and ascent could

FIGURE 5.13a. The scene just before the launch of *Explorer* on July 28, 1934 from the Stratobowl. NGS.

not have been smoother, nor the air clearer. With Anderson handling the pneumatic valving and ballast, the craft was brought to equilibrium at 12,400 meters to check all systems and allow the scientific instruments to cycle through a set of observations. At about noon, they started their ascent again by closing the pneumatic valve. Kepner, Stevens, and Anderson were still ascending to their maximum equilibrium height when, around 1 p.m., they realized that something was terribly wrong. They found that the bottom of the bag was ripped in several places, and the

FIGURE 5.13b. *Explorer* in ascent. AAC/NASM.

rips were growing larger, although they had not yet reached the expanding gas bubble within the bag. Prudently, the crew decided to leave the stratosphere immediately and land as soon as possible.[95] But the bag began to rupture at about 5 kilometers during their descent, and the balloon that had given them lift was now nothing more than a flimsy and rapidly disintegrating parachute. With the gondola falling evermore rapidly, the three bailed out in a terrible rush which was made all the more harrowing when one of them got caught in the escape hatch. Only scant moments after they were free of the gondola, the hydrogen-filled balloon shell, contaminated with atmospheric oxygen, exploded and sent the gondola crashing on the plains near Holdrege, Nebraska.

The crew parachuted to safety, but most of the photographic records of the scientific instruments were destroyed in the crash. Two of O'Brien's three spectrographs were completely destroyed. O'Brien was in a two-seat Army plane following the flight when the balloon exploded. He did not see the crash but landed at the crash site only minutes later. There he heard the account of what had happened. The three aeronauts waited until the gondola had descended to the point where they could open it without the threat of explosive decompression, and then Kepner climbed

FIGURE 5.14. *Explorer* at an altitude of 2,000 meters and descending at 200 meters per minute. The ruptured balloon temporarily acts as a parachute. AAC/NGS/NASM.

out onto the support ring above the gondola to assist the other two. According to O'Brien: "Stevens had walked on his chute and it was unfortunately all bunched up . . . Because of that they decided that they had better get him out first. He had to jump out backwards with the chute in his arms."[96] Kepner then cut O'Brien's hanging spectrograph loose

FIGURE 5.15. Explorer gondola and balloon fragments in free fall an instant before crashing to earth. Note two parachutes visible in the debris trail of the balloon. AAC/NGS/NASM.

"right on the dot" as previously planned, "and it came down as it should [by parachute] and landed perfectly on the Kapok cushion. But everything else was wrecked," as O'Brien recalls. He was able to secure the help of a local farmer to carry his one surviving spectrograph to a local town, where it was packed and shipped back to Rochester after the film was removed.

Mohler was excited when he heard that some of O'Brien's spectra survived. He hurried to supply O'Brien with the needed barograph records;

FIGURE 5.16. Remnants of two of O'Brien's three spectrographs. NGS.

Pekeris's negative assessment of Götz and Dobson's ozone studies had just appeared in print, and the field was wide open to respond to Regener, who had just conducted ozone studies with a tiny automated balloon-sonde spectrograph. Mohler wished O'Brien well: "I trust you will soon know the actual facts about [the] ozone distribution."[97] But it took quite some time for O'Brien and his staff to analyze the spectra they obtained, which, O'Brien speculated, might provide 90 percent of what they hoped for, despite the crash and the smashed cameras.[98] However, the special microdensitometer prepared to reduce the photographic spectra did not work properly, and they had to perform the analysis visually. This procedure became especially tedious as O'Brien insisted on checking every spectrum for the slightest evidence of solar absorption in the 210-nanometer ozone window, in addition to determining continuum intensities for ozone abundance.[99] Then O'Brien took ill in November. His was the last paper outstanding, and it threatened to delay the National Geographic's technical publication schedule. McKnew therefore asked Briggs to get something out of O'Brien. Briggs tried to convince O'Brien that Regener's differing results required a prompt answer, but O'Brien insisted that because there were differences, he had to be very careful with the quantitative reductions.[100] The tenacious O'Brien finally completed his reductions and submitted his manuscript in March 1935, five months behind the society's schedule. The society could not wait, however, and so O'Brien's paper did not appear. He later incorporated it into his combined results from the 1935 flight.

FIGURE 5.17. O'Brien's direct sunlight spectrograph survived and provided data. It had been hanging outside the gondola, was released by Kepner just before he jumped, floated to earth under its own parachute, and was retrieved several hundred meters away from the crash site. NGS.

Swann had the ironic satisfaction to know that at least one of his instruments survived: his Stösse chamber that had been removed from the flight at the last moment. His apparatus for determining the directionality of cosmic radiation was completely destroyed, as were its records, but he did manage to retrieve some information on the average number of counts per second as a function of altitude, and was also able to find some evidence of a variation in cosmic-ray intensity with variations in zenith distance.[101] Swann first saw his photographic records on

August 23 and was delighted that more than expected had survived, even though much of the photographic remains had been fogged owing to light leaks. However, he was more concerned that Stevens apparently forgot to tend to his experiment near maximum altitude, as cosmic-ray counts during descent were missing from the photographic reel. Swann surmised that Stevens failed to switch on the thyratron circuit since the photographic reel had been exposed, but lacked these data.[102]

Millikan's shielded electroscopes, with their films, were completely destroyed in the crash.[103] But he did retrieve some data from his unshielded instrument, which allowed him to reconfirm his earlier Fordney–Settle results, specifically the "peculiar shape of the altitude-ionization curve"

FIGURE 5.18a. Westover and Stevens ponder test photographic records from Millikan's unshielded electroscope prior to flight. NGS.

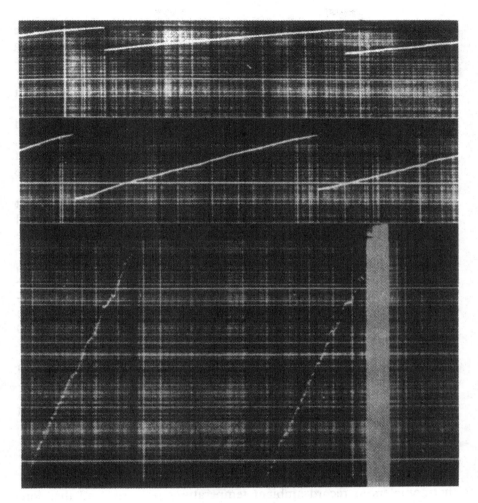

FIGURE 5.18b. Three actual records from sea level (top), 12,000 meters (middle), and 18,000 meters (bottom) show the increasing slope of the cosmic-ray trace, indicating an increasing rate of discharge, and hence a higher incidence of cosmic-ray energy. NGS.

constructed from the Fordney–Settle data taken at 9,300 meters, which differed from the data obtained by Millikan and Bowen in the 1920s. The Millikan–Bowen data were taken in flights at 42 degrees, the Fordney–Settle data at 53 degrees geomagnetic latitude. Millikan therefore attributed the difference to the latitude effect, which he now recognized and rationalized in terms of his own cosmic-ray photon model. These results allowed Millikan to proclaim, "The flight was 50% effective so far as our objectives were concerned." But losing the shielded electroscope data prevented them from determining the penetrating power, and hence the energy spectrum of cosmic rays. Millikan also regretted not obtaining readings at elevations in excess of 23,000 meters, which might have detected cosmic-ray primaries. In a report to Stevens, Millikan added: "May I congratulate you and the other members of the crew on the magnificent display of poise, intelligence, and courage under trying conditions."[104]

W. G. Brombacher of the National Bureau of Standards had placed his barographs in heavily insulated balsa wood boxes outside the gondola, and the balsa deadened the impact of the crash. In addition, some of Stevens's aerial photographs survived, so that he was able to calibrate the barometric altitude measurements against stereoscopic photographic pairs. He also used photograph pairs to derive wind directions from the drift of ground patterns. The combined results of the barographs and aerial photographs gave his Wright Field photogrammetrist colleagues some insight into the variations of barometric pressure as a function of altitude and temperature. To Stevens's delight, the official thermometer records and the all-important official F.A.I. barographs survived. All in all, *Explorer* did obtain some useful temperature/altitude records, which revealed an "S-shape curve instead of a straight line," as Stevens noted to Fordney in November 1934.[105] The temperature between 3,000 and 15,000 meters dropped faster than expected, and then rose above that height:

From 50,000 feet up, it begins to increase slowly and I dare say it keeps on increasing, at least in daylight, to figures that may be astonishing. One scientist, Dr. Humphreys [of the National Bureau of Standards], says that the air at 25 miles elevation may be as hot as that of the Sahara. Other scientists have ventured even greater temperatures.[106]

This fascinating result, new to Stevens, validated his belief in the utilization of heavy manned balloons. He argued to Fordney that all previous sondes rose too fast through the 3,000–15,000-meter region to allow their thermometers to record ambient temperatures properly, and therefore they failed to detect the nonlinear S-shaped drop. On the other hand, he admitted that if the rate of climb was too slow, radiation from the balloon could influence temperature measurements. Only in a manned flight could the rate be properly controlled to ensure that the ambient temperature was recorded. Although this was quite correct, the rate of ascent of a manned system was constrained by the maximum allowable flight time, determined by the respiration requirements of the crew, and the most efficient combination of ballasting and venting. Further, the maximum allowable ascent rate for accurate temperature measurements was still far greater than what was required for the cosmic-ray observations. Stevens did not discuss this last point.

Stevens was generally encouraged by the performance of the balloon, before the rip and subsequent explosion. His photogrammetrist colleagues at Wright Field were able to perform a preliminary calibration of the relation between altitude and barometric pressure and during the ascent, all had gone as planned, including aerial photographs at near-maximum elevation. Stevens therefore declared the mission a general success, since some scientific results were obtained, even though the balloon had failed to equal the altitude record set by Settle and Fordney. However, they had reached a maximum altitude only 180 meters lower than the official international record. Thus Stevens claimed with confidence, "Had the rip not occurred, it is probable that the balloon would have risen at least an additional 15,000 feet." He concluded:

Our most cheering thought of the recent ascent is that we feel we have successfully solved the problems of living and working efficiently in the stratosphere. It is gratifying to state that not a single piece of scientific equipment attached to the gondola failed us during the flight; every instrument worked exactly as planned. As for the balloon, we think another can be built that will go to its calculated maximum elevation without mishap.[107]

Briggs, more reserved than Stevens, was satisfied that the project stimulated cooperation between his bureau, the military, and the National Geographic Society, each contributing from their strengths. He added that another flight could be made if the design of the balloon was altered and helium was substituted for hydrogen. The crash had taken no lives. In fact, it presented Briggs with a real opportunity: to demonstrate the

FIGURE 5.19a. Three-dimensional flight profile of *Explorer*, from Rapid City, South Dakota, to Holdrege, Nebraska. Vertical scale is 8 times the horizontal. Swann Papers, APS Library.

FIGURE 5.19b. Vertical view from 19,000 meters. The thin dark line is the out-of-focus image of the rope leading to the hanging spectrograph. NGS/NASM.

value of the National Bureau of Standards in its assessment of what really happened to the balloon.

Between July and September, a Joint Board of Review was organized and convened to investigate the causes of the balloon failure. An informal inspection had been made of the shards of the balloon the day after the flight; board members also collected all available records on the construction, storage, and handling of the balloon prior to the flight. Briggs and the society's McKnew, with Hugh L. Dryden and L. B. Tuckerman of the bureau, inspected the salvaged balloon shards in August on several occasions, and later in August they examined reports from the Goodyear-Zeppelin Corporation. The bureau became a clearinghouse for technical information on the fate of the balloon. All remains, minus shards carried away by a few lucky treasure hunters, were sent there for analysis. Bureau personnel scrutinized fabrication, storage, inspection, and flight records and concluded that nothing was amiss. The flight crew reported that the balloon opened unevenly and, after direct on-site inspection, interrogations of witnesses, scrutiny of the records, and microscopic analyses of the remains, the Joint Board of Review concluded that the initial tears occurred radially in the fabric itself as the balloon expanded. During flight, folds in the balloon bag had not fully opened, and adhesion between the folds caused "excessive stresses ... through the shear resistance of the adhesions. These stresses produced multiple radial tears."[108] A new type of folding procedure had been used for the huge balloon, and during the month of waiting for good weather, sufficient adhesion had occurred between different parts of the cloth on opposite sides of the folds to keep them stuck together even as the balloon expanded.[109] The second failure, the explosion of the balloon, was relatively easy to reconstruct by the review board: clearly, once the hydrogen in the bag was mixed with sufficient oxygen in the atmosphere, an explosion was inevitable.

The board submitted its report in mid-September 1934, arguing that the experience of the flight and its analysis "form a distinct contribution to the science of stratosphere ballooning and will add to the security of future stratosphere flights."[110] No mention was made of a future flight for *Explorer*.

Notes

1. Kepner and Scrivner (1971), p. 124; A. W. Stevens to R. A. Millikan, 21 July 1933, RAM/CIT, R23 F22.5.
2. Stevens to The Chief of the Air Corps, 18 December 1933, and Karl Arnstein to A. W. Stevens, 10 May 1933, RAM/CIT, R23 F22.5.
3. Kepner and Scrivner (1971), p. 124.
4. Stevens was identified as a "reconnaissance zealot" by Burrows (1986), p. 40.
5. [News Letter Correspondent], "Record Long Distance Photograph," Air Corps News Letter XIII #15 (9 November 1929), NASM Technical files.
6. Albert Stevens to Chairman, Board of Trustees of the National Geographic Society, p. 3, 25 July 1933, NGS.
7. See Stevens to L. M. Mott-Smith, 24 December 1933, RAM/CIT, R23 F22.5 p. 4. See also Gilbert Grosvenor to Albert W. Stevens, 2 October 1931, NGS.
8. William Beebe (New York Zoological Society) was also supported by the National Geographic Society for deep sea investigations near Bermuda in his bathysphere in 1934, his third season, where he reached just over 930 meters. His two-man bathysphere was of cast steel, 1.4 meters in diameter, and contained cameras, a searchlight, and a telephone. The sphere was constructed by Otis Barton of MIT who accompanied Beebe on some of the dives. See Oehser (1975), pp. 17–18.

9. A. W. Stevens to R. A. Millikan, 21 July 1933, RAM/CIT, R23 F22.5; Stevens to Chairman of the Board of Trustees, 25 July 1933, NGS.
10. Stevens to Chairman of the Board of Trustees, 25 July 1933, NGS.
11. Stevens to Chairman of the Board of Trustees, 25 July 1933, NGS, p. 3. See also Stevens to Millikan, 21 July 1933, NGS.
12. Stevens to Grosvenor, 20 July 1933, NGS.
13. Stevens to Millikan, 21 July 1933, RAM/CIT, R23 F22.5.
14. Goetzmann (1986), p. 420. For a discussion of the motives of the founders of the National Geographic, see Pauly (1979), pp. 517–532; and Abramson (1987).
15. Abramson (1987), pp. 119, 149.
16. Pauly (1979), p. 518.
17. Abramson (1987), p. 137.
18. Stevens to Chairman of the Board, 25 July 1933, NGS, p. 4; Stevens to Millikan, 21 July 1933, NGS.
19. Ibid.
20. Stevens to Millikan, 21 July 1933, pp. 6–7, RAM/CIT, also NGS.
21. Millikan to Stevens, 5 August 1933, RAM/CIT, R23 F22.5.
22. R. A. Millikan, "Report to the Carnegie Institution of Washington," 13 September 1933, RAM/CIT, R23 F22.5.
23. See R. A. Millikan to Isaiah Bowman, 9 November 1933, with attached report addressed to Henry A. Wallace, 13 November 1933, "Report of the Special Committee on the Weather Bureau of the Science Advisory Board of the National Academy of Sciences, appointed by President Roosevelt, Executive Order No. 6238"; R. A. Millikan to Gen. B. D. Foulois, 4 December 1933, RAM/CIT, R40 F35.19. On the Science Advisory Board, see Kevles (1978), chap. 17; Kargon and Hodes (1985), pp. 301–318.
24. On Millikan's contacts and continued need for military services, see R. A. Millikan to William R. Blair, 22 July 1933; Millikan to Gen. B. D. Foulois, Chief of the Air Corps, 3 August 1933; J. H. Newton to Millikan, 20 July 1933; Millikan to Comdr. F. W. Reichelderfer, 5 September 1933, RAM/CIT, R33 F30.5. See also Bates and Fuller (1986), pp. 34–35.
25. See NBS/NARA General Correspondence, 1933 INA folder; 1934 INA folder. F.A.I. was the international body for validating aviation records.
26. Albert Stevens to W. G. Brombacher, 1 August 1933, and W. G. Brombacher to A. W. Stevens, 8 August 1933, General Correspondence, 1933 INA Folder, NBS/NARA.
27. Lyman J. Briggs to Gilbert Grosvenor, 10 August 1933, NGS.
28. Kevles (1978), chap. 16, "Revolt Against Science," pp. 236–251, especially pp. 250–251; Cochrane (1966), pp. 308–311, 345–350.
29. The perception of duplication of effort is noted in Bilstein (1984), p. 125. Roland (1985) examines how National Bureau of Standards aeronautical research programs kept the NACA from developing its own research in several key directions, but the effect of this avoidance of duplication on the bureau, or the bureau's perception of NACA as a competitor, has not been addressed, although the bureau did seem to have the upper hand. See Roland (1985), p. 135, pp. 139–140, p. 162.
30. Cochrane (1966), p. 323.
31. Ibid., p. 316.

32. Dr. John Oliver LaGorce to Gilbert Grosvenor ("G.H.G. from J.O.L."), 13 December 1933, NGS.
33. The society proudly advertised that it supported Byrd's Antarctic expeditions to the tune of $100,000, and gave $25,000 to help preserve the giant Sequoia in California, over $50,000 for a series of expeditions to Peru to unearth traces of the Inca civilization, and $65,000 to the Smithsonian Institution to establish a solar monitoring station for six years in Southwest Africa. See Endpapers of the *National Geographic Society Stratosphere Series Publications #1 and #2*, 1935–1936.
34. Stevens to Millikan, "Contributions . . . ," n.d., RAM/CIT; Stevens to L. M. Mott-Smith, 24 December 1933, RAM/CIT, R23 F22.5.
35. Stevens to The Chief of the Air Corps, 18 December 1933, RAM/CIT, R23 F22.5, p. 3.
36. O. Westover to Chief, Materiel Division, 9 January 1934, Orvil Anderson Collection, MAFB.
37. W. F. G. Swann to Lyman Briggs, 13 January 1934; Briggs to Swann, 16 January 1934, Swann Papers, APS. Donald Piccard recalls that his parents visited Grosvenor asking for support before the Explorer flight was announced, and that Grosvenor did not mention his own plans. Donald Piccard to the author, 18 July 1988. Surviving correspondence indicates only that the Piccards met with Grosvenor just at the time of the announcement. See G. Grosvenor to J. Piccard, 19 January 1934, PFP/LC.
38. Bates and Fuller (1986), p. 43–44.
39. Stevens to Millikan, 21 July 1933, RAM/CIT, p. 7.
40. This is confirmed in Thomas Settle to Lewis M. Mott-Smith, 24 December 1933, RAM/CIT, R23 F22.5.
41. A. W. Stevens to The Chief of the Air Corps, 18 December 1933, RAM/CIT, R23 F22.5, p. 3.
42. Lyman Briggs to Gaertner Scientific, 17 February 1934, folder VI-6/INA-678-C, NBS/NARA, RG 167.
43. F. L. Mohler to Daniel Mann and to A. W. Stevens, 23 March 1934, and to R. W. Wood, 27 March 1934, folder VI-6/INA-678-C, NBS/NARA, RG 167.
44. Brian O'Brien oral history interview, 9 March 1987, NASM.
45. See O'Brien (1925), pp. 486–491; O'Brien (1930), p. 381 (abstract); and O'Brien (1931), p. 471.
46. Mohler to R. W. Wood, 27 March 1934, folder VI-6/INA-678-C, NBS/NARA, RG 167.
47. O'Brien oral history interview, 9 March 1987, NASM.
48. Stevens to L. M. Mott-Smith, 24 December 1933, RAM/CIT, R23 F22.5.
49. Stevens to Millikan, 15 January 1934, RAM/CIT, R23 F22.6.
50. Quoted from *Monthly Evening Sky Map* (October 1934), in H. G. Phair to J. A. Fleming, 24 October 1934, RAM/CIT, R23 F22.9. Phair, an engineer in the Signal Corps and specialist in radio transmission, wrote to volunteer his services to Compton's project.
51. Compton (1934a), p. 8.
52. In February, Compton responded more positively to an inquiry from Swann on Stevens's behalf, indicating that he was willing to share the experience he had gained, and possibly a detector, if desired. But clearly he did not

want to participate in an active manner. See A. H. Compton to W. F. G. Swann, 15 February 1934, Swann Papers, APS.

53. Compton later argued that these expenses, along with shipping and travel costs arising from his World Survey and other activities, had to be somehow reimbursed by the University, possibly by setting up an account into which Compton might deposit revenue he had gathered from consulting for General Electric and other interests. See A. H. Compton to Frederick A. Woodward, December 4, 1934. A. H. Compton Papers, Regenstein Library, University of Chicago.

54. Stevens to Millikan, 15 January 1934, RAM/CIT, R23 F22.6.

55. Millikan to Stevens, 19 January 1934, RAM/CIT, R23 F22.6.

56. Stevens to Millikan, 23 January 1934, and Millikan to Stevens, 8 February 1934, RAM/CIT, R23 F22.6.

57. Minutes of Advisory Committee, 21 February 1934, NGS, pp. 1–2.

58. Bernson (June 1934), pp. 131–134.

59. Grosvenor to Stevens, 8 February 1934, NGS.

60. T. W. McKnew to G. G. Brooder, 23 May 1934. See also "Research Committee Grants Stratosphere Flight I," folder NGS.

61. "Minutes of a Meeting of the Executive Committee of the Scientific Advisory Committee . . ," 13 May 1934, pp. 9–10, NGS.

62. Ibid., p. 13, NGS.

63. Millikan to McKnew, 22 May 1934, RAM/CIT, R23 F22.7.

64. "Minutes of a Meeting Held Wednesday Afternoon, February 21, 1934 . . ," pp. 1–2. National Geographic Stratosphere Flight I File, NGS. The minutes recorded that the actual cost to Goodyear was over $35,000.

65. Arnstein (1935), p. 96.

66. Ibid., pp. 96, 105, and fig. 6, p. 104.

67. At first Kepner was to take the lead, but as he became involved with the highly controversial military takeover of the airmail service at Wright Field, Stevens took charge. See A. W. Stevens to A. W. Winston, 25 March 1934, Stevens Folder, Swann Papers, APS.

68. See, for instance, Stevens to I. S. Bowen and H. V. Neher, 5 May 1934, Stevens Folder, Swann Papers, APS.

69. Stevens to McKnew, 21 April 1934, NGS.

70. Winston (1935), p. 110.

71. Newspaper fragment, "Temperature Little Changed Inside Gondola, Says Fordney" (November 22, 1933), n.p., "Century of Progress" Associated Press wire story, NASM Technical Files.

72. Stevens (1934), pp. 401, 432–433.

73. Bonham (1936), pp. 260–261.

74. G. Grosvenor to J. Piccard, 19 January 1934, PFP/LC.

75. Jean later reversed his opinion about the use of ropes for controlling the rip valve on the balloon, became more interested in Stevens's methods for air filtration, and also sent various motors and other devices to Stevens at Wright Field for testing in their high-altitude chamber. See Jean Piccard to A. W. Stevens 27 February 1934, Stevens Folder, Swann Papers, APS.

76. See letters between Jean Piccard and A. W. Stevens, February 1934 to August 1934, boxes 42 and 53, PFP/LC.

77. "Radio Speech of Gilbert Grosvenor," 23 April 1934, NGS.

78. Ibid.
79. "Radio Address by Brigadier General Oscar Westover . . ," 6 May 1934, NGS.
80. Ibid.
81. Grosvenor to LaGorce, 29 June 1934, NGS.
82. Transcript, "Broadcast of Dr. John Oliver La Gorce . . ," 14 May 1934, NGS.
83. "Minutes of a meeting . . . " 13 May 1934, p. 16. NGS.
84. News Release, "From National Geographic Society," 26 June 1934, NGS.
85. Kepner (1935), p. 27.
86. Swann to Jean Piccard, 5 July 1934, PFP/LC. See also Swann (1933c), pp. 1025–1027.
87. Untitled and undated transcript of radio broadcast, NGS.
88. Ibid., p. 4.
89. Ibid., p. 5.
90. O'Brien oral history interview, 9 March 1987, NASM.
91. Untitled and undated transcript of radio broadcast, NGS, p. 5.
92. Ibid.
93. News Release, McFall Kerby, "Most Ambitious Stratosphere Flight Yet Planned . . ," 17 June 1934, NGS.
94. The official record conforming to international regulations was set at 60,613 feet, or 18,790 meters, but the barometers indicated 62,720 feet and the photographic measurements 62,100. See Briggs (1935a), pp. 5–6.
95. Briggs (1936a), p. 5; Stevens (1934), pp. 397; 410–417.
96. O'Brien oral history interview, 9 March 1987, NASM.
97. F. L. Mohler to B. O'Brien, 4 August 1934, folder VI-6/INA-678-C, NBS/NARA, RG 167.
98. Brian O'Brien to Lyman Briggs, 3 August 1934, D/INA-678-C, O'Brien folder, NBS/NARA, RG 167.
99. O'Brien to Briggs, 12 November 1934, D/INA-678-C, O'Brien folder, NBS/NARA, RG 167.
100. Briggs to O'Brien, 20 November 1934; O'Brien to Briggs, 22 January 1935, D/INA-678-C, O'Brien folder, NBS/NARA, RG 167.
101. Swann and Locher (1935), pp. 12–13.
102. Swann to Briggs, 23 August 1934, Swann Papers, APS.
103. Millikan (1935), pp. 24–25.
104. Millikan to Stevens, 29 August 1934, RAM/CIT, R23 F22.8. He was also more generous in his assessment of the flight at this time, reporting to J. A. Fleming that the flight "gave us two-thirds of the important data which we hoped to get." See Millikan to John A. Fleming, 28 August 1934, RAM/CIT, R23 F22.4.
105. Stevens to Fordney, 19 November 1934, PFP/LC.
106. Ibid.
107. Stevens (1934), pp. 397–434; see also pp. 425; 434. On Stevens's calculations of altitude, see Stevens to Fordney, 19 November 1934, PFP/LC.
108. [Briggs] (1935a), p. 80.
109. Briggs (1935b), p. 304.
110. Ibid.

CHAPTER 6

The Flight of *Explorer II*

FIGURE 6.1. Dignitaries at Stratocamp. Left to right: John Oliver La Gorce, vice president of the National Geographic Society; R. L. Bronson, secretary of the Rapid City Chamber of Commerce; Albert Stevens; Gilbert Grosvenor, president of the National Geographic Society; Orvil Anderson; Lyman Briggs, director of the National Bureau of Standards; George Hutchinson, secretary of the National Geographic Society. NGS/NASM.

In an NBC radio talk on Sunday, November 11, 1934, when asked if he wanted to return to the stratosphere, Stevens replied: "Certainly!.. We feel that we learned so much from the last flight that we could make another stratosphere flight to a much higher elevation."[1] A return flight would provide more information on the structure of the stratosphere, which Stevens felt was a practical necessity "if airplane travel develops to the point where planes may travel in the stratosphere, all these things [air pressure, temperature, mass distribution, wind velocities, and streaming] have immediate value."[2] Stevens also believed that stratospheric flight would be the ultimate man could possibly attain. When asked, "Is stratosphere exploration the first step to the stars?" he replied: "No, I wouldn't say that. Man can never successfully leave the earth. Also, I do not believe that he could exist anywhere but on earth."[3] To Stevens, as to Piccard, the stratosphere was a new realm ripe for exploitation. It was accessible through available technology and was within the reach of any willing to journey there. It was an appropriate and practical national goal.

On December 10, Gilbert Grosvenor reconvened the Scientific Advisory Committee to read a letter he had just received from Gen. Foulois, Chief of the Air Corps.[4] Grosvenor had been prepared for another flight even before *Explorer* first flew, but had thought of it as an economical reuse of original equipment. In late October he turned first to his old friend George H. Dern of the War Department, and then to Foulois for support. Foulois knew that Grosvenor and Dern were close and that Dern was sympathetic to *Explorer*, and so gave his endorsement as well. With the endorsement of the Army, and with the comforting knowledge that the balloon, gondola, and instruments of *Explorer* had been insured by Lloyd's of London, there was nothing in the way of another try.[5]

Grosvenor's continuing patronage of *Explorer*, beyond his obvious need for corrective publicity, came from a sincere fascination with high flight. He later expressed to Stevens his great personal interest in the human experience of the flight and the fact that life could be sustained in such an alien environment.[6] Without Grosvenor's support and his efforts to keep Dern and Foulois interested, the proceedings on December 10 would have been far different. Before the committee met, hydrogen was still the lifting gas of choice. Briggs thought helium was too expensive, but along with other committee members wanted to find a way to streamline launch procedures so as to minimize the contact time between hydrogen and the sensitive rubberized balloon skin.[7] But after the meeting, the advisory committee decided that helium should be the buoyant agent in what was now to be called *Explorer II*.[8] The committee also decided that women were not to be allowed entry to the Stratobowl, especially the tent city and working areas called Stratocamp, for the 1935 launch. Details were to be worked out by the principals.

Within a week after the committee meeting, Stevens and McKnew traveled to Akron to work out the details with Winston of Dow and

Arnstein of Goodyear. There they decided that the gondola should be larger and its skin thickness reduced to maintain weight limits. They also discussed how to incorporate the use of helium into the new balloon design to maintain lift, and how a resulting balloon bag, possibly as large as 105,000 cubic meters, should be prepared and handled. The Akron meeting identified most of the major design details and flight operations requirements for a return to the stratosphere. One major flight parameter that changed several times, however, was the size of the crew. Foulois would not let Kepner fly again; he was too valuable to the Materiel Division as a procurer of new aircraft. Foulois also recited an obscure policy that precluded Kepner's continued participation. Kepner confided to Briggs: "There are many breaks in this life where one is unable to do as he pleases." Kepner certainly would have flown again, if given the chance.[9] Without Kepner, the planners considered a two-man crew, instead of three, but the size of the crew changed back and forth several times after this point. In the spring of 1935, the flight roster included Stevens as commander, with Anderson as pilot and Capt. Randolph Williams as observer. But this would soon change again.[10]

Most of the new balloon fabric was ready by the spring, and although Dow was slightly behind schedule, the larger, but lighter, gondola was undergoing both water and air pressure tests.[11] The new helium balloon was redesigned with heavier rubberized cloth for the lower section below the catenary; the portion of the balloon that had ruptured now had the heaviest weight cloth. Also, reacting to the lessons caused by the failure of the folds in the cloth during the first ascent, the Goodyear engineers reduced the stickiness of the inner cloth surface by treating it first with a smooth rubber compound and then dusting it with powder to make it dry and nontacky.[12]

The new gondola, still constructed of Dow's magnesium/aluminum alloy Dowmetal, was larger by 20 centimeters; it was now Stevens's original 2.8-meter sphere, except that it weighed in at 290 kilograms, 28 kilograms lighter than the Explorer I gondola. The most significant alteration in the design of the Explorer II gondola was that the manholes were enlarged by 5 centimeters in each dimension, since Kepner, Anderson, and Stevens had serious problems getting through the manholes in their hasty exit from *Explorer I*. Other minor alterations in the gondola included more secure strapping and bolt fittings to tie down all the equipment directly to the walls of the gondola, thereby eliminating the shelves. The connecting system between the gondola and load ring was also redesigned to reduce its weight, and the working floor was enlarged by 10 centimeters. Dow conducted the same types of stress tests as it did on *Explorer I*, but this time air pressure tests were continued both at Wright Field and at the Stratocamp during the integration and installation phases, when the gondola was hooked to the load ring and the instruments were all installed with sealed bolts.[13]

Publicity, which had been quieted during the fall and winter while the society and the Air Corps regrouped, resumed in the spring. On April 20, the National Broadcasting Company's Blue Network ran a repeat series with Grosvenor, Dern, and Stevens speaking of the new patriotic venture, sandwiched between renderings by the Army Band of "Americans We," "Stars and Stripes Forever," "Men of Valor," and "The Star Spangled Banner." Grosvenor spoke of the stratosphere as "lifeless, lonely and desolate. Yet only there may some emanations from outer space be recorded in their full intensity. Such are cosmic rays."[14] He also emphasized that the new Explorer would use nonflammable helium, "a wonder gas which once was almost as rare and costly as radium."

Anderson, Stevens, and Williams were constantly showered with letters and requests from well wishers. Most common were autograph requests, although many asked if self-addressed letters might be taken up in the gondola. Even welfare society balls were held in the name of the flight; in May, a "Stratosphere Ball" to raise funds for children was held in Evanston, Illinois.[15]

A First Attempt in July

By late May, 1935, Stevens had been officially designated as commander, scientific observer and photographer, Anderson was pilot and Williams was ground officer and alternate pilot. The new gondola arrived in Rapid City in May, as did the balloon. The June flight date slipped because of poor weather; while the tension mounted, Stevens and Briggs would take long walks in the hills and with Swann would anxiously check and recheck all life-sustaining systems aboard the gondola at each delay. When the daily weather forecast would cancel possibilities for launch the next day, Stevens would try to ease tension that night by keeping in close, cordial contact with the scientists, and possibly even singing the refrain from a locally inspired ballad "Home in the Bowl," sung to the tune of "Home on the Range:"

> Home—home in the Bowl
>   Where the Soldiers and Scientists play
> Where ever is heard the discouraging word
>   There'll be no inflation to-day.[16]

To try and cheer up the crew, O'Brien, who was in Rochester, relayed his good wishes to Anderson and, recalling the humorous saga of the fruit flies in Settle's flight, which had become institutionalized by then, telegrammed them saying: "BEST WISHES DON'T BE LIKE THOSE FRUIT FLIES YOU ARE ALL RIGHT AS YOU ARE."[17]

On July 10, a stable high-pressure condition finally made it possible to launch *Explorer II*. Everyone moved quickly to prepare the balloon for inflation and ready the gondola, instruments, and passengers. The gondola was opened, and with ground crew properly stationed all around, the crew entered and were sealed in the gondola. Then the inflation began late in the evening of the 11th and continued through the first dark hours of the 12th.

During the inflation and roping process, as the balloon swelled over the camp and was being connected to the gondola, it suddenly collapsed, burying the gondola and three ground workers in only seven seconds. O'Brien, who was in Rochester, recalls stories about the enormous gas bag, which momentarily hovered and then fell to the ground. In those few moments, no one knew just where the acres of heavy cloth were going to land, and just who or what was going to be buried. Fabric holding the upper end of the rip panel had torn, and the entire upper portion of the balloon had failed.[18]

As attending soldiers rescued the three workers, who escaped serious injury, near panic set in as Goodyear-Zeppelin at first accused the ground crew of overinflating the balloon. Members of the advisory committee quickly stepped in to ease tensions and asked for an immediate inspection of the evidence, and then reminded everyone that they all were in the same boat, as far as the rest of the world was concerned.[19] Much more than a summer launch was lost now. Goodyear-Zeppelin had to reconstruct and strengthen the balloon, and launch procedures had to be reevaluated. These were hardly sanguine days for the project. Soon after the July failure, the advisory committee met to assess the situation. Every aspect of the project was now in question, even the use of the Rapid City site. La Gorce sensed the feelings of those assembled: "In the hearts of all of us is the feeling of combativeness in that we are not satisfied with terminating the effort but it would be a very serious matter to have a third failure."[20] Everyone realized that Rapid City was still probably the best site, and that they should push ahead as fast as possible, if for no other reason than the fact that Westover, a strong Army ally, was about to be reassigned. Heavily depressed, Stevens apologized to the group "for this whole business because if it had not been, perhaps, for my getting such an ambitious thing started in the first place it would have saved a lot of trouble."[21] But, he added, since "we have started it . . . we are in a position where we certainly have to finish it." Stevens pledged another $1000 to the enterprise, which moved La Gorce to say that the society was not "sorry that you brought the project to us. We are very proud that you did and proud of the two failures. They were steps forward in the realm of science that no fair-minded person can deny was a pioneering effort."[22]

Stevens was by now four and a half years into the project, and his incessant drive and critical attention to every detail was beginning to wear him down seriously. His intimate involvement with every collaborator and assistant, through long and exhaustive correspondence, and his constant traveling to every center of activity over the past two years had already caused Lyman Briggs to become concerned about Stevens's health: "He is driving himself unmercifully, is not getting enough sleep, and I have been told that he seemed to be getting very nervous."[23] But those who knew Stevens better, such as the society's La Gorce, assured Briggs that this is how Stevens lived. As a bachelor, his "only interest in his life is his work." La Gorce agreed that they had to be vigilant to make certain that Stevens did not exceed his limits. By July 1936, he was very close to the limit, but the project always came first.[24]

Brian O'Brien followed the proceedings that summer with more than casual interest. He had the impression that La Gorce and McKnew indeed responded very positively to the crisis, and that there was no hesitation on the part of the society to continue with the project. To O'Brien, the society was a staunch supporter: "it was such an obvious thing to continue. The National Geographic did not understand science . . . but they recognized science and they had a number of very fine advisers."[25] O'Brien understood and appreciated the image the society wished to project. In August, a few members of the advisory board learned differently, when Grosvenor had to draw in the reigns.

Millikan Pulls Out

At their July 29 meeting, the Scientific Advisory Committee decided to be prudent about press coverage; this time, no buildup program would be mounted because they had little idea when or if they would actually launch again. No longer were the NBC broadcasts needed to keep the millions interested.[26] The July disaster had an extremely sobering influence on every member of the project, but the society's concerns over image had already intensified as a result of derogatory public comments on manned ballooning the month before by none other than Robert A. Millikan, who never encouraged manned expeditions, but was compelled for many reasons to cooperate when invited. But Millikan had finally lost all patience with the manned effort.

As he did with most of the principal scientists, Stevens always kept Millikan informed of progress in detailed and encouraging personal reports. After the 1934 crash of *Explorer*, Millikan had been compensated for his destroyed detectors by the society's insurance account; although the insurance did not cover the total cost, Millikan could at least produce another set of shielded electroscopes.[27] But this was little compensation, as Millikan's colleagues had planned to reuse the Explorer instruments immediately.[28] Although contentious about the whole project, Millikan continued to cooperate. He remained preoccupied with obtaining support for equatorial flights and continued to remind Grosvenor and others that he would rather see *Explorer* fly from a more southerly site. Grosvenor held firm arguing that the extent of their support was limited to *Explorer*

II, which was consuming "a substantial portion of our research budget, the flight being among the largest and most expensive wholly scientific expeditions sent into the field in recent years."[29]

But in the spring of 1935, as Millikan gained Army participation to fly his unmanned balloonsondes in equatorial regions as well as Weather Bureau support to continue domestic flights, his attention was drawn more and more in that direction. He had to be sure—when opportunities arose to exploit available manpower at little or no cost—that he had resources ready to take advantage of them. He therefore had to ensure that his operation was not stretched too thin: his budget for both instruments and technical assistance, although comparatively large at the time, was small compared with the enormous budget of *Explorer II*. The irony could not have been lost on Millikan, but *Explorer* was his only chance to send a very heavy payload to a great height.

Contention developed into contempt during March, when Millikan and Neher learned that the ionization chambers they had been developing for *Explorer II* were too large; scale models that Stevens had requested simply would not fit even in the larger gondola space. Stevens asked them to provide smaller chambers, which gave them problems.[30] By May, Millikan and Neher were very busy coordinating mountain, aircraft, and balloon experiments, and tried to get a straight answer from Stevens as to when they—probably only Neher—would be required at the launch site.[31] But when Stevens did not respond, Millikan decided to pull out of the Explorer flight and use his electroscopes and ionization chambers elsewhere, especially in south Texas and the Philippines.

On June 21, before the July failure but during the frustrating period of waiting for good weather, Millikan gave a lecture at the University of Iowa, where, according to an Associated Press release titled "Stratosphere Jaunts Just a Fad For Adventurers, Says Millikan," he said:

Stratosphere flights in manned balloons are a fad which will cease when they no longer are regarded as spectacular . . . it is chiefly the spirit of adventure which has caused men to reach new heights in balloons.. Such flights probably have the same scientific importance as the Byrd expedition to the South Pole . . . but it is the spirit of adventure which plays the most important part. When the flights are no longer spectacular it will be impossible to finance them.

Millikan's indictment and clear slap at the National Geographic Society's support of Byrd, which was even greater than that given over to *Explorer*, continued in a familiar vein: "I would rather have my data obtained at a height of nineteen miles in a small unmanned balloon than all that can be obtained in the larger balloons such as the Explorer."[32] He added that his own wish was to send balloons aloft in equatorial regions, where they could do more good and at far lower cost—$1,000 to $2,000 per flight as opposed to some $60,000. Millikan's strong criticism should not have come as a complete surprise to the National Geographic

Society. Since May, Millikan had clearly lost enthusiasm for the flights, and in his correspondence with Stevens, he made clear that he wanted to withdraw his participation since his forces were spread so thinly already.[33] In fact, as early as January 5, 1935 Millikan hinted to Grosvenor that he no longer sought National Geographic Society support for his experiments since he was fully supported by the Carnegie Institution, which allowed him to engage in observations of cosmic-ray phenomena as he saw fit.[34]

Even though Millikan's statements were understandable, they certainly were imprudent; publicity such as this could well ruin everything Stevens had worked for. The abortive flight attempt in July brought matters to a head. Millikan's instruments were still on the roster even after his Iowa speech. But after the July disaster, Millikan pulled out. Swann's group took over the general cosmic-ray program, and Millikan's instruments were returned to Caltech in August 1935.[35] Now there was room for Swann's large Stösse chamber, removed from *Explorer* at the last moment.

Stevens, in a general state of despair, willingly and without hesitation returned Millikan's apparatus. He had been running between Rapid City and Akron and Dayton trying to save the program, and admitted:

It is very much regretted that the collapse of the balloon put an end to the July flight. Whether another flight may be made is still in question. I wish to thank you and Dr. Neher for your interest and only wish that we could have been of service in getting these instruments into the stratosphere.[36]

These words were written while Stevens was experiencing a loss of confidence in the mission, and his highest state of stress. Millikan's attitude added considerable injury and may have helped to place the entire effort in jeopardy as now the National Geographic Society was turning sour on *Explorer II*.

Cutting Back the Science

The greatest stress on Stevens was the doubt about the continuation of the project itself. Even through August, internal decisions at the National Geographic Society kept the future unclear. Lyman Briggs and Oscar Westover argued that *Explorer II* should be supported, but Gilbert Grosvenor remained "rather doubtful about the wisdom of undertaking another flight."[37] Notwithstanding the results of the July 29 advisory committee meeting, Grosvenor still could not make up his mind. Safety and image now were uppermost in his mind, as Briggs told Swann in confidence:

Dr. Grosvenor very properly insists that safety must be a prime requisite in another flight. He also is anxious from the standpoint of the Society to see a new altitude record established. Dr. La Gorce is perfectly frank in saying that from the Society's standpoint these are the two main considerations now.[38]

To ensure the safety of the crew, Grosvenor recommended they not be given too much to do while in flight. He therefore: "urged that the amount of scientific equipment be reduced with particular reference to equipment requiring unusual attention or work on the part of the flight personnel."[39] "Banned" were Millikan's devices which no longer were in Rapid City anyway, because they had required too much attention to discharge the lead ballast shielding; one of the high-frequency radio experiments of the National Bureau of Standards; Swann's cloud chamber, "because it requires a good deal of attention" and because it was not yet operational; and O'Brien's suspended spectrograph because of difficulties in launching, and "because Anderson feels that in case of the failure of the balloon the gondola might strike it on the way down and be punctured."[40] On the other hand, Swann's Stösse chamber survived the cut; although it was quite large, it was not too heavy and it was automatic. It was also a high-priority item for Swann.

To Briggs, these were devastating cutbacks in the scientific complement, amounting to 50 percent, and were made with priorities far different than those he held dear when he offered to reduce the instrumentation for the first flight. Aware that Millikan had already pulled out, Briggs lamented in particular the loss of the horizon spectrograph; he confided to Swann, "I don't want to see the whole ozone program go by the board." Briggs had to admit that the National Geographic Society had spent more than it planned to, and would be unwilling to put more than an additional $15,000 into the project for another flight. Before Grosvenor committed anything extra, he had to be assured by the reconvened board of review that the problems of the July abort had been solved fully. Goodyear-Zeppelin was cooperating completely, at no cost to the society, and Briggs hoped that local South Dakota interests, such as the "Homestake Mine and other good citizens" would be able to raise another $10,000 to cover additional expenses the society would not pay for. Thomas McKnew had argued that a flight from Scott Field would cost $10,000 less and be better controlled than one from the Stratobowl. But if Rapid City would pay the extra cost, the Stratobowl was still acceptable. The society favored Scott Field for a time, but despite the costs, the flight crew and remaining scientists still were committed to the Stratobowl.

Gilbert Grosvenor, who was under considerable self-imposed pressure, was rightly concerned for the safety of the crew, if not for the public image of the National Geographic Society supporting an expedition that might end in disaster. The January 1934 crash of the ill-fated "Ossoaviakhim-1" had been a nightmare that was no doubt remembered all too well as successive balloons failed—from the crash of *Explorer I* through the previous July's abort, to the belated announcement on August 9 of a near fatal incident in a second Soviet balloon launch. Grosvenor decided that he had to step in and obtain assurance that the same fate would not meet *Explorer II*.[41]

On June 26, a crew of three Soviet balloonists including the aeronaut K. I. Zille, physicist Alexander Verigo, and engineer J. G. Prilutsky took a battery of Wilson Cloud chambers and other instruments beyond 16,000 meters on the *U.S.S.R.1-Bis*. First reports indicated complete success, but on August 9 the weekly *Science-Supplement* revealed that the balloon had developed a large rip like the one that "caused the failure of the Army Air Corps-National Geographic Society balloon *Explorer II* just before it left the ground." The details of the Soviet experience became an object lesson for Explorer: "They offer a picture of what might well have happened to the *Explorer II* if its accident had occurred in the stratosphere instead of on the ground."[42] Prilutsky and Verigo bailed out, and Zille was then able to maneuver the lighter gondola under its weakened balloon to a safe landing. The Soviets had suffered a traumatic experience that was certainly not one Grosvenor wanted to have associated with the National Geographic Society.

Yet for all his concern, Grosvenor still demanded a new altitude record in the same breath. This is what killed the January 1934 Russian crew, according to a detailed failure analysis performed by G. Prokofiev for the Soviet Academy of Military Aeronautics, part of which was published in *Pravda*.[43] Saving weight by canceling half the scientific program was one thing, but the crew was soon to be reduced to two as well. Clearly, saving the weight of one crewman was not going to make life any easier or safer for the remaining two, save for the fact that there would be more room in the gondola. Attainable altitude was the only real motive.

For the rest of his life Stevens would remember the stress he experienced that summer. In the late 1940s, in ill-health and within a year of his death, Stevens recalled the strain of the wait for good flying weather and the summer of 1935. He wrote Swann remembering how they tried to reduce the stress by long walks in the hills; but Stevens also remembered keenly how he did his best not to show his feelings in front of the Stratobowl crew. Briggs was sometimes his companion on these walks, and they, with Swann, would sometimes spend the evenings together over a drink to settle their nerves "after a day or days of suspense because of weather uncertainties."[44]

Grosvenor eventually agreed to another flight. During September the balloon and the rip panel sections were strengthened by Goodyear-Zeppelin in Akron. The Goodyear-Zeppelin people were on the spot: They decided to take no chances and so strengthened everything they could think of in the balloon structure to ensure success, and at the last moment decided to replace the rip panel completely by a heavy, uniform upper panel. The function of the rip panel was taken over by a steel wire ripping device.[45] In late September they returned to Rapid City with a newly repaired and slightly larger balloon, and the camp quickly and enthusiastically re-formed for another attempt. In early October, Oscar Westover wrote to the three crew members urging them to keep their spirits up:

The Stratosphere flight has been a bold plan and adventure so far, and naturally we have had to learn through sad experience some of the practical difficulties to be whipped before success could be achieved, but I confidently feel that the difficulties so far experienced, as well as the experience gained by all personnel connected therewith, will surely result in a successful third attempt.[46]

Westover added that the new rip panel design furnished by Goodyear-Zeppelin had his complete approval and greatly improved the reliability of the balloon. All that was now required, he hoped, was good weather.

At some point during October, Captain Williams was dropped from the flight roster, leaving Stevens and Anderson. All was in uneasy readiness. The ground crew and Goodyear engineers were on constant watch for problems—they found additional small rips due to the rapidly cooling days, and countless small technical problems created delay after delay but at least kept everyone busy. Nevertheless, it became clear that rubberized cloth balloons were almost as uncontrollable as the weather.

Lift Off

The weather finally cleared up in the second week of November. Well before dawn on November 11, the balloon crew suited up while Goodyear and Army personnel fought back the cold to inflate the balloon. But the cold was a threat from the very start: A 5-meter rip opened as soon as the bag billowed with the first helium charges. Goodyear people swarmed around the ripped bag. They took two hours to repair it and to recheck all visible folds. When they finished, filling started again, and just before sunrise, the gondola was attached to the great balloon. In the worklights of Stratocamp, both remained vertically poised like a giant octopus sheltered by the pocket of cold air trapped in the Stratobowl. *Explorer II*

FIGURE 6.2. The inflation of the Explorer II balloon. Helium is being fed from tanks to the right of the illuminated portion of the compound. NGS/NASM.

FIGURE 6.3. The gondola is connected to the balloon by Goodyear technicians. Once the balloon was inflated, the gondola was rolled in and all connections made while the crew worked inside. Stevens is visible in the facing port. NGS/NASM.

strained at its tethering tentacles as the ground crew waited for the signal to launch.

The helium-filled Explorer balloon carrying Stevens and Anderson slowly lifted from the Stratobowl at 7:01 a.m. Mountain Standard Time under what were thought to be ideal weather conditions. Small sounding balloons released just minutes before had revealed no currents. But just after liftoff, in the dim light of dawn, and with the huge balloon still well within the Stratobowl, warmer winds aloft penetrated the bowl and caught the balloon and drove it toward the cliffs. It barely missed the trees on the edge of the limestone escarpment as Anderson raced to dump ballast, which landed all over the assembled crowd at the rim of the bowl.[47]

With its 6,800-kilogram payload, which included almost 1,000 kilograms of scientific instrumentation, the balloon rose rapidly because Anderson had thrown out a great deal of ballast. Anderson then countered

FIGURE 6.4. *Explorer II* rising out of the Stratobowl, drifting towards the cliffs where spectators are massed. Note parked automobiles in the background. NGS/ NASM.

the quick rise by venting some helium, and reached the first plateau at 5,300 meters in 29 minutes, where they stopped to inspect all systems. Then they started a slow rise of 60 meters per minute, increasing to 180 meters per minute until they reached the pressure equilibrium height of 20 kilometers at 10:35 a.m. At this point, they released ballast so they could rise to maximum height, and reached 22 kilometers at 10:50, where they remained until 12:20 p.m.

Most of the observations were taken at peak altitude, although some scientists, for example Swann, had planned to have their instruments work on the ascent, at the peak, and during the descent. Some of the automatic devices were turned on only when the balloon was close to maximum altitude: Swann's Stösse chamber was turned on at peak altitude and continued throughout the descent, as planned, but the crew did not turn on Swann's Geiger counter arrays until about an hour before they reached maximum altitude.[48] Stevens's aerial cameras and O'Brien's spectrographs worked continuously during the flight and were only shut off when the crew had to release the heavy batteries before landing.

The descent started slowly at 12:20 p.m. when the crew opened the balloon valve, and increased from 90 meters per minute to over 220 meters per minute, with the balloon reaching 5600 meters by 2:30 p.m. The crew then checked their rate of fall by releasing more ballast, along with the batteries, and opened the gondola at around 5,000 meters. With ballast to spare, the balloon and gondola descended to a float level at 600 meters above the ground and then, under perfect control, descended very slowly as it was being frantically followed by hordes of automobiles and light aircraft. Stevens and Anderson let out the drag ropes when they found a completely clear field near a road, and the gondola gently touched down at 3:14 p.m. near White Lake, South Dakota. At the instant of landing, Anderson pulled the wire ripping device to open the giant gas bag completely so it would collapse and not drag the gondola across the prairie.

Both the scientific equipment and the gondola were unscathed and virtually everything worked perfectly, according to Stevens.[49] Later reports admitted that a number of items did not work properly, such as the valving and a small fan mounted outside the gondola, which was supposed to rotate the gondola for Swann's study of the directionality of cosmic rays.[50] Still, *Explorer II* was a great success overall, ironically for Millikan. The Army Air Corps and the National Geographic could boast a new world's altitude record for manned flight and a significant return of scientific data about the stratosphere and cosmic radiation. Stevens's dream had come true, and the society was delighted, if not considerably relieved.

The Scientific Complement and Results

Briggs believed that the Explorer series "constituted a striking example of experimental cooperation."[51] Altogether, seven institutions provided apparatus for about two dozen explicit experiments, each of which had to be integrated into the gondola, or suspended from the gondola, and also had to be manageable by the balloonists during flight. The cooperative nature of each experiment extended from conception, design and integration to the reduction of the data and their elucidation. And although Stevens above all had been the coordinator and the glue, it was Briggs's institutional fabric that ultimately held it all together and gave it substance. Accordingly, on December 21, 1935, the National Geographic Society sponsored a coordinating conference at the National Bu-

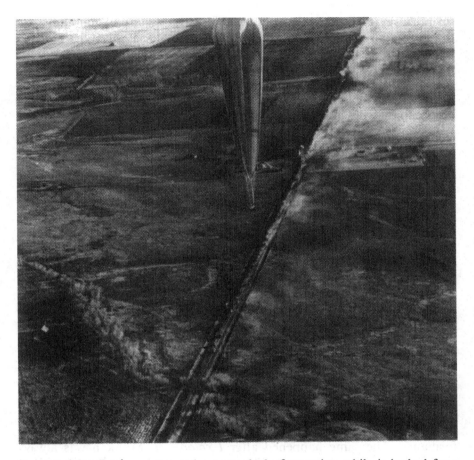

FIGURE 6.5a. *Explorer* attracted a great deal of attention while it looked for a good landing spot. The flight trajectory was also known through radio broadcasts. NGS/NASM.

reau of Standards to discuss how the data from *Explorer II* were to be analyzed and distributed. The bureau made available housekeeping data on pressure and temperature, compass readings, and timing records so that the participating scientists could calibrate and reduce their instrument data. Both the bureau and the society wanted to increase contact between the participants, avoid duplication or confusion over the methods of reducing the flight and environmental data, and decide how the results would be announced. Stevens and his staff at Wright Field were still reducing altitude data from their vertical aerial reconnaissance photographs, but most other data were now available for analysis.

At this meeting, the society announced that it would produce a second collection of "Technical Papers" containing tabulated data along with scientific reviews and analyses. To satisfy its far vaster popular audience, the society also asked Lyman Briggs to coordinate short reports by each scientist; Briggs therefore requested that each principal scientist submit preliminary results, and what photographs they might possess, to the society before March 15 so that it could prepare the article as quickly as possible.[52] Most of the scientists made the deadline, but were obviously preoccupied with their own detailed analyses.

Cosmic Rays

The Bartol group constructed a new set of shielded and unshielded cosmic-ray incidence-measuring and intensity-measuring experiments including an array of Geiger–Müller tubes, a Wilson Cloud chamber, Swann's old Stösse chamber, and photographic plate stacks for recording cosmic-ray tracks. Swann, G. L. Locher, W. E. Danforth, C. G. and D. D. Montgomery flew all but the cloud chamber to determine the vertical component of the cosmic-ray flux and the spatial orientation of the cosmic-ray spectrum to obtain a refined altitude/intensity profile, and to search for the existence of very large cosmic-ray bursts and what caused them. By excluding the cloud chamber, they simplified the duties of the crew, but the amount of "look-time" available was also curtailed since the smaller crew complement now had their hands full with housekeeping tasks. The plate-stack experiments were, in fact, a simplified replacement for the heavy cloud chamber. Even without the cloud chamber, *Explorer II* was a most exciting flight for Swann.

As we saw in Chapter 4, Swann was able to fly his Stösse chamber on Piccard's Detroit flight, but afterwards became very quiet about its performance. The same happened after *Explorer II*. Millikan was always skeptical about the practicality of making Stösse observations in a short balloon flight, arguing that the chamber was better suited for long-term observations, as was shown by the work of the Montgomerys on Pike's Peak. Nevertheless, Swann was determined to use the bulky chamber in the stratosphere, and hoped that with it he could verify his theory of non-ionizing primaries. But sadly, Swann's Stösse chamber remained silent during the entire trip, just as it had on the Piccard flight. In order to record only the strongest bursts, which were the ones Swann was most interested in, the chamber detectors were restrained by a discriminator circuit. Based upon extrapolations from the Montgomery Pike's Peak data, Swann had expected that at least 52 events would be recorded during the period of time the balloon was planned to remain at maximum altitude.[53] The apparent null result left Swann uncharacteristically silent. But he did gather interesting data from the other Bartol experiments that allowed him to speak and speculate at length.

Swann's staff detected a significant cosmic-ray flux attributed to neutrons. Using photographic emulsions carried in a nitrogen-filled box car-

FIGURE 6.5b. The balloon was quickly ruptured to prevent it from dragging the gondola across the field. Helium escaping out of the collapsing balloon carried with it soapstone dust that had lined the interior of the balloon skin. NGS/NASM.

ried within the gondola, L. H. Rumbaugh and G. L. Locher found complete low-energy recoil tracks from protons, which they surmised were produced by neutrons in the stratosphere. This led the Bartol scientists to conclude, following in Millikan's footsteps, that they were observing secondaries, the nature of which indicated that "heavy particles can produce only an insignificant fraction of the total cosmic-ray ionization in the region of the stratosphere reached by *Explorer II*; consequently, most of the cosmic-ray energy must be carried by photons and by positive and

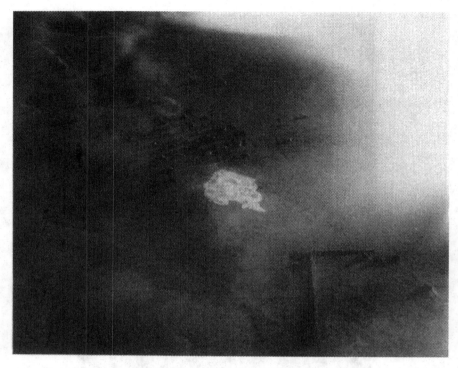

FIGURE 6.5c. The gondola was quickly surrounded by automobiles when it landed on the plains near White Lake, South Dakota. AAC/NASM.

negative electrons."[54] Rumbaugh and Locher believed the presence of neutrons suggested that high-energy primary cosmic rays had already collided with particles in the upper atmosphere or in space long before encountering the detectors on *Explorer II*. What the detectors were sensing were not the primary rays, but products of these collisions: the neutron flux and electrons. Reversing the process, they surmised that a significant amount of the parent radiation causing the observed neutron flux had to be in the form of gamma radiation. This was an unfortunate conclusion to reach at a time when the Compton–Millikan debate was still lingering because their detection of neutrons was ignored; Serge Korff recalls that a "distinguished physicist" declared, well after *Explorer II*, that "no evidence whatsoever" existed for neutrons in cosmic radiation.[55]

T. R. Wilkins, another Swann associate from the University of Rochester, flew photographic plate stacks outside the Explorer II gondola. His analysis of the particle tracks recorded on his plates revealed direct evidence for the existence of alpha particles, confirming the hint Compton had obtained from his Fordney–Settle flight data, and greatly increasing their acceptance.[56] Swann, however, was more interested in another track

FIGURE 6.5d. After the gondola was opened, an inventory had to be taken immediately. Lt. B. B. Talley (in goggles) confers with Stevens, visible in the open manhole. O'Brien's spectrograph was just removed from the top of the gondola and the National Bureau of Standards temperature sensor is on the ground to the right of the gondola. NGS/NASM.

that seemed to come from a heavy particle with incident energy of over 100 million electron volts, in the range that his theory predicted for non-ionizing primaries.[57]

Owing to the short look time and problems with gondola rotation, Swann and Johnson's large Geiger counter arrays failed to detect the east–west effect found originally by Johnson and Street, and by Alvarez. The Geiger counter data did confirm the already well-known vertical distribution of particle tracks for that geomagnetic latitude, and showed in addition a noticeable dropoff in the rate of increase in intensity of the vertical component of the cosmic-ray flux at the highest elevations. Thus Swann confirmed the recent detection by Pfotzer and Regener in July 1935 of the sought-after region in which a maximum incidence of vertical cosmic rays existed.[58] Swann wanted dearly to discover this maximum intensity region for himself; not only would it vindicate the manned mission, but it would indicate whether his instruments had entered a region in which he believed his non-ionizing primaries might be found.

Millikan had previously argued that he had detected the region in which the intensity was maximized, and that the level at which this observation was made confirmed his photon model.[59] Now, however, most physicists believed that the level at which the maximum was found by Regener and Pfotzer confirmed the charged particle model. Swann, somewhere in-between, believed that there had to come a height at which the production of secondaries was maximized: above this height one moved into a realm in which non-ionizing primaries dominated. Knowing the exact altitude where the maximum appeared, corrected for seasonal temperature, density, and atmospheric composition, could as well provide critical information on where, Swann thought, Stösse production would also be at maximum. Thus, finding where this maximum occurred became one of many scientific prizes sought by scientific balloonists such as Swann.

Beyond the fact that the level found by Regener, Pfotzer, and by Swann did not convince anyone that non-ionizing particles existed, the irony of coming in second in the race to detect the cosmic-ray maximum with his Explorer II instruments stayed with Swann for many years. Although he was silent about the failure of his Stösse device and of the fate of his theory, Swann never forgot that if *Explorer II* had flown in July 1935, when first planned, he would have had the pleasure of being the one to discover the Regener–Pfotzer maximum, as he recalled to Stevens in 1944:

We were practically simultaneous with Regener in getting first evidence of a maximum in the intensity altitude curve. Indeed, we should have been ahead of him had the balloon not suffered the accident during inflation on the first attempt of the flight of Explorer II. In fact, the apparatus was not removed from the gondola for the final flight, so that it was in exactly the same condition.[60]

Reviewers commonly cite Georg Pfotzer's July 25, 1935 observation of the cosmic-ray maximum as the first definite detection of the maximum, with confirmations by Millikan's colleagues Neher and W. Pickering. *Explorer II* reached 22 kilometers, well below Pfotzer's maximum altitude of 29 kilometers, yet high enough for Swann to catch the beginnings of the drop in vertical intensity, which was all that Swann claimed to have seen. Swann was beaten by a tiny 6-kilogram instrument that included a triple-coincidence counter array, three-stage amplifier and mechanical counter, and a 1,000-volt dry battery. Hugh Carmichael, who visited Regener's Stuttgart laboratory in 1936 to learn about ballooning techniques, was deeply impressed with the experimental elegance and efficiency of these sondes, and remarked in his notebook that Pfotzer's discovery was "probably the finest cosmic-ray experiment of the year."[61]

Swann, denied the joy of discovering the secondary maximum, turned his attention to its details: He was able to elaborate on the character of the maximum, since his counter telescope records could distinguish fluxes at selected zenith distances. Thus a tiny bump that Pfotzer had detected on his curve was explored by Swann in greater detail. At times, Swann spoke of two maxima: the main one detected by Regener and Pfotzer and a barely detected "secondary" maximum (in the mathematical sense) at a somewhat lower elevation detected in his Explorer II data. The visibility of this second maximum was greatest for observations close to the vertical direction, and it decreased with a resultant flattening of the entire profile as the vertical angle, hence air mass, increased, as Swann expected it would.[62]

Swann was convinced of the overall value of the manned ascent, and of the need to have people present to perform accurate scientific observations. But in December 1936 he admitted that he was disappointed with the way that the cosmic-ray observations had been constrained on *Explorer II.* He had hoped that continuous observations on the ascent and descent would have been possible, but "the processes concerned with the navigation of the flight itself" made this impossible.[63]

After the Explorer experience, Swann relied on Army Air Corps support for a series of lower-altitude airplane flights to fill in the profile, and with junior colleagues continued with a full agenda of monitoring cosmic-ray showers from the Bartol laboratory in Swarthmore as well as in deep mines and on shipboard around the world. He also encouraged his staff, particularly Thomas Johnson and Jean Piccard, to develop improved balloonsondes. Justifying this latter activity to his Bartol patrons in December 1936, Swann admitted "Unmanned balloons of this type have the advantage that they can be sent up at all latitudes and from places where the ascent of a larger balloon would be impracticable. Moreover, the cost of such an apparatus is negligible compared with that of a normal stratospheric flight." Swann added, however, that balloonsondes could not return data that would be as precise as those obtained from a larger capacity balloon flight, "but one may repeat the measurement several times" with a sonde.[64] Swann continued to think of piloted flights as "normal stratospheric flight" and as balloonsondes as experimental. In the next year he felt more positive about balloonsondes, especially since Johnson had demonstrated a recent success with radio transmission of data. But even then, Swann remained ardent about the advantages of large piloted balloon flights:

It is true that if one is concerned with cosmic-ray investigations alone, the whole program of sending up many unmanned balloons with apparatus so as to secure a large number of results which in the aggregate would provide the accuracy desired could be carried out at far less expense than for a single manned stratospheric flight. However, a manned stratospheric flight is usually carried out in the interests of a wide range of scientific projects; and, when such a flight is made, students of cosmic-rays should take the fullest advantage of it.[65]

For Swann, the advantages of manned ballooning outweighed its disadvantages. He quieted his own frustrations and looked forward to further flights, although he never initiated one. For Swann, who had the continued backing of the Bartol Research Foundation, such opportunities were all part of playing ball with his fellow administrators of scientific organizations. He felt close to Lyman Briggs, freely devoting a share of the pages of the *Journal of the Franklin Institute* to spread the good work of Briggs's National Bureau of Standards, and maintained close relations with the National Geographic Society as well. And with the Bartol at his beck and call, mustering support to build instruments for balloon ventures was a straightforward matter. To Swann, it was fate that denied him the chance of detecting the cosmic-ray maximum first found by Pfotzer and Regener, and was not a condemnation of the mode of conveyance. In correspondence with colleagues and in lectures in following years, he frequently recalled that his Explorer observations were uniquely valuable to his long career in cosmic-ray research.

Swann was never as central to the study of cosmic rays as Millikan or Compton; indeed, his own Bartol subordinates, especially Johnson, were cited by more colleagues than he ever was. Although Swann's contributions to cosmic-ray physics are cited here and there, no significant discussion of his Explorer experiments have been found beyond those published by the National Geographic Society and in his own publications and later commentary. His scientific reputation was established early by his contributions to problems in magnetic induction and earth currents, by his fine record of teaching and promoting the research of his students, and by his long and productive directorship of Bartol.[66] Yet through his Explorer activities he established the high-altitude balloon program at Bartol, and brought many of his younger colleagues, whom he left free to pursue major observing programs with balloonsondes, into the field.

Millikan incorporated his *Explorer I* observations into a general compilation of his intensity/altitude results from his many expeditions and flights since 1932. His unshielded Fordney–Settle flight data from 1933, previously published within a general review of their observations using aircraft and balloons in 1934,[67] and his July 1934 data from the Kepner–Stevens–Anderson flight provided "fairly reliable" complementary evidence for an "unambiguous determination of the absorption of the atmosphere for electrons."[68] His total observations led him to admit that cosmic rays even of the highest energies cannot penetrate more than the upper one-tenth of the earth's atmosphere before they are in equilibrium with their daughter products. Still, he argued that the incoming rays, "whatever their nature act just like x-rays or gamma-rays" from their absorption characteristics.[69] None of the manned observations changed Millikan's mind on anything. They merely added points on a curve, which was reproduced frequently by reviewers of cosmic-ray research.[70] Only in this manner did his results from *Explorer I* become part of the cosmic-ray literature.

Aerial Photography

Stevens's highest priority was photographic aerial reconnaissance. His aerial mapping cameras produced thousands of images for study and analysis of barometric altitude calibration and long-range reconnaissance techniques. Both he and the National Geographic Society delighted in displaying a spectacular montage of the Bad Lands and the Black Hills which showed a definitely curved horizon over 500 kilometers distant, encompassing a number of states. Reproduced both in the technical reports as a eight-section foldout and in the society's magazine, its title proclaimed its significance: "The First Photograph Ever Made Showing the Division Between the Troposphere and the Stratosphere and also the Actual Curvature of the Earth—Photographed from an Elevation of 72,395 feet, the Highest Point Ever Reached by Man."[71] Stevens and the society also proudly displayed vertical images of the earth taken from peak altitude.

Stevens fully deserved the high praise the Army bestowed on him for his design and application of the aerial camera system. But the pain-

FIGURE 6.6a. Vertical aerial photograph taken at the peak of *Explorer II* ascent, 22,066 meters above sea level, at 11:41 a.m. November 11, 1935. The photograph covers 250 square kilometers surrounding the south fork of the White River. A suspended meteorograph is visible as an irregular white shape at the top of the photograph. NGS.

staking reductions themselves and their analysis were the work of his photoreconnaissance assistant, Lieutenant B. B. Talley, Corps of Engineers, at Wright Field. Talley's reports from both Explorer flights provided Stevens with detailed analyses of the path of the balloon in three dimensions, time/altitude curves, time/orientation, sun brightness, wind velocities, and planimetric maps that revealed broad expanses of the Great Plains and the nature of the terrain photographed during the overflight.[72]

The main camera Stevens used was a standard issue Fairchild K-type single-lens system with a 235-millimeter focal length, modified for automatic operation. It used Eastman Kodak Supersensitive Panchromatic

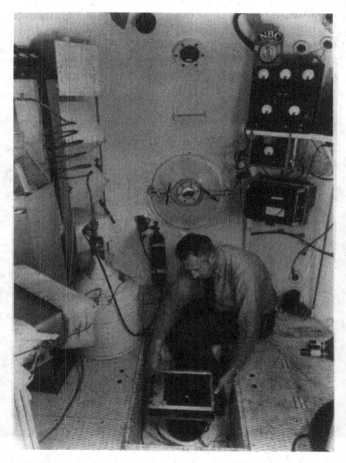

FIGURE 6.6b. Fitting the Fairchild aerial camera into the vertical observation port in the floor of the gondola. This camera photographed the earth automatically once every 90 seconds. The large white flask and coil to the left were part of the modified air conditioning system developed by Stevens. NGS.

film behind a red filter and in this configuration was an accepted part of the Army Air Corps program for rapid reconnaissance, photographic mapping, and spotting missions.[73] Stevens's second camera, designed for maximum atmospheric penetrating power, was not standard. It used near-infrared sensitive Kodak Type R film and was mounted to photograph the distant horizon. Each camera was fitted with an electrical exposure and film advance mechanism devised by Stevens and built at Wright Field. Operating at F/6.8, they took 0.04 second exposures every 1.5 minutes without manual supervision. Both cameras were extremely rugged: Even on the 1934 flight, with the camera magazines badly smashed, 163 exposures out of 200 were still "printable" and were used in Talley's analysis.[74]

Stevens's pictorial images of the curvature of the earth were splashed around the world by the *National Geographic Society* magazine. Technically, they were useful studies of horizontal air transparency in the stratosphere, helped Stevens calibrate and extend the all-important barometric/altitude relationship used in photoreconnaissance, and demonstrated to the Army the potential of high-altitude long-range reconnaissance from manned balloons.

Ozone

For shear complexity and technical sophistication, none of the other Explorer II instruments could compare to the ultraviolet spectrographs designed and used by Brian O'Brien and H.S. Stewart, Jr., of the University of Rochester Institute of Optics. Assisted by Fred L. Mohler of the National Bureau of Standards and O'Brien's numerous commercial industry colleagues, the Rochester team built three types of instruments to study the ozone profile and search for an extreme ultraviolet atmospheric window.

The University of Rochester/NBS group, with Bausch and Lomb assistance, modified two Bausch and Lomb quartz Cornu-type laboratory spectrographs for the Explorer I 1934 flight to study the vertical distribution of ozone in the atmosphere. These designs were duplicated in 1935. A large automatic spectrograph examined direct sunlight, and a smaller instrument observed diffuse scattered sunlight from the horizontal direction through a quartz porthole within the gondola.[75]

The direct-sunlight spectrograph was designed to cover the spectral range between 200 and 330 nanometers, although O'Brien's group hoped in March 1934 that it might be possible to reach deeper into the far ultraviolet.[76] The main obstacle lay in obtaining maximum sensitivity in the far ultraviolet, a particularly difficult goal because the scattered visible and parasitic light would have to be reduced with the aid of diaphragms and specially designed optical components. The spectrograph had to have a free-flowing ventilation system, to minimize thermal hysteresis, but also had to remain completely light tight, so the ultimate design was a compromise between the two requirements. O'Brien and his group were concerned about the mechanical design of a film holder and film-advance system that would be reliable in the harsh stratosphere. They considered standard commercial plate holders, film rolls and exotic rotating film drums that could accept a series of flat film strips. O'Brien decided to leave the details up to Bausch and Lomb, including designs for appropriate shutter mechanisms and automatic timing devices that would both advance the film and record the time of each exposure.

One of the most persistent design problems was how to get direct sunlight into the automatic direct sunlight spectrograph. O'Brien's group first considered using a conventional heliostat as a feeder, but they soon decided that this was impossible to do reliably from a balloon.[77] Nor could they use Regener's simple technique of having the spectrograph face a gray scatter plate, because this was an inefficient use of sunlight. Instead, they decided sometime after March to capture direct sunlight by a wide-field quartz diffusing sphere. Both O'Brien and Gustave Fassim of Bausch and Lomb had suggested such a device, and Fassim produced it through the Thermal Syndicate Ltd. of Brooklyn, New York, which specialized in making unusual optical components.[78] O'Brien remembers Fassim, a Belgian who had a joint appointment at his institute and at Bausch and Lomb, as a "brilliant designer of instruments" who always searched for creative solutions.[79] The 3-inch sphere he designed for O'Brien was able to capture a great amount of sunlight. The inside of the sphere was coated with magnesium oxide, and provided the diffusion characteristics needed to fully and evenly illuminate the slit of the spectrograph.

O'Brien recalls that the stigmatic spectrograph optics provided very clear images of the slit, which made it possible to use the images for accurate photographic sensitometry, his primary interest. He designed a special step wedge to cut the spectral lines into 10 segments with a 100:1 variation over the length of the wedge. Not only did the wedge provide very accurate sensitometry using only predetermined exposure times over the course of the flight, but it improved the chances of at least one portion of the spectrum having optimal exposure for reliable sensitometry.[80] But making the step wedge was a problem in itself, and O'Brien looks back with pride at the results of the effort. Aluminum, unlike silver, had almost total reflectance in the photographic ultraviolet, and so was an excellent material to use in a variable density filter such as the step wedge. However, techniques for depositing aluminum onto an optical substrate (quartz or glass) were just then being perfected by laboratory specialists in experimental optics, such as John Strong at Caltech and Robley Williams at Cornell. Thus it was even more difficult to attempt a variable density coating such as O'Brien desired.[81] O'Brien's solution for progressive evap-

FIGURE 6.7. The housing for the suspended sunlight spectrograph. The 3-inch quartz sphere that collected sunlight sits just to the right of the top of the box. NGS/NASM.

oration was eventually patented and applied to split-image aerial photography systems used by the Army Air Corps during World War II.[82]

Henry F. Kurtz, Fassim's supervisor, was responsible for the overall design and construction of the spectrographs. He suggested that the direct sunlight spectrograph be suspended several hundred meters below the gondola and be equipped with a large parachute that could be released from the gondola. When the balloon was at maximum inflation, it would still not obscure the summer sun or the zenith sky. From the gondola, both would happen. Also, the spectrograph could be released from the gondola for safe retrieval independent of the fate of the rest of the experiments. This proved to be an especially farsighted design decision, considering the fate of the 1934 flight.

The large spectrograph, with its solar light collector, thick quartz Cornu prisms, and associated optics turned out to be transparent only to 270 nanometers. It could not penetrate as far into the ultraviolet as O'Brien desired, but it went far enough for him to look at the long wavelength end of the ozone absorption region. To get at the shorter wavelength regime, O'Brien's team also designed and flew a small double-dispersing spectrograph in 1934 that could detect the atmospheric window near 220 nanometers. This complex system was especially immune to scattered light, but was equally cantankerous and inefficient. Both the predisperser and the main spectrograph used quartz prisms transparent in the range of 180 to 330 nanometers. The system was limited only by the transmission of the small quartz Cornu prisms and the mechanical baffling of the instrument, and by the sensitivity of specially formulated fluorescing photographic emulsions prepared by Kodak. This spectrograph was also designed to be suspended below the gondola and retrieved by parachute.[83] It did not detect the atmospheric window during the 1934 flight; O'Brien noted that it did not fly again in 1935 because their results from 1934 indicated that "optical and mechanical changes" were required that were "too extensive to be undertaken."[84]

The double-dispersing system was far more sophisticated than the slightly modified sun and horizon spectrographs. Reflected sunlight was sent through a series of mechanical baffles that acted as a very broad slit, and then into the predisperser, which illuminated a second narrower slit that in turn isolated the 185 to 330 nanometer band. This second slit was covered by a collecting lens and an aluminized quartz densitometric wedge. The slit then became the object for the second spectrograph that imaged the solar spectrum onto the fluorescing film. The use of a predisperser instead of an entrance baffle or a filter to limit the spectral range of the instrument and to reduce scattered sunlight was also a part of the original design of the hanging sunlight spectrograph in March. The predisperser was soon dropped, however, presumably because it greatly reduced the efficiency of the instrument.

Beyond the fact that the double-dispersing spectrograph did not work properly in 1934, Briggs used it as a pawn in negotiating with Grosvenor to save what he could of the ozone experiments for the third flight attempt. At first, Grosvenor wanted to cancel the suspended direct-sunlight spectrograph. However, Briggs convinced him to keep that one and to drop the double-dispersing instrument. He also removed the direct-sunlight spectrograph from its suspended position to a location above the gondola, which did not sacrifice its effectiveness because the winter sun was lower in the sky and would not be obscured by the balloon.[85] O'Brien apparently was not directly involved in Briggs's negotiations with Grosvenor.

O'Brien recalls nothing but the finest cooperation from Bausch and Lomb and Kodak. Both companies had been major underwriters of the Institute of Optics during its formative years in the 1920s and early 1930s, and still enjoyed very close contact with its staff; many at the institute, including O'Brien, acted as industry consultants. O'Brien's expertise in precision photographic sensitometry and his study of the problems of reciprocity failure and intermittancy effects also made him useful to Kodak's C. E. Kenneth Mees, another member of Briggs' Scientific Advisory Committee for *Explorer*. Thus both companies provided all the support he required; to O'Brien, it was an "ideal setup."[86]

The ozone results of the 1934 flight contradicted earlier findings. In particular, the distribution O'Brien found differed from what Götz, Meetham, and Dobson had found by indirect ground-based observations. To a lesser extent, O'Brien's results also differed from the recent results of Regener. But O'Brien remembers that it was Regener's work that compelled him to stick with *Explorer* for a second year. Regener, again using a simple automatically recording spectrograph in a balloonsonde, had found that the ozone distribution was lower than previously measured or inferred.[87] Apparently, O'Brien learned through Mohler that C. L. Pekeris of MIT had found that the Götz and Dobson ozone results based upon the Umkehr effect were incorrect, which left only Regener's observations to answer.[88]

O'Brien tried all but the double-dispersing system again on *Explorer II*. The original horizon spectrograph had to be completely rebuilt, but the direct sunlight (or "sun" spectrograph, as it was subsequently called by O'Brien's group) spectrograph from 1934 was in excellent shape, and so was reused with only a few changes to reduce scattered light. Hundreds of spectra were captured on some 40 meters of film which was exposed throughout the flight, but primarily during the two hours *Explorer II* remained at its 22-kilometer ceiling. Exposures were electrically timed from an automatic clock within the gondola, and the photographic film was advanced by a small automatic electric timing circuit designed by the Folmer-Graflex Corporation. All of O'Brien's spectrographs had a specially designed segmented slit that produced a series of solar spectra simultaneously, each behind a different part of a graduated optical wedge. The varying optical density of the wedge along the slit provided both for intensity calibration and a range of effective exposures to sunlight, which increased the chances of obtaining a proper exposure.

O'Brien's spectra reached down only to 296 nanometers, penetrating just beyond the atmospheric cutoff, and showed no evidence of the sus-

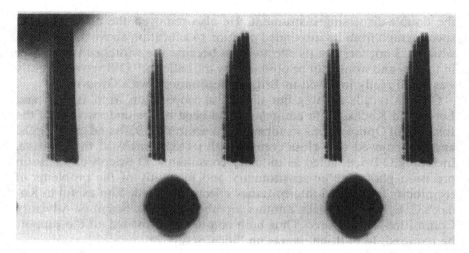

FIGURE 6.8. Example of the spectra taken with O'Brien's "sun" spectrograph. The solar spectrum was exposed through a segmented slit and variable density wedge filter to increase the chances of obtaining a proper photographic density. The circular images are the overexposed face of a timing watch. NGS.

pected window at 210 nanometers after considerable scrutiny. Thus his measurements of ozone absorption as a function of height were confined to bands beyond 300 nanometers. Employing supplemental indirect techniques similar to those used by Regener, Götz, and others, O'Brien extrapolated his photographic data to discuss the vertical distribution of ozone up to about 35 kilometers, or 13 kilometers beyond the heights attained by *Explorer II*. Like Regener before him, O'Brien found that the ratio of ozone to air as a function of height increased rapidly beyond 18 kilometers. But O'Brien found the first maximum at 22 kilometers whereas Regener had detected the first maximum at an altitude of some 27 kilometers. O'Brien's and Regener's extrapolated observations hinted that more than one maximum existed; O'Brien concluded that "the ozone concentration must have a second maximum at a height indefinitely above 30 kilometers."[89]

O'Brien was the only one to have measured ozone concentrations from a manned flight. This seemed to give the measurement special value in some reviews. In any case, most treated the differing results of Regener, O'Brien, and post-war observers not as contradictory, but as an indication that the vertical distribution of ozone was constantly changing. Since Regener had taken more observations than anyone else, and had flown the highest, his were usually considered the best available observations prior to the rocket era.[90]

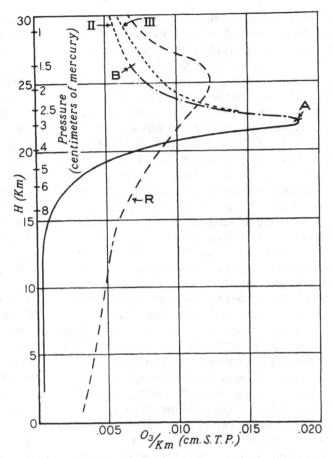

FIGURE 6.9. O'Brien's summary of his ozone distribution results, compared to Regener's. O'Brien concluded that his data supported Curve II reliably from point A to point B above the highest altitude obtained by *Explorer II*, but beyond B (at 26 kilometers) he had little faith in his results. Marked differences with Regener's distribution (R) are both the points of maximum concentration and the sharpness of the distribution. NGS/NASM.

Other Scientific Experiments

Explorer II contained a wealth of instruments that collected data useful to studies in aeronomy, meteorology, biology, and radio propagation in the high atmosphere. The National Bureau of Standards, the Carnegie Institution of Washington, and the Department of Agriculture, among other groups, carried out detailed experiments on the electrical conductivity of the air, the composition of the high atmosphere, the variation of sky brightness with altitude, wind direction and velocity changes with height, the distribution of micro-organisms at various levels in the atmosphere, and the effect of ultraviolet radiation and general stratospheric extremes upon spores and fruit flies.

The electrical conductivity experiments used a charged insulated rod set within a cylindrical condenser that carefully channeled air around the rod to try and determine the electrical potential in the Kennelly–Heaviside layer. The discharge of the rod was a measure of the electrical conductivity of the air. O. H. Gish and K. L. Sherman of the Carnegie Institution developed this experiment as a problem in geophysics and found that their observations revealed inconsistencies compared with laboratory data using vacuum chambers that indicated that the water vapor content of the high atmosphere might be far different than expected. They acknowledged the value of being able to conduct such an experiment from a manned balloon that carried with it a "full line of auxiliary and control data," but they also stated that the same experiments would be feasible using the same equipment on unmanned sondes.[91] Gish highlighted his Explorer experiment in a later review of atmospheric electricity, edited by Carnegie's Fleming, but noted that varying results between the ascent and descent phases may well have been caused by local electrical "pollution" of the atmosphere by the enormous intruding balloon.[92]

The air composition experiments consisted of three large collection flasks inside the gondola that were filled at peak altitude by remote valving systems in the gondola skin. Martin Shepherd of the National Bureau of Standards wished to determine the composition ratios of oxygen and carbon dioxide and the degree of convective mixing in the upper atmosphere. Two of the three valves worked, and the two resulting samples were analyzed in laboratories at the bureau for helium, oxygen, and carbon dioxide. Shepherd found that his results for the diffusive separation of oxygen fit a model suggested by Sydney Chapman's "best guess, which seems to have been a happy one," but that results for other elements were not clearly in agreement with the model. Shepherd tentatively concluded that, in general, atmospheric composition did change greatly with altitude, but that convective mixing also extended above the isothermal boundary between the troposphere and stratosphere. Shepherd called for an "extended" study, including additional unmanned observations at "great heights,"[93] although his precise measurement of the carbon dioxide concentration at 21.5 kilometers aboard *Explorer II* remained a standard through the 1950s.[94]

The remaining experiments conducted on *Explorer II* were described in the National Geographic's technical publications. 10 particles were captured in a micro-organism collector between altitudes of 11,000 and 22,000 meters that produced active cultures when incubated in the laboratory. This indicated to the experimenters that spores rising to such heights might explain global migration of plant and animal diseases. Al-

FIGURE 6.10. The contents of the stratosphere air flasks removed from the interior of *Explorer II* being analyzed at the National Bureau of Standards by Martin Shepherd. NGS/NASM.

though terrestrial spores carried aloft were subjected to very low humidity, low temperature, high concentrations of ozone, and ultraviolet radiation for some four hours, six of the seven fungi survived. This in turn supported the stratospheric model of disease transmission indicated from the results of the micro-organism collectors. Furthermore, exposed *Drosophila* showed a fivefold increase in mutation rate, which was far higher than expected if the only cause was the incidence of cosmic rays. Low temperature had to play a role as well, but to Victor Jollos of the University of Wisconsin's Agricultural Experiment Station, one stratospheric flight did not constitute a conclusive test. He therefore continued to expose his stocks to conditions found atop 4,340-meter Pike's Peak.

The Significance of the Explorer Series

Both Explorer flights garnered a great deal of publicity in the popular press. The National Geographic Society publications and wire service releases enjoyed wide distribution; even though the society was virtually the sole source of photographic material (Army Air Corps photography was carefully orchestrated through society channels), it freely distributed text and illustrations through the wire and news services.

The National Geographic Society devoted much attention to the Explorer series of flights in eight issues of its own magazine between April 1934 and March 1937. With a membership of over one million at the time, which meant a readership of about five million, the *Geographic* made *Explorer* a household conversation piece.[95] These issues contained announcements and reports of the Explorer flights, written by Gilbert Grosvenor, Albert Stevens, or Geographic staff. The earlier articles emphasized the importance of science, while the later articles hailed the record-breaking successes of *Explorer II*. The first announcement, in the April 1934 issue, stated that the society was supporting *Explorer* "To increase scientific knowledge of the upper air" and went on to describe some of the planned experiments.[96]

After the crash of *Explorer I*, in his official announcement to continue with the mission, Gilbert Grosvenor reviewed the scientific data that were saved, especially the high-altitude photographs, and pointed out in the February 1935 issue of the magazine that the accident was a learning experience that "initiated studies which form a distinct contribution to the science of stratosphere ballooning and will add to future balloon flights."[97] A June 1935 report emphasized flight safety and the decision to use helium, and finally, in January 1936, the magazine heralded the success of the venture in an article by Stevens entitled "Man's Farthest Aloft." The society also ran full-page advertisements soliciting new members by displaying its support for *Explorer*, and companies that had participated in the mission also bought advertisement space to commemorate the successful flight. (See Fig. 5.7a.)

Stevens reviewed "The Scientific Results of the World-Record Stratosphere Flight," in May 1936, crediting the participation of the many scientists. At the end of the article, he identified 13 technical and scientific "firsts" associated with setting the world's altitude record. To Stevens, the altitude record now became the raison d'être of the flight. The last word on the Explorer series in the pages of the *National Geographic* came

𝔑𝔞𝔱𝔦𝔬𝔫𝔞𝔩 ⑥𝔢𝔬𝔤𝔯𝔞𝔭𝔥𝔦𝔠 𝔖𝔬𝔠𝔦𝔢𝔱𝔶

WASHINGTON, D. C.

GILBERT GROSVENOR, PRESIDENT
JOHN JOY EDSON, TREASURER

June 1, 1935

JOHN OLIVER LA GORCE, VICE PRESIDENT
GEORGE W. HUTCHISON, SECRETARY

To Members of the
National Geographic Society:

The new National Geographic Society–U. S. Army Air Corps
Stratosphere Flight, foretold in this issue of The Magazine, is
a notable example of your Society's constant efforts to increase
our knowledge of the world about us. Your Society's income is
used to support such important researches, world–wide explorations, and photographic surveys, and to publish the results of
these activities in The National Geographic Magazine.

Only as The Society grows can this useful work be expanded.

By helping to enlarge the membership, you enable The
Society to extend its educational services and to return to you
a magazine of ever–increasing interest and value.

Sincerely yours,

George Hutchison.

Secretary

DETACH HERE—OR NOMINATE BY LETTER IF YOU PREFER NOT TO CUT YOUR MAGAZINE

Nomination for Membership

Secretary, National Geographic Society,   _____1935
    Sixteenth and M Streets N. W., Washington, D. C.

I nominate for membership in the National Geographic Society:

(1) Name_____
    Address_____
                            _____(Occupation)_____

DUES
Annual membership in U. S., $3.00; Canada, $3.50; abroad, $4.00; life membership, $100. Please make remittances payable to the National Geographic Society. Please remit by check, draft, postal or express order. The membership fee includes annual subscription to The National Geographic Magazine.

(2) Name_____
    Address_____
                            _____(Occupation)_____

(3) Name_____
    Address_____
                            _____(Occupation)_____

(6)   [*Name and Address of Nominating Member*]

FIGURE 6.11. National Geographic Society membership invitation, appearing in
the *National Geographic 67*, No. 6 (June 1935). NGS/NASM.

in March 1937 with a note proudly displaying the fourteenth "first": the
only color photograph (of the underside of the balloon) taken in the
stratosphere.

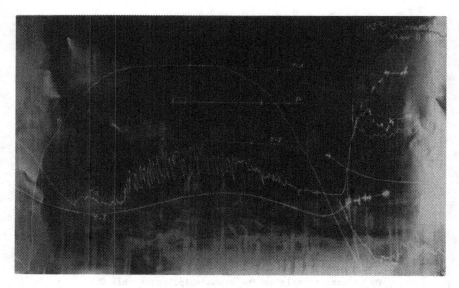

FIGURE 6.12. The official record from Brombacher's instrument that established the temperature (T-A = air temperature; T-I = instrument temperature) and pressure (P) profiles of the Explorer II flight. NGS.

Both Explorer flights made Stevens, Kepner, and Anderson public heroes. Telegrams, cards, and letters flowed in for months after the second flight. Many were from well-wishers, but some were also from people wanting to express what this success meant to them personally and to the world. The secretary of the American Association of Engineers hoped that the Explorer success would serve "to keep America aware that we are still a nation of pioneers."[98] Six-year old Jim Johnson of Diamond, Missouri, poured through the pages of the *National Geographic* and eagerly listened to reports on the radio of the flight. He wrote to Anderson inviting the *Explorer* to land on their farm, enclosing his drawing of their balloon and gondola.[99] In April 1936, a House bill was drafted for consideration by the 74th Congress, in its second session, to "recognize and reward the accomplishments of the pilots of the stratosphere balloon Explorer II," through giving them both a promotion.[100] Even President Roosevelt recognized them. Stevens had found a way into not only the stratosphere, but into the hearts of the nation and the most rarefied realms of official Washington.

The flyers were not the only ones to be honored. Also in April 1936, the master balloon tailor and designer for Goodyear-Zeppelin, J. Frank Cooper, who must have spent many sleepless nights after the fateful crash of *Explorer I* and the aborted *Explorer II* attempt, received a company award for his efforts. A member of Goodyear's flying squadron since 1913,

FIGURE 6.13. Stevens and Anderson with Franklin D. Roosevelt. Standing at center left is Gen. Malin Craig, Chief of Staff of the Army, and at center right Brig. Gen. Oscar Westover, Acting Chief of the Army Air Corps. Acme (UPI) photograph, courtesy Dow.

Cooper had designed balloons during World War I and rose to general foreman of the Balloon Department. He was a balloon racer, and was well known locally for designing and constructing the "grotesque figures for the Macy parades."[101] Arnstein's counterpart at the Dow Chemical Company, Arthur W. Winston, won similar recognition.

The scientific return, although gratifying, did not compare with that obtained from the far more numerous unmanned flights conducted continuously throughout the decade. Regener's flights, for example, are more often cited in scientific reviews than all of those from manned flights. Yet, certain characteristics of the Explorer series made it a significant event: It grappled with the technical obstacles that had to be overcome in order to address the scientific problems that access to the high atmosphere made possible.

First, the Explorer series was a cooperative venture between government, military, and civilian scientific interests, working through a private institution that existed in the name of scientific exploration. Each had his own agenda: glory through exploration for the society; re-establishing credibility and scientific veracity for the Bureau of Standards; and establishing both a competitive edge and a capability for the military. Each scientist too had specific motives, ranging from O'Brien's interest in the physiological effects of ozone, which helped to showcase the products of his institutional patrons, to Millikan's desire to please Foulois and Westover, for political reasons. The military possessed the manpower and training, and contacts within Goodyear; Dow Chemical was anxious to establish the superior character of Dowmetal; the Bureau of Standards knew how to make and test the proper type of equipment; the society wished to promote exploration through sales of its publications; and Stevens wanted to take pictures of the earth at "Man's Farthest Aloft."

Although there was little explicit discussion of military interests in the National Geographic "Stratospheric Series," the experiments had direct military application. Both Dern and Westover made this clear during meetings of the advisory committee and, as we have seen, in their radio talks. Aerial mapping and photoreconnaissance benefited in particular: The Army gained valuable experience in the technique of manned flight out of reach of conventional aircraft or of any means of ground defense, and conducted experiments in aerial photography that simulated an operational mission. In addition, cosmic-ray and biological studies provided information for assessing the survivability of high flight, something that both the Army and Navy became concerned about in the ensuing years. Much was learned about wind currents, and the radio experiments tested systems of effective communication; during the flight, the Explorer crew spoke with National Broadcasting Company radio personalities, with distant ground stations, and with the *China Clipper*, airborne over the Pacific at the time. Although they were done primarily for publicity, some of these demonstrations in radio propagation were of interest to the Signal Corps. Finally, the military gained valuable practical experience in coordinating a launch and all the operational details during the flight.

Second, looking again at the science, but now within the context of manned ballooning, the Explorer II flight was highly successful when compared with all other attempts prior to World War II. As already pointed out, much good science was done. The question remains, however, was it a significant scientific enterprise? One way to answer this question would be to look at the amount of support provided for analyzing and disseminating the scientific results. The National Geographic Society did provide such support: True to its original mandate, the society provided a coherent forum for the dissemination of the results obtained from the scientific experiments and the engineering studies that made up the program. The society sponsored an initial private conference to coordinate the use of housekeeping data, and then supported a limited circulation publication: the two-volume 400-page *Stratospheric Series*.

The two volumes contained 16 purely scientific articles, 19 technical articles on the design and operation of the scientific and flight-support equipment, 3 flight reports by the crew, and 2 reprints of popular accounts by Stevens from the *National Geographic*. Although not widely distributed—copies were difficult to obtain even by Swann within several years of their printing—the series stands as the most extensive report of any scientific ballooning program before the NASA era.[102] Lyman Briggs also organized a special "Symposium on the Results of the 1935 Stratospheric Flight" at the National Bureau of Standards during the April 1936 meetings of the American Physical Society in Washington. In many respects a cosponsor, the bureau was a logical site for the symposium. Most of the principal investigators were featured speakers.[103] Within six months, members of Swann's Bartol group as well as Brian O'Brien published short articles or abstracts of talks based on their Explorer observations in journals such as the *Physical Review*. Most significantly, the society, the bureau and Stevens worked feverishly to reduce all the housekeeping data for the entire flight profile. These data were then made available to all the participants who required information on altitude, temperature, and pressure to reduce their own observations. In many respects, this was the clearest indication that the sponsors of the flight expected to provide continuing support for the scientific activity.

Third, manned scientific ballooning—*Explorer II* in particular—stimulated a general interest in using ballooning to solve scientific problems. Brian O'Brien's introduction to ballooning was similar to that of both Compton and Swann as well as his colleagues at Bartol. All were already working on scientific questions that could be addressed through ballooning, all came to ballooning via the manned flights, and, as we shall see, all promoted ballooning studies using unmanned sondes. Swann, in particular, encouraged his Bartol staff to develop unmanned balloon radiosondes for cosmic-ray studies. Before Swann entered ballooning, T. H. Johnson and other Swann associates spanned the globe from South America to Fort Churchill, Manitoba, conducting ground-based and mountain observations. But after Swann's initiation into scientific ballooning, through his work with Piccard and Stevens, Johnson turned to multiballoon chains using cellophane, and also pursued radiotelemetry, while other Bartol staff developed more efficient radio transmitters and smaller lightweight counters and cloud chambers suitable for balloon flight. Bartol staff continued to send counters around the world on ships, and Swann promoted observations from aircraft. Johnson collaborated with Serge A. Korff (who had worked closely with Millikan throughout the 1930s and then moved to Bartol) to produce Bartol's first fully operational unmanned balloon systems by 1938.[104] As a result of his experience with manned flight, Swann added balloonsonde development to the agenda of the experimental laboratory at Bartol.

Manned ballooning was, of course, an expensive way to be introduced to ballooning. Several reviewers of research in the upper atmosphere decried the great expense and dangers associated with manned ascents. We have already reviewed Millikan's attitude, and need only mention here that Regener's 1935 discovery of the potentially harmful effects of ozone on rubberized cotton balloons in the lower stratosphere gave rise to an added concern among some balloonists.[105] Accordingly, the physicist Pierre Auger observed in 1941 that the costs involved, especially in trying to reduce the inherent dangers, and the manpower required to handle the logistics, prevented manned ascents from becoming regular scientific pursuits. Auger felt that too many flights had ended in disaster, and Auguste Piccard's original success could only be accounted for by his "presence of mind and sportsmanship."[106] Louis Leprince-Ringuet concluded in the 1940s that the Explorer series had demonstrated that manned flights "were still very difficult and that many accidents could happen despite the greatest precautions." He added that the cosmic-ray experiments were "satisfactory," but that the "length of the ascent was too short and conditions too difficult to obtain any momentous data."[107]

The scientific community also criticized Grosvenor's decision to reduce both the flight crew and the scientific agenda. Although this decision helped to achieve Grosvenor's goals—a new altitude record and a safe return—a British commentator writing in *Nature* in 1938 argued that leaving out the cloud chamber seemed "a great pity and in the opinion of the reviewer more useful information would have been obtained by taking up a cloud chamber than a wireless transmitting set."[108] The *Nature* reviewer admitted that the plate stacks that were flown worked nicely, but didn't require the presence of man. In the spirit of Millikan's criticisms in June 1935, he hoped "that this work will be continued with unmanned balloons."[109]

The *Nature* review only said in public what Briggs and others expressed in private. In 1934, Briggs wanted to aircraft-test a new instrument Swann developed to measure the electrical potential gradient in the atmosphere. He easily obtained Stevens's approval to fly it in *Explorer*, but did not see much hope for it because "it looks as if the omnipresent broadcasting equipment would preclude the potential gradient apparatus."[110] While it is true that Grosvenor reduced the scientific agenda, there was good reason to do so. All previous flights—those of Auguste Piccard; Settle and Fordney; Jean and Jeannette Piccard; and Stevens, Kepner and Anderson—demonstrated that the crew was overloaded with work. Of course, Grosvenor also reduced the crew, and many tasks remained (such as the NBC radio communications) that further compressed the time available to do fruitful research. The same *Nature* reviewer admitted that "a considerable amount of very useful data was obtained on this flight," and commended the aeronauts for their heroism and the ground staff for their good planning, but he added: "Nevertheless, one is inclined to wonder whether, from a purely scientific point of view, the money would not have been better spent in sending up a large number of unmanned balloons."[111]

Virtually all of the instruments carried aloft on Explorer were automated, or were amenable to automation, as Regener had demonstrated. At the same time, the great weight and housekeeping requirements added by the human presence limited the height the balloon could attain, as well as the total time aloft. Yet, the *Nature* reviewer appreciated the priorities of patronage: "On the other hand, the popular appeal was far greater, which doubtless made it much easier to collect the necessary funds."

There is no question that popular appeal created patronage and that man had to be present to create the appeal. Another critic of manned ascents warned that Regener's observations of higher ozone concentrations at lower altitudes meant that manned balloon flights using ozone-sensitive rubberized cloth balloons were potentially dangerous and unnecessarily costly. He also argued that Regener's important contributions had not been given proper credit by those advocating manned flights, "because, maybe, of the crew's longing for dramatic moments of adventure."[112] Drama and adventure attracted visibility and patronage, as they do now. But patronage came also from the military and from the aircraft industry, which linked the study of cosmic radiation with practical applications. For example, it could be used to better understand the physiological problems encountered in high flight, and even disclose untold sources of atomic power, as some news accounts proclaimed. But man had to be present—aircraft technology was improved for transportation and strategic advantage only with man in mind.

Finally, the Explorer flight agenda and experience hinted at the central scientific, technical, and administrative problems that would arise in the postwar era when balloons, aircraft, and rockets were all used as vehicles in upper air research. The questions facing the Explorer scientists, like those facing the scientists who placed experiments on captured German V-2 rockets, revolved around cosmic rays, the ultraviolet spectrum of the Sun, the atmospheric distribution of ozone, air sampling, atmospheric electricity, biological studies, and long-range radio propagation. O'Brien's application of a spherical sunlight collector in his direct sunlight spectrograph, his use of photographic film, and his preliminary conclusion that the ozone distribution contained a double maximum all reappeared in the work of both German and American groups who planned to exploit the V-2 rocket for scientific research. The technical problems faced by O'Brien, Swann, and some of the others during the development of Explorer reappeared both during and after the war in both ballooning and rocketry. These included electronics stability and survivability; temperature hysteresis; adequate ventilation and pressure equalization; how to get enough sunlight into a spinning spectrograph; how to maximize the chances of retrieval; and how to design photographic plate holders and automatic photographic film advancement mechanisms and use double-dispersing spectroscopic instrumentation to delve deep into the ultraviolet. The question of how to design the most efficient cosmic-ray and geophysical experiments dominated early rocketsonde projects, as they had the Explorer flights. The only significant departures were the increased use of telemetry after the war, increased attention to ionospheric studies, and, above all, an expanded interest by the military in certain aspects of the research. The military's heightened awareness of the value and applicability of high-altitude studies after the war stimulated new organizational arrangements. As before, there would be cooperation between military agencies and civilian scientific groups in the conduct of high-altitude experiments. This was necessary in virtually all postwar efforts in scientific rocketry and high altitude ballooning. But in the postwar era, as we shall see, the military took an ever-stronger role, both as patron and as director of the flight agenda.[113]

A major technical lesson learned during the Explorer series was that the rubberized cloth balloon had been taken to its practical limit. A substitute had to be found if man wanted to rise higher under a balloon. Even so, some still hoped that yet another manned flight into the stratosphere might be possible with Explorer.

Epilog: Stevens's Hopes for an Explorer III

One immediate effect of the Explorer success was that its chief proponent, after experiencing the rush of high flight, wanted to return as soon as possible, whether for science or adventure. Albert Stevens, now an Air Corps Major, advocated a third flight for Explorer. But after *Explorer II*, General Westover quietly ordered Stevens and other Army spokesmen to make it known that no plans existed for another flight. Stevens complied in 1935 and 1936, especially in his official report to the National Geographic Society, in which he noted that no future flights would be considered until a justification was found.[114] But he was constantly asked if a return was being planned as he toured the country giving lectures on the Explorer flights. By 1937, his convictions reemerged, and he diplomatically approached Westover again:

It is possible to get 10,000 feet higher, with a full load of practically the same kind of apparatus, but . . . it is not the policy of the War Department to engage in such projects too frequently, even though the training to the Air Corps and to other Army personnel is recognized as of considerable value.[115]

At first, Stevens lectured to repay political debts, but soon he took temporary leave from the Air Corps to lecture to raise funds for a third Explorer flight, "to go to 82,000 feet, or possibly 85,000 feet." During these tours, he made many contacts and by April 1937 had acquired a "nest egg of $20,000, which I believe I could expand in a few months to $50,000," as he claimed to Westover in his renewed plea for support.[116] He required approval from Westover before he could formally approach the National Geographic Society again.

Stevens wanted one more flight. He was now 51 years old and knew that before long, his hopes for another ascent would be lost. But, for the moment, the original organizational structure was still intact, as was the gondola, and the whole project could be simplified because of the experience already gained. Another "positively . . . last expedition" would not be an experimental one, but would be designed to return "maximum altitude and . . . maximum scientific work."[117] Both Swann and O'Brien, Stevens argued, would likely be willing to prepare experiments for the flight. And, Stevens told Westover, the balloon volume could be increased by another 25 percent, according to Goodyear, which would ensure a new altitude record:

We are one jump ahead at the present time, both in altitude and in useful work accomplished. As a matter of pride in accomplishment by the Army Air Corps, I would like to help set up new marks for altitude and work accomplished that will maintain the lead we already have.[118]

Stevens exploited the fierce interservice competition between the Army and Navy as well as national pride in his appeals to Westover. Stevens believed the best tactic to take was to ask for one more chance to "maintain the lead" rather than emphasize the importance of establishing a meaningful scientific capability.[119] Stevens thought his best shot was also to appeal to national prestige when he found out what the Polish were planning to do.

In the fall of 1938, Stevens was invited by Captain Burzynski of the Polish Army to attend the first flight of the silk balloon, *Star of Poland*. This was, in fact, Auguste Piccard's final effort to construct a balloon that would reach 30,000 meters.[120] Burzynski had visited the United States in July 1938 to procure helium for the flight, and during a meeting with Stevens showed him a sample of the silk for their 65-meter-diameter, 142,000-cubic-meter balloon, which was "grown, spun and woven in Poland."[121] Stevens found that, although the rubberized silk balloon material ripped quite easily, the entire balloon weighed only 60 percent of the Explorer balloon.

After this meeting, Stevens tried to find a way to provide Burzynski with helium because it might stimulate another return to the stratosphere in the United States. He admitted to Anderson:

Part of my motive was selfish, for I figured if we make another flight in this country, we must use helium, and therefore, the two flights would be on an equal basis. If we force them to use hydrogen, it would necessitate us building a still bigger balloon to equal their proposed performance.[122]

Stevens did try to help Burzynski obtain the costly and rare helium, but not enough cylinders were available to store and ship the gas, and so Burzynski settled for Polish hydrogen. Stevens traveled to Europe to see the *Star of Poland* carry Burzynski in his 2.5-meter aluminum gondola

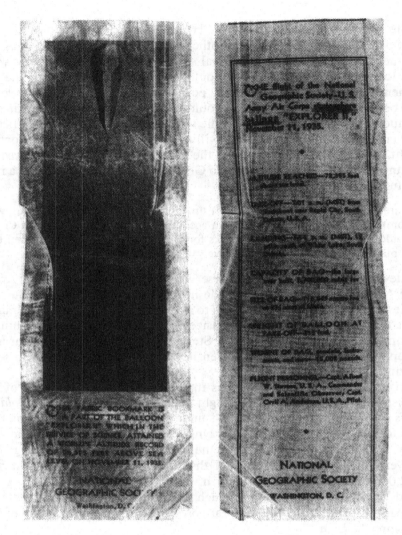

FIGURE 6.14. *Explorer II* bookmark, back and front, made from portions of the balloon. NASM Collection.

into the stratosphere. But whatever hopes Stevens still harbored that the Poles would rekindle stratospheric fever in Westover were dashed in December when the great balloon burned in a fire caused by static electricity.[123]

Stevens's hopes did not materialize. The huge three-ton rubberized cotton balloon bag for *Explorer II* was a very inefficient container and had poor structural properties when subjected to extreme stratospheric conditions. Meanwhile, rubberized silk was too expensive to use and was even less resistant to rupture and fire, as the Polish had demonstrated.

In the late 1930s, the general opinion was that 24 kilometers was the practical limit set by present technology, and Stevens had met that limit. Even though the Explorer II balloon had returned to earth intact, the National Geographic Society sent a clear message to one and all that it had reached its limit; it duly cut up the balloon into one million bookmarks for its loyal members.

Notes

1. Albert W. Stevens, "Draft of Radio Talk," NBC, Sunday 11 November 1934, NGS.
2. Ibid., p. 2.
3. Ibid., pp. 4–5.
4. "Minutes of a meeting of the Advisory Committee . . ," 10 December 1934, NGS.
5. Stevens (1934), p. 433.
6. G. Grosvenor to A. Stevens, 4 February 1935, NGS.
7. Briggs to Kepner, 5 December 1934, D/INA-678-C, ! ·pner folder, NBS/ NARA, RG 167.
8. "Minutes of Advisory Committee December 11, 1934," Research Committee Grants Stratosphere Flight #1, NGS.
9. Kepner to Briggs, 24 November 1934, D/INA-678-C, Kepner folder, NBS/ NARA, RG 167; B. D. Foulois to Gilbert Grosvenor, 10 December 1934 [original date of letter November 17, 1934], NGS; see also: Kepner and Scrivner (September 1971), p. 128; Stevens to Anderson, 16 December 1934, Orvil Anderson Collection, MAFB.
10. Stevens to Anderson, 16 December 1934, Orvil Anderson Collection, MAFB.
11. Stevens to Thomas McKnew, 10 March 1935, NGS.
12. Arnstein and Swann (1936), pp. 240–245. For an interim design profile, see Karl Arnstein, "Project for National Geographic Society Stratosphere Balloon," 24 January 1935, Goodyear-Zeppelin Corp. Technical Memorandum, Orvil Anderson Collection, MAFB, frames 817–827.
13. Winston (1936), pp. 248–250.
14. "Continuity of Radio Program," 20 April 1935, NGS.
15. See, for instance "Baz" Bagby to O. A. Anderson, 25 April 1935; L. M. Paulson to Anderson, 7 June 1935, Orvil Anderson Collection, MAFB.
16. Letters, Stevens and Swann, 21, 27 July 1948, Stevens Folder, Swann Papers, APS; "Home in the Bowl," (1936 Version), Orvil Anderson Collection, MAFB.
17. Brian O'Brien to Orvil Anderson, 11 July 1935, Orvil Anderson Collection, MAFB. This was sent on the day of the aborted flight; O'Brien was apparently unaware of its fate.
18. Stevens (1936), pp. 160–161; Brian O'Brien oral history interview, 9 March 1987, NASM.
19. "Minutes of a Meeting of the Scientific Advisory Committee, July 29, 1935," NGS, p. 12.
20. Ibid., p. 10.
21. Ibid., p. 11.

22. Ibid., p. 12.
23. Briggs to Oliver La Gorce, 16 April 1934, Blue Folder Files 1903–1952, Folder 678-INA, NBS/NARA, RG 167.
24. LaGorce to Briggs, 17 April 1934, Blue Folder Files 1903–1952, Folder 678-INA, NBS/NARA, RG 167.
25. Brian O'Brien oral history interview, 9 March 1987, NASM.
26. "Minutes of a Meeting of the Scientific Advisory Committee July 29, 1935," NGS, p. 13. On the value of the broadcasts, see: LaGorce to Briggs, 17 April 1934, Blue Folder Files 1903–1952, Folder 678-INA, NBS/NARA, RG 167.
27. Grosvenor to Millikan, 20 December 1934; Millikan to Grosvenor, 14 January 1935, RAM/CIT, R23 F22.10.
28. See Victor Neher to John Fleming, 25 October 1934, RAM/CIT, F22.9.
29. Millikan to Grosvenor, 5 January 1935; Grosvenor to Millikan, 14 January 1935; Millikan to Stevens 14 January 1935, RAM/CIT, R23 F22.10.
30. See Stevens correspondence with Neher and Millikan in RAM/CIT, F22.11 circa March, 1935.
31. Millikan to Stevens 13 May 1935, RAM/CIT, R23 F22.11.
32. Associated Press Release, "Stratosphere Jaunts Just a Fad For Adventurers, Says Millikan," *The Baltimore Sun*, 22 June 1935, NGS.
33. See letters between Millikan and Stevens, May through July 1935; see also Millikan to John C. Merriam, 4 October 1935, RAM/CIT, for a general review of Millikan's activities.
34. See Millikan to Merriam, 5 January 1935, RAM/CIT.
35. Millikan to Stevens, 23 July 1935; Stevens to Millikan, 1 August 1935, RAM/CIT, R23 F22.12.
36. Stevens to Millikan, 1 August 1935, RAM/CIT, R23 F22.12.
37. Briggs to Swann, 13 August 1935, Swann Papers, APS.
38. Ibid.
39. Ibid.
40. Ibid.
41. Bernson (1934), p. 1. Ossoaviakhim was the Soviet "Society for Air and Chemical Defense." See "Soviet Balloons for Exploring the Stratosphere," *Science-Supplement, 81,* (17 May 1935), p. 6.
42. See "The Soviet Stratosphere Flight," *Science-Supplement 82*, no. 2118 (9 August 1935), pp. 9–10; see also "Ascent of the Russian Stratosphere Balloon," *Science-Supplement 82*, no. 2114 (5 July 1935), p. 5.
43. Bernson (1934), p. 1.
44. Stevens to Swann, 21 July 1948, Swann Papers, APS.
45. Stevens (1936), p. 161; Arnstein and Swan (1936), pp. 242–244.
46. Westover to Stevens, 2 October 1935, Orvil Anderson Collection, MAFB.
47. F. C. Brockett to Orvil A. Anderson, 6 December 1935; C. C. Curran to Stevens and Anderson, 24 January 1936, Orvil Anderson Collection, MAFB. See also Stevens (1936), p. 190.
48. Stevens (1936), p. 198; Swann (1936a), pp. 678–679; Swann (1937), pp. 443–445.
49. For a diary of the flight, see Stevens (1936), p. 173ff.; and on the preparations, see A. W. Stevens, "Report of the Commanding Officer . . ," 13 December 1935, p. 9, Orvil Anderson Collection, MAFB.
50. Swann, Locher, and Danforth (1936), p. 24.

51. Briggs (1936a), p. 6.
52. Briggs to Swann, 28 December 1935, Swann Papers, APS, with appended "Minutes of Conference."
53. Swann, Montgomery, and Montgomery (1936), p. 29.
54. Rumbaugh and Locher (1936), p. 36.
55. The observation of a neutron component in the cosmic-ray spectrum was later verified by Serge Korff, who flew boron trifluoride proportional counters in radio balloonsondes in the late 1930s. See Korff (1985), p. 257.
56. Identified in Pfotzer (1974), p. 221; Wilkins and H. St. Helens (1935), p. 855.
57. Briggs (1936a), p. 11; Wilkins (1936), pp. 37–48.
58. Cited in Pfotzer (1974): Regener (1935), p. 718; Pfotzer (1935), p. 400; Pfotzer (1936), p. 23.
59. Galison (1987), p. 93.
60. Swann to Stevens, 3 April 1944, Swann Papers, APS.
61. Carmichael (1985), pp. 106–107.
62. Swann (1936a), pp. 672–675.
63. Ibid., p. 678–679; Swann (1937), pp. 443–445.
64. Swann (1936a), p. 680.
65. Swann (1937), p. 449.
66. Based on citations in scientific review literature prior to 1940. See, for example, Chapman and Bartels (1940), Chaps. 21–22.
67. Bowen, Millikan, and Neher (1934), pp. 645–646.
68. Bowen, Millikan, and Neher (1937), pp. 82–83.
69. Ibid., pp. 84, 88. This seemed to mitigate his new idea that two types of radiation had to be at play: field-sensitive and non-field-sensitive. The former were electrons, which he argued were responsible for his original observations of X-ray and gamma-ray absorption.
70. See, for instance, Braddick's reproduction of Bowen, Millikan, and Neher's latitude curves, in Braddick (1939), p. 22.
71. National Geographic Society Stratospheric Series II (1936), folded map after last page, "Photographic Supplement."
72. Talley (1934).
73. Lutz (1936), p. 276.
74. Talley (1934), p. 2.
75. O'Brien (1936), p. 51; see also H. Kurtz, "Memorandum on Spectrographs for Stratosphere Flight," 23 March 1934, Stevens File, Swann Papers, APS.
76. Kurtz, Ibid.
77. O'Brien (1936), p. 57.
78. At first, O'Brien was to develop the diffusing device, see: H. Kurtz, "Memorandum on Spectrographs for Stratosphere Flight," 23 March 1934, p. 7. On the Thermal Syndicate, see Swann Papers, "Thermal Syndicate" folder, APS.
79. Brian O'Brien oral history interview, 9 March 1987, NASM.
80. Ibid.
81. Ibid. On the early application of vacuum evaporation techniques to the aluminization of astronomical optics, see DeVorkin (1987b).
82. Brian O'Brien oral history interview. O'Brien performed this work as part of his responsibilities as chief of Section 16.2, "Illumination and Vision" of the OSRD. See Stewart (1948), p. 93.

83. Fassim and Kurtz (1935), pp. 112–118.
84. O'Brien, Mohler, and Stewart (1936), p. 72.
85. Briggs to Swann, 13 August 1935, Swann Papers, APS.
86. Brian O'Brien oral history interview, 9 March 1987, NASM.
87. Ibid. See also Regener and Regener (1934b), p. 380.
88. F. L. Mohler to C. L. Pekeris (MIT), 9 June 1934, Folder VI-6/INA-678-C; Briggs to O'Brien, 20 November 1934; O'Brien to Briggs, 22 January 1935, D/INA-678-C, O'Brien folder, NBS/NARA, RG 167. On the Umkehr effect, see Mitra (1948), pp. 101–102.
89. O'Brien, Mohler, and Stewart (1936), p. 92.
90. Goody (1954), p. 95; Mitra (1948), pp. 95–97; Hulburt (1939), p. 570.
91. Gish and Sherman (1936), p. 116.
92. Gish (1939), pp. 204–206.
93. Shepherd (1936), p. 133.
94. Goody (1954), pp. 70–72.
95. On circulation figures, see Abramson (1987), pp. 149–150.
96. "Your Society Sponsors an Expedition to Explore the Stratosphere," *National Geographic*, 65 (April 1934), pp. 528–530.
97. Grosvenor (February 1935), p. 265.
98. M. E. McIver to Orvil A. Anderson, 23 November 1935, Orvil Anderson Collection, MAFB.
99. Jim Johnson to Orvil Anderson, n.d., Orvil Anderson Collection, MAFB.
100. *H. R. 12394*, House of Representatives, 20 April 1936, Orvil Anderson Collection, MAFB. The bill erroneously stated that their flight took place on 12 July 1935.
101. Goodyear News Service, n.d., "Master Balloon Tailor To Get Award For Stratosphere Work," Orvil Anderson Collection, MAFB.
102. The society sent out several hundred complimentary copies of each volume. Additional copies could be purchased for $1.50 each. See enclosed card dated 4 January 1937 in *The National Geographic Society—U. S. Army Air Corps Stratospheric Flight of 1935 in the Balloon "Explorer II"*, (1936).
103. See [Briggs] (1936b), pp. 23–24.
104. In order to codify what was known about the new technique of radio telemetry of scientific information, Korff and E. T. Clarke reviewed a large portion of the field when the war put a halt to their activity. See Clarke and Korff (1941), pp. 217–355.
105. "Professor Regener Cautions Piccard Ozone in Stratosphere Dangerous to Balloon Fabric," newspaper fragment, n.d., attached to routing slip, Stevens to Swann, 5 May 1935, Stevens Folder, Swann Papers, APS.
106. Auger (1945), p. 98.
107. LePrince-Ringuet (1950), p. 115.
108. [R.T.P.W.] (1938), p. 271.
109. Ibid.
110. Briggs to Swann, 1 November 1934, Blue Folder Files 1903–1952, Folder 678-INA, NBS/NARA, RG 167.
111. [R.T.P.W.] (1938), p. 274.
112. "Professor Regener Cautions Piccard Ozone in Stratosphere Dangerous to Balloon Fabric," newspaper fragment, n.d., attached to routing slip, Stevens to Swann, 5 May 1935, Stevens Folder, Swann Papers, APS.

113. DeVorkin (1987a).

114. Stevens (1936).

115. Stevens to Westover, 4 April 1937, NGS.

116. Ibid.

117. Ibid.

118. Ibid., p. 2.

119. Ibid. See also Stevens Folder, Swann Papers, APS.

120. Philp (1937), p. 200.

121. Stevens to Anderson, 18 August 1938, Orvil Anderson Collection, MAFB.

122. Ibid.

123. Stevens to Kepner, 12 December 1938, Orvil Anderson Collection, MAFB.

CHAPTER 7

Jean Piccard and Progress in Balloonsonde Development

FIGURE 7.1. Jean Piccard building a cellophane balloon in Minneapolis, June 1936. NASM.

Plastic Balloon Research at Bartol

Explorer was the last of the grand balloon projects of the 1930s. Its record for "Man's Farthest Aloft" was not broken until well into the postwar era and stood as a challenge to all who dreamed of high flight. Even after his 1934 flight, Jean Piccard remained one of those who harbored the dream.

Jean actually had two main goals at the time: to return to the stratosphere and to find a professional position that fit his temperament and talent. These goals were to lead him to Minnesota and into aeronautical research, where he continued for a while in balloonsonde technology. The path in this direction began at Bartol, where W. F. G. Swann's continuing patronage and friendship eventually brought some stability to the Piccards' life in the form of a real academic position. But until that happened, Swann was willing to let Jean continue in his research position to pursue things that interested him and that were useful to Bartol.

In early 1935, T. H. Johnson suggested that he and Jean collaborate on developing cellophane balloons. Rubber balloons, still not plentiful, had the unfortunate habit of bursting at high altitudes, and some thought that cellophane would be less resistive than rubber to high altitude conditions.[1] Regener had argued in 1934 that the thin rubber skins of his balloons were degraded by the enriched ozone in the stratosphere, and therefore he experimented with cellophane cells with volumes ranging up to 100 cubic meters.[2] Johnson and other members of Bartol wanted to improve balloonsonde capabilities as well; Jean Piccard recalls that Johnson once asked him about how one might use cellophane as a rubber substitute, and in return he designed their first cells.[3] Following Regener's lead, they decided that the two most useful ways to go would be to substitute plastic cells for rubber balloons and develop miniature radio transmission devices.[4] Much of the progress in balloonsonde research in the next decade was made in these two areas, and Piccard, though not a leader, participated actively in both, first at Bartol and then at Minneapolis. As we follow his fortunes, we can also keep track of how others were progressing along these fronts in the United States.

By the end of the year, Johnson and Piccard had constructed open-neck cellophane plastic cells as large as 70 cubic meters to lift 5- and 10-kilogram payloads. They taped together gores of the light clear plastic substance and reinforced the top of the hemispherical cell with additional strips of cellophane tape. The lower section of the cell was shaped into a truncated cone, which Johnson and Piccard believed would form an extensible surface of minimum stress. The open-neck cell could be partly inflated with hydrogen at launch, and would then rise to an altitude set by the constant volume of the cell when fully extended.[5] The open-neck design provided for a degree of altitude control not possible with expandable rubber superpressure balloons.

Through 1936, as *Explorer II* flew to a new world's record, Johnson was busy improving their plastic cells, while other Bartol colleagues were developing miniature radio transmitters, power supplies, and Geiger counter arrays. Toward the end of the year, Johnson, assisted at first by Piccard, began sending up complete Geiger counter telescope arrays to test each component. Weight restrictions were a significant obstacle; Johnson sought creative solutions by replacing the heavy power supplies with a weight-driven static electricity generator (where the weight was the payload itself) and bulky cameras by miniature radio transmitters.[6] Although Piccard helped with this work in the first part of 1936, he soon became preoccupied with securing a real position.

Stability at Last

After their Detroit flight, Jean and Jeannette felt the time was ripe to exploit their success to obtain for Jean and possibly Jeannette real paying positions. Jeannette wrote to Swann asking if Jean might be hired on as a staff member in chemistry, and offered her own services as well, but Swann had to turn both down, although he was willing to help Jean look elsewhere.[7] Through 1935, as Jean worked with Johnson at Bartol, no offers were forthcoming, although both Jean and Jeannette were now frequently sought out as lecturers because of their popularity. They both wrote constantly to dozens of universities and colleges, and received just as many rejections. While the market for faculty was terrible, Jean and Jeannette often set their sights quite high, looking for college presidencies as well as academic professorships. While they looked, Jean never faltered in his enthusiasm for high flight. In his speeches, Jean often mentioned his continued thirst for exploration; at times he linked stratospheric exploration with polar exploration to establish its broadest and most appealing context. He saw the stratosphere as the next frontier ripe for exploitation.[8] Jean also remained occupied with his inventions and patent applications, which, as before, were attended by dispute and failure.

During one lecture tour, they passed through Minneapolis. After meeting the Mayor, Jean and Jeannette were shown around town by John D. Akerman, professor and head of the Department of Aeronautical Engineering in the College of Engineering and Architecture at the University of Minnesota.[9] Akerman was convinced that Jean Piccard would be an asset to the faculty and tried to obtain a professorship for him in the department. But the Dean would not agree to such an arrangement until such a time when the college's other staff requirements were met. Akerman did arrange for Piccard to return to give a short lecture series in 1936. Relaying the bad news to Piccard, Akerman vowed to keep trying. Swann, as it turned out, was involved in the matter, as he still enjoyed some influence at the University of Minnesota. During the 1920s, he had been in the physics department, and still had good contacts there with John T. Tate and probably with Akerman.

Akerman's continued advocacy and Swann's good wishes brought Jean to Minnesota again. He began teaching part-time in the spring quarter of 1936 and seemed quite happy with the assignment. In early June, Swann warned his impetuous friend that a real appointment might soon be opening at the university, and that he had better "prepare your reactions accordingly. In other words this is the time to see all things in the most favorable light, and get in well with as many as possible of the people who matter."[10] Jean took the hint and behaved himself, and as a result was appointed to the faculty in July with a contract that ran to the end of the year. As the Piccards settled into their new home, Jean reported to Auguste that they both were very happy in Minnesota. There was hope for rapid advancement, and it was also a nice place to live.[11]

Research at Minnesota

True to his word, Akerman helped Jean establish his research at Minnesota and the two collaborated on a number of projects related to balloon technology. Through Akerman, Jean received $900 from the National Advisory Committee for Aeronautics (NACA) to continue work on high-altitude sounding balloons that he had started at Bartol in 1935 in collaboration with Johnson. Akerman also helped Jean keep abreast of recent advances in aeronautical design. Sometime after Johnson suggested the use of cellophane to Piccard and they both started work on open-necked balloons made of the material, Piccard visited Goodyear in Akron and was shown a new rubber-derived plastic substance called Pliofilm, which its staff thought was superior to cellophane. At the time, Mr. Jacobs, a Goodyear engineer, suggested that Piccard try out the new Goodyear product. Jacobs, however, never sent Piccard any samples, even though Piccard had requested them on numerous occasions. At Minnesota, Akerman showed Piccard a sample of Pliofilm that he had been experimenting with. In Minneapolis, Pliofilm "was on the market," according to Jean, and he even heard rumors that Goodyear was about to use the material to make open-necked balloons. Worried that his Akron visit had moved Goodyear to action, Jean asked Swann to convince Johnson to issue a claim to their open-neck concept, and with it to apply for a patent on behalf of Jean and the Franklin Institute.[12] Swann agreed, and the matter was settled, but it redoubled Jean's efforts to develop plastic cells at Minnesota.

Jean's activities now ranged from the design of propellers to the development of cellophane and plastic sondes and searching for support for new manned ascents. In this latter activity Swann continued to be highly supportive, seeking out possible patrons in New York and Philadelphia, but nothing came of these efforts.[13] What did develop were the plastic sondes, and with Akerman's collaboration, radiotelemetry.

Advancing Radiotelemetry

The first successful application of radiotelemetry to balloons appeared in France, Germany, and the Soviet Union between 1927 and 1930. Scientists there employed electromechanical time-cycle devices based on nineteenth-century design concepts.[14] Compton was among the first in the United States to support the development of radiosondes for cosmic-ray research; J. M. Benade and R. L. Doan of the University of Chicago's Ryerson Physical Laboratory in 1934 developed and flew a test radio set weighing 4.5 kilograms that transmitted successfully from a height of 15 kilometers. Their radio transmitter was designed to relay the output of a small ionization chamber amplified by a cumbersome electro-optical system.[15]

Although radiosondes showed great promise for cosmic-ray research, they were developed largely for meteorological purposes. The Blue Hill Meteorological Observatory, asscooiated loosely with Harvard University, built lightweight short-wave transmitters for radiosondes starting in 1934 that could transmit data on temperature, pressure, and humidity using rotating electromechanical contacts attached to aneroid barometers, bimetallic strips, and hair hygrometers.[16] By 1936 they had produced a serviceable radiosonde design that they licensed for manufacture.[17]

Blue Hill's early efforts were soon overshadowed by the National Bureau of Standards, which had already built devices for the radiotelemetry of data from inaccessible or dangerous places and was deeply involved with radio propagation research by the early 1930s.[18] The NBS concentrated on improving the technology itself and establishing a standard design that could be mass produced. Responding to requests from the Weather Bureau and the Navy Bureau of Aeronautics for a reliable radio meteorograph, as well as for the establishment of frequency standards, the bureau's interest and involvement grew as well from experiments that its staff developed for the Explorer flights. Not only did the bureau provide Explorer experimenters with development and testing services in 1934, but its staff also performed a number of detailed studies, especially in high-frequency radio propagation.[19]

The Weather Bureau request was turned over to Leon Curtiss and Allen V. Astin of the electrical division in 1935, and early in the new year, the Navy's Bureau of Aeronautics meteorological sonde was taken up in the bureau's radio laboratory by Harry Diamond, Wilbur S. Hinman, and Francis W. Dunmore.[20] The Navy's sonde had to weigh less than .5 kilograms, cost around $25, be amenable to mass production, operate accurately over a wide range of temperature, and lend itself to radio direction finding.[21]

Before Diamond and his staff responded to the Navy's needs, they had already become involved with the Explorer flights at the Stratobowl. Diamond and G. H. Lester flew relatively large UHF transmitters from the Stratobowl to an elevation of 13,000 meters to test reception in anticipation of the experiments on *Explorer II*. As we have seen, the high-frequency experiments planned by the bureau were canceled by Grosvenor in early August 1935, but Diamond was so impressed with the potential of radiosondes for meteorological measurements that he continued to develop ultralight transmitters. By the end of 1935, his laboratory had designed an RC oscillator that provided a frequency modulated signal to a small radio transmitter that was well within the severe weight restrictions set by the Navy and the Weather Bureau.[22]

The efforts of the Weather Bureau and the National Bureau of Standards set an example for able students of radio. The NBS design was not put into production until 1938, but even before then, graduate students in engineering schools such as Minnesota's had little trouble adapting radio transmitters that were on the market for the production of "micrometeorographs." The first ones made at Minnesota for Piccard and Akerman weighed over two kilograms, but eventually were miniaturized to a tiny 41 grams.

The Minnesota Balloonsonde

Akerman, like Piccard, was something of a showman. He had an avid interest in balloon research and engaged in many experiments in balloon design, such as a hot-air barrage balloon. When he first got to Minnesota, however, Piccard turned his nose up at Akerman's hot-air balloon, which did not win Jean many points.[23] Still, the two collaborated in balloon research and with their students went on to more practical projects, such as their long-lived balloonsonde.

Piccard and Akerman designed their Minnesota balloonsonde to remain aloft at a constant altitude for gathering data rather than taking measurements only during ascent and descent. Akerman wished to develop a capability distinct from other efforts by trying to produce a simple sonde that had greater lifting power and transmitting range than those of the Weather Bureau, and one that could stay aloft for long periods of time, rather than one that would "establish new records for altitude."[24] After a summer of flying single cells, Piccard and Akerman linked a series of open-necked, partly inflated, cellophane balloons in both pairs and chains for flights in November 1936, but soon switched to standard superpressure latex balloons when the cellophane cells cracked in the cold dry Minnesota winter air. The small latex balloons came from the Dewey and Almy Chemical Company of Cambridge, Massachusetts, which during the 1930s was the primary American provider of such balloons for unmanned sondes.

After trials with single Dewey and Almy balloons, Piccard and Akerman found they needed greater lift to carry a relatively heavy but powerful radio transmitter designed by a student, Robert M. Silliman, and so turned to clusters of four and six balloons. Silliman had specifically designed his device to extend the reception area of the radiometeorograph, choosing the accessible amateur radio frequency of 62 kilocycles (5-meter band) for his 1.5-watt signal.[25] The radio transmitter's weight lay in its batteries and in an attached clockwork-driven drum and time cam system for controlling data transmission in a four-minute broadcast sequence of 1-minute intervals. The number of "pips" received per minute from the mercury barometer, bimetallic thermometer, and hair hygrometer corresponded to the amount of pressure, temperature, and humidity that each were experiencing.

Piccard and Akerman's first publication on their sonde consisted of two distinct parts. The first section was written by Akerman, the second by Piccard. Whereas Akerman concentrated on the design of the system and its instruments, Piccard discussed the development of the plastic balloons, temperature control, and ballast systems. He noted that this work was being done "without, however, giving up the hope of making another stratospheric flight." At every turn, he wished to remind his audience of his legacy; his plastic balloon cell, "which looks and behaves like the manned balloon," was reliable enough so that "the author would not hesitate, if necessary, to go up on a calm summer morning in such a balloon, slightly larger than the ones made up to the present."[26]

Despite his ultimate goal, Piccard and his colleagues at Minnesota gained useful experience flying many different combinations of balloons with thermally protected meteorological payloads. After trial flights with their prototype 30-foot plastic cells and Dewey and Almy latex balloons, their first full test flights in 1937 contained instruments that telemetered data on the popular 5-meter amateur radio band.[27] The payloads included standard automatic Weather Bureau meteorographs to check the radio-telemetered data.[28] Demonstrating the reliability of the radio data in test flights during 1936 and 1937, Piccard and Akerman continued with meteorological flights, using radio telemetry both to gather data from the sonde and to provide location information in order to retrieve the expensive payload and determine detailed wind patterns at the 16-kilometer level. Even though Piccard had asked Swann and Johnson to lend small cosmic-ray instruments for these flights, the request was not followed up.[29]

Piccard and Akerman's balloonsonde development is rather typical of university-based efforts of the time. For example, Brian O'Brien continued with his ozone studies after his Explorer experience; he recalls that he knew then that manned flights could not go high enough to penetrate through the ozone layer and that Regener's methods held more promise.[30] O'Brien's continued solar radiation and ozone studies from unmanned balloon sondes were supported in part by the Smithsonian Institution's

FIGURE 7.2. A cellophane balloon is launched from the University of Minnesota campus in June 1936. From Akerman and Piccard (1937). NASM.

Charles Abbot and by Francis W. Reichelderfer, chief of the Weather Bureau. O'Brien's first sondes, developed in 1937, before Smithsonian support was provided, contained small photocells that fed signals to a condenser circuit and radio transmitter that telemetered data to the ground from altitudes up to 20 kilometers.[31] O'Brien's corps of graduate students, including L. T. Steadman and Harold S. Stewart, provided optical and electronics expertise, as well as manpower for launching, optical tracking, and data reduction. Although he had no interest in developing balloons,

O'Brien had to address the problem of radiotelemetry, but left it to electronically inclined graduate assistants at the University of Rochester. Balloons could be purchased, but radiotelemetry had to be designed for the specific project in mind.

Akerman wanted to develop the capabilities of sondes themselves, as an engineering project, and O'Brien continued to study ozone. Both were interested in systems; the former wished to develop them and the latter to use them. And both also relied upon younger people versed in electronics and radio to provide the telemetry. Finally, both demonstrated by their examples that effective programs of research with balloonsondes could be maintained on university budgets, supplemented to some extent by outside sources. This model was already well in place in Millikan's and Compton's rather larger groups, and at Bartol.

But none of these people shared Jean Piccard's burning desire to fly once again. As we have just seen, his balloonsonde research was merely an interlude until he could build a manned stratospheric system again.

Jean Piccard's *Pleiades*

While Jean Piccard and his Minnesota colleagues were studying the lifting power of sounding balloons during the spring and summer of 1936, it occurred to him that "if one balloon can lift one pound of useful load to an altitude of 17 to 20 miles, then, under proper conditions a cluster of 3,000 sounding balloons should lift to the same height, scientific instruments, aeronautical equipment and crew in a stratosphere gondola weighing 3,000 lbs."[32] It was the availability of Dewey and Almy balloons in large quantity that first gave Jean his idea, as he told Auguste in July, 1936.[33] They were efficient and cheap, and were a good way to reach the stratosphere. He called his balloon project *Pleiades*, after the tiny celestial asterism containing seven visible stars. At various times, he thought of using 1,000 to 1,500 balloon cells. But the proper geometry of such a huge cluster, and other characteristics such as control, eluded him. Each rubber balloon cost only about $2.00 and the cluster concept had many advantages: There would be no need for the feared rip cord or for a net, as each balloon would be suspended on its own wire. And, according to Jean, Dewey and Almy was now providing new synthetic rubber balloons that could be even cheaper, larger, and available in greater quantity. Jean also hoped that the new synthetic substances might allow him to develop drum-shaped balloons that would nest more efficiently in a cluster, and possibly be more controllable. Auguste liked this idea, but cautioned that it posed new technical problems. He added that Jean should not advertise his *Pleiades* as an original concept: It had been already suggested by others for manned flight.[34]

Through the summer, Jean continued to develop the idea, and stoked his dreams of returning to the stratosphere, which were inflamed partly by an unfulfilled promise of financial support from the *Minneapolis Star*. To minimize the geometry problem, he decided on a lightweight, open, one-man gondola and a double cluster of 92 standard small Dewey and Almy latex balloons to test the balloon cluster idea. By August, Jean had sponsorship from the local Kiwanis Club of Rochester, Minnesota; and with the help of the university shops, he developed an explosive device for rapidly releasing the upper balloon cluster just before landing. He also had a system for releasing a controlled number of balloons during flight, but worried that releasing balloons during or after the flight might disqualify it for any records.[35]

His first and only flight, on July 18, 1937, was intended to be a low-altitude float. 92 Dewey and Almy 350-gram balloons were grouped into two clusters to carry aloft his light aluminum carriage, a radio set transmitting to a local Rochester, Minnesota, station (KROC), and himself. All went well at first, but during the flight, his balloon release system seemed not to be working right. As he released single balloons from the lower cluster, he feared that they became captured by the upper cluster. They were still providing lift, therefore, but could easily be dislodged by a crosswind. So he pulled balloons down directly by their ropes and burst them with a knife. When he decided to land after several hours, after traveling southeast along the Mississippi, he pulled down and burst more balloons from the lower cluster, but also reportedly used a revolver to shoot the balloons.[36]

Although nothing came of this last effort at manned ballooning before World War II, Jean and Jeannette Piccard's continued dreams, as with the Detroit flight highly publicized affairs (The Piccards prepared brochures and souvenir packets to bring attention to the flight. One was titled "Who Said We Couldn't Do It."),[37] had a lasting effect on at least one young aeronautical engineering student at the time. Robert A. Gilruth, who headed the Mercury Program for NASA and became the first director of the Johnson Spaceflight Center in Houston, recalls with poignant clarity Jean Piccard's fascinating and prescient efforts. In particular, Gilruth remembers flamboyant demonstrations of a small pressure valve that he helped Piccard prepare and that impressed everyone, including people from nearby General Mills.[38] Piccard had a strong influence on the young engineering student: "I learned many things from him, ways of looking at problems. He had a way of simplifying things, in talking about [how his devices worked]."[39] Gilruth feels that he gained a sense of what was required to build a gondola safe for manned flight that could be automatically regulated, and that this was a valuable experience, "learning about the gondola and how he managed it, [which was] just 20 years ahead of designing the Mercury capsule."[40]

Throughout the remainder of the prewar years, Jean Piccard continued to write and preach about the many reasons for returning to the stratosphere. The driving forces remained the same: the need to develop new

FIGURE 7.3. Night launch of the *Pleiades*. The lower cluster is easily visible, with the upper cluster dimly illuminated. NASM.

flight technologies for safer high flight, the need to refine pressurized cabins for aircraft, the need to extend the human presence into near-space, and the study of the mysterious cosmic ray.[41] Clearly, his emphasis at Minnesota shifted more and more toward the value of balloon gondolas as testing grounds for aircraft technology.

Through 1939, Jean continued to experiment to some extent with radiosondes, following Johnson's lead, and, with students such as Harold Larsen, developed smaller radiosondes that could transmit in standard Morse code.[42] Jean acknowledged the many possibilities that improved radiosondes promised, such as greatly extended reception range and attainable altitude,[43] and he knew as well that by the latter part of the 1930s, advances in both balloon technology and in radiotelemetry indicated a bright future for the study of the high atmosphere. Reliable, cheap, and lightweight radiotelemetry systems were readily available from the National Bureau of Standards and their liscencees, balloons were easily available in quantity, and hence more and more university groups turned to them for access to the high atmosphere. But Jean Piccard never lost sight of his own goals; to remain active he studied physiological problems of stratospheric flight, convinced that the lifting capacity of a manned mission was not going to be matched by unmanned automata in the near future.

Unlike manned ascents, unmanned flights were inexpensive, common ventures that accordingly attracted little popular attention. The public only became interested when it saw the starlike reflection of the sun from a high-altitude balloon, or when a payload landed in someone's field and a small reward was offered for its recovery. Despite its low profile, unmanned scientific ballooning made considerable progress in the 1930s, and presented scientists with a real choice in how they would gain access to the stratosphere, as we shall see in Chapter 8.

Notes

1. Auguste to Jean Piccard, 12 August 1936, PFP/LC.
2. See Ibid., and Pfotzer (1974), pp. 225–226.
3. Auguste to Jean Piccard, 12 August 1936; Jean to Auguste Piccard, 29 August 1936, PFP/LC.
4. Swann (1936a), pp. 679–680.
5. See "Statement," Jean Piccard and T. H. Johnson, witnessed by W. F. G. Swann, 18 May 1936, Swann Papers, APS; Johnson (1937), pp. 339–354.
6. Swann (1936a), p. 680.
7. Swann to Mrs. Piccard, 14 December 1934, PFP/LC.
8. See speech and writing file, PFP/LC, "The Influence of Stratospheric Research on Tomorrow" (1935); "Why Do We Go To The Stratosphere?" (n.d.); "Eleven Miles High and Why?" n.d.
9. John D. Akerman to Jean Piccard, 28 May 1935, PFP/LC.
10. Swann to Jean Piccard, 1 June 1936, Swann Papers, APS.

11. Jean to Auguste Piccard, 26 June 1936, folder 1934–1947, PFP/LC.
12. Jean Piccard to Swann, 16 May 1934, Swann Papers, APS; "Statement," Jean Piccard and T. H. Johnson, witnessed by W. F. G. Swann, 18 May 1936, Swann Papers, APS.
13. See letters, Swann to Piccard, 1 June 1936 and 18 August 1936, Swann Papers, APS.
14. See: Middleton (1969a), Chaps. 9 and 10; Middleton (1969b), Chap. 10; Mayo-Wells (1963), pp. 499–514; Pfotzer (1974), p. 224; Vernov, Grigorov, et al. (1985), p. 357. Charles Ziegler, John DuBois and Robert Multhauf are currently reviewing in detail the history of radiosonde technology. See also: Vernov (1934), p. 822; Vernov (1935), pp. 1072–1073.
15. Benade and Doan (1935), p. 198.
16. "Sounding Balloons and Weather Conditions," *Science News (Science Supplement)*, 81 (3 May 1935), p. 8; Middleton (1969a), pp. 333–335.
17. Middleton (1969b), p. 109, fig. 104.
18. Cochrane (1966), p. 352, n. 156.
19. Snyder and Bragaw (1986), pp. 122–124.
20. Cochrane (1966), p. 353; Snyder and Bragaw (1986), p. 124.
21. Middleton (1969a), pp. 337–338.
22. Snyder (1986); Middleton (1969a), pp. 338–343.
23. Robert Gilruth oral history interview no. 1, 21 March 1986, GWS/NASM, p. 55rd.; no. 2, p. 72rd. Robert Gilruth was a student in the department at the time, and recalls that Akerman and Piccard did not get along too well when Piccard first arrived because of this type of criticism.
24. Akerman and Piccard (1937), pp. 332–337; p. 332.
25. Robert Gilruth oral history interview no. 1, 21 March 1986, GWS/NASM, p. 77rd; Akerman and Piccard (1937), p. 334, fig. 4.
26. Akerman and Piccard (1937), p. 336.
27. Jean Piccard to "All Scientific Collaborators," Project Helios File 1946, PFP/LC.
28. Akerman and Piccard (1937), p. 335.
29. Piccard to Swann, 16 May 1936, Swann Papers, APS; Jean to Auguste Piccard, 21 July 1936, PFP/LC. Translated by Renata Rutledge.
30. Brian O'Brien oral history interview, 9 March 1987, NASM.
31. O'Brien, Steadman, and Stewart (1937), p. 445. This prototype transmitted reproducible data from altitudes in excess of 20 kilometers, and over distances greater than 80 kilometers. See also Abbot (1939a), p. 62; Abbot (1939b), pp. 102–103; O'Brien oral history interview, 9 March 1987, NASM.
32. See "The Voyage of the Pleiades, 1937," attachment to Jean Piccard to Lt. Commander D. F. Rex, 11 April 1946, "Proposition for Pleiades II Flight," Project Helios File, ONR. For a slightly different version of this story, see: Crouch (1983), pp. 636–637.
33. Jean to Auguste Piccard, 21 July 1936, PFP/LC. Translated by Renata Rutledge.
34. Auguste to Jean Piccard, 12 August 1936; Jean to Auguste Piccard, 29 August 1936, PFP/LC.
35. Jean to Auguste Piccard, 29 August 1936, PFP/LC.
36. "Voyage of the Pleiades, 1937," Project Helios file, ONR; see also "Who Said We Couldn't Do It," n.d., PFP/LC, folder "Ballooning." Donald Piccard states that Jean Piccard never used the revolver.

37. Ibid.
38. Robert Gilruth oral history interview no. 2, 14 May 1986, GWS/NASM, pp. 73–75rd.
39. Ibid., p. 79.
40. Ibid., p. 80.
41. Jean Piccard (1937), p. 31, PFP/LC.
42. Piccard and Larsen (1939).
43. Jean to Auguste Piccard, 25 April 1939, folder 1934–1947, PFP/LC.

CHAPTER 8

Epilog to the 1930s: Advancing Balloonsonde Technology

FIGURE 8.1. R. A. Millikan with double chamber set within wire frame to hold cellophane "hothouse" for thermal protection, circa 1935. Photograph by W. Connell, permission of the California Institute of Technology Archives.

Manned or Unmanned Ballooning?

In the late 1920s, Auguste Piccard argued that the instrumental limitations of automated sondes justified his proposals to the FNRS for manned
stratospheric flight. After he flew, he traveled across the United States
advocating manned ascents, and continued to do so after returning to
Europe. Jean Piccard and Swann, too, emphasized the advantages of
manned ascents, but as automated balloonsonde technology began to
improve in the middle of the decade, they both supported balloonsonde
development. Toward the end of the 1930s, even Jean Piccard admitted
that radiotelemetry and plastic cells opened up new possibilities. Although William Swann argued after *Explorer II* that laboratory instruments sent with human operators were still preferable to unmanned sondes
and were the normal way to conduct observations in that realm whenever
possible, he admitted that the sondes were cheaper, could be flown almost
anywhere if physical retrieval was not necessary, and were safer. Accordingly, Swann encouraged the development of unmanned sondes at Bartol.

The Piccards were now almost the only scientists advocating manned
flight for research in the high atmosphere. Even their most sympathetic
allies, such as Swann, felt that any means available should be exploited
to study cosmic rays or the solar ultraviolet. The scientific issues had to
be considered first, not the means of transportation to the site. We have
already noted in Chapter 6 how Brian O'Brien, A. H. Compton, and
Swann and his colleagues at Bartol all brought preexisting scientific interests to the manned flights which they carried over to unmanned sondes.
After Harry Diamond developed radiotelemetry devices for the National
Bureau of Standards in connection with the Explorer flights, the Bureau
competed for both Weather Bureau and Navy contracts to build radiosondes. And Johnson at Bartol, along with Serge A. Korff and E. T. Clarke,
refined their radiosonde designs to produce operational systems, which
by 1938 were their primary mode of high-altitude research.[1] The conversion to unmanned ballooning was mainly a technical shift, although
for some scientists, it required a shift in attitude as well, as in Swann's
case.[2]

Plastic Balloons

To increase both the lifting capacity and passive controllability of balloonsondes, Jean Piccard and T. H. Johnson had tried to make balloons
out of cellophane, and both continued to work on this independently
after Piccard moved to Minneapolis. At first, cellophane balloons looked
highly promising—they were thought to be more resistive than natural
rubber to ozone exposure—but they were susceptible to cracking at low
temperature and humidity. As a result, most balloonsonde flights were

conducted with sealed rubber balloons since the economies of scale imposed by the Weather Bureau and military, and the introduction of synthetics, brought the Dewey and Almy balloons to levels well within university budgets. The miniaturized meteorographs and radiotelemetry devices developed by the National Bureau of Standards to meet government and military needs were also used by numerous university groups. It was, indeed, the problem of data retrieval that remained most pressing at the outset of the 1930s.

The Retrieval of Data

Ever since balloonsondes were first used in the 1890s, some means was necessary to record data automatically during flight and to preserve that data until the sonde was retrieved. Scribe markers on smoked surfaces, ink pens drawn over rotating paper drums, maximum and minimum recording thermometers and barometers, all gradually were replaced by photographic recording of faces of dials or gauges, and, as in the case of cosmic rays, of the changing positions of tiny quartz fibers. But the photographic record had to survive the flight and be recovered. When Bowen and Millikan required equatorial data, they chose heavily populated areas like Madras, India, to increase the possibility of retrieval.[3] Methods of detection, recording, surviving and recovery were interrelated. Photographic recording required a highly controlled physical environment, natural and artificial sources of illumination, ruggedized film cassettes, automatic exposure control, and film advance mechanisms. The most prolific early proponent of balloonsondes, Erich Regener, flew some 51 times between the years 1928 and 1936, and amassed abundant data useful to cosmic-ray, meteorological, and atmospheric composition research. In particular he was also responsible for improving the technique of photographic recording on unmanned balloonsondes. As a gifted experimentalist (largely as a result of his tenacity), Regener was able to overcome many of the problems self-recording devices posed. His most successful early cosmic-ray detector was a photographic ionization chamber weighing only 1.5 kilograms.[4] Regener remained wedded to photographic recording through the 1930s and even in the 1940s as he designed instrument packages for V-2 rockets. His students, however, turned readily to radiotelemetry when it was possible to do so.

The fortunes of balloonsonde research also depended on the accuracy and reliability of the instruments and detectors. Piccard, Swann and the FNRS had stressed that laboratory instruments in manned chambers were far more reliable and precise than tiny automata. Ionization chambers fitted with fiber electrometers were the most common form of balloon cosmic-ray detector. Designs by Millikan and Bowen, by Regener, and others were all amenable to photographic recording, but the fibers were very small and highly sensitive to temperature and vibration, and had to be carefully illuminated and thermally stabilized. Both ionization chambers and Geiger counters could be attached to electrometers to record intensity data. Geiger counters could provide directional information as well if more than one tube were placed in line, in the manner suggested by Bothe and Kohlhörster in 1929, or by Bruno Rossi, who in 1930 used electrical coincidence circuits with far higher time resolution.[5] A coincidence circuit was first successfully used in an unmanned balloonsonde on a July 25, 1935 flight designed by Regener's student Georg Pfotzer. This instrument, which consisted of a triple-coincidence amplified counter and photographically recorded meter, detected the vertical cosmic-ray maximum.[6]

But in the 1930s Geiger counters were not the easiest devices to make; nor could they be relied upon to function continuously. The introduction of vacuum-tube amplification into the Geiger–Müller circuit in 1927 made the device extremely sensitive, but such tubes required constant attention and careful calibration. A 1941 review of the Geiger–Müller tube called it the most sensitive detector known for the measurement of high-energy radiation, but building one during the early 1930s depended on trial and error laboratory art: "Although simple in appearance [their construction] used to be more hazardous a game with unpredictable results.."[7] A. H. Compton used high-pressure argon ionization chambers with Lindemann electrometers instead of Geiger tubes in his worldwide survey in 1933 because he found the former gave more reproducible results and lower statistical errors.[8] Millikan continued to use ionization chambers primarily, asking younger members of his Caltech group to develop Geiger counters as experimental additions to their detector arrays later in the 1930s. And when Bruno Rossi first began making tubes in 1930, he felt the process was "a kind of witchcraft."[9] Again, Regener and his students led the way, using photographic recording of cosmic-ray events, and turning to Geiger counters only when the conditions were most appropriate. Until Geiger counters could be relied on to work consistently for hours unattended during flight, they were not commonly used on balloonsondes. On the other hand, Geiger counters were smaller and lighter than typical ionization chambers and, properly designed in pairs, provided directional information. As a result, both types of detectors were used through the late 1930s in groups where the expertise was available to build them.

Even the best-planned scenarios for payload survival and recovery created many frustrations and questions for Regener and all the others. The payload might float for many hours and even days, and might land hundreds of kilometers away. It might be days or weeks before the sonde was retrieved; but the exposed photographic emulsion had to be protected from light and extremes in humidity and temperature or the record would be ruined. Finally, analysis had to await physical recovery, so that the

weather data gathered were always out of date, and this further complicated the reduction of time-dependent cosmic-ray phenomena.

The faster that the data were retrieved—or the more direct the route—
the more likely that the experiment would succeed. If a way could be
found that did not require immediate physical recovery, if at all, there
would be less chance that something could influence or destroy the recorded data. With instantaneous data retrieval, airborne recording would
not be necessary and the payload could be simplified. If metered data
could be electrically or electro-mechanically sensed and transmitted by
radio, many steps in the photographic method could be eliminated. For
many reasons physicists and meteorologists desired an alternative to photographic recording and found it in radiotelemetry.

Stimulated by its own development of radiotelemetry, the National
Bureau of Standards supported two distinct projects to design and use
sondes for a wide range of aerial phenomena, including ozone and cosmic-
ray research. Diamond's radiosonde made possible the ozone studies of
W. W. Coblentz and R. Stair of the bureau's Radiometry Section. William
W. Coblentz, the "founder of modern radiometry,"[10] was well known for
his application of photoconductive and electrothermal devices to the study
of heat radiation from the sun and stars and for his laboratory standardization techniques. In 1937, with Stair, Coblentz applied photoelectric techniques to radiosondes to measure ozone distribution in the stratosphere. They developed an "ultraviolet-intensity meter" that used a
cadmium cathode photocell sensitive to a range of 260–325 nanometers.[11]
The cell itself had been used in a light-intensity meter Coblentz and Stair
had developed for ground stations. They replaced their amplifier and
microammeter with the small audiofrequency generator, amplifier, and
radio transmitter developed by Diamond's team.

Coblentz and Stair conducted six flights in June and July 1937, supported in part by the National Academy of Sciences. In ascents during
1938 to 27 kilometers with improved amplifiers and transmitters, they
determined that the sondes had penetrated 65 percent of the ozone layer,
that the ultraviolet flux was ten times greater than it was at sea level, and
that the ozone layer extended from 18 through 27 kilometers, with a
maximum concentration at 25 kilometers.[12] Coblentz and Stair continued
their unmanned ascents through 1940, and confirmed their 1938
observations.

Leon F. Curtiss and Allen V. Astin of the electrical division of the NBS
turned their attention to cosmic-ray research as well as meteorological
sensing when the Weather Bureau decided that it would adopt the competing Diamond design. Curtiss had a long-standing interest in cosmic-
ray studies and in developing Geiger counters, and shared the NBS's
interests in radio propagation studies. Cosmic-ray events were interesting
to the bureau as well because they could interfere with radio reception.[13]
Curtiss and Astin's radiosondes used simple electromechanical recorders

connected to automatic telegraph circuits based on the venerable Olland time-cycle design.[14]

Curtiss and Astin enclosed their balloonsondes in a cellophane "hothouse," following what was by now standard procedure, to maintain ambient temperature at 35° Centigrade during initial experimental ascents with a barograph. They were concerned that the clockwork would freeze and that the batteries used for the transmitter would die in the -60° Centigrade environment expected at operational altitudes. Compton's group, as well as Johnson and Korff, routinely used cellophane hothouses for the same reason, and all were delighted with how well these simple expedients, first used by Regener, worked. Curtiss and Astin's first meteorological sondes (0.7 kilogram radio-barographs) flew as high as 38.7 kilometers under single 1-meter (uninflated diameter) rubber balloons. By 1938 they devised special Geiger counters for their sondes, and lofted them to 20 kilometers for cosmic-ray measurements. After 18 flights, they found that most of the cosmic-ray phenomena they were recording were from secondaries produced in the earth's atmosphere, which Pfotzer had shown was at a maximum in that region.[15]

The National Bureau of Standards staff worked closely with Blue Hill Meteorological Observatory staff and Bartol's T. H. Johnson, all of whom were engaged in similar tasks: the automatic gathering and telemetering of data. Each enjoyed sharing the experience and talents of the other. Johnson was closest to problems in cosmic-ray physics, and shared his discipline-related knowledge with Curtiss and Astin, while the latter two freely shared the facilities and technical support of the bureau.

All the groups knew only too well that some types of data are more amenable to radiotelemetry than others. Transmitting the record of an electrical pulse from a Geiger counter or a photoelectric cell is trivial compared with transmitting a dispersed ultraviolet spectrum of the sun, or the attitude of a quartz fiber in an electroscope without proper amplification, or information on atmospheric composition, temperature, pressure and humidity, unless the devices are carefully designed for these purposes. Thus these balloonsonde researchers faced common problems and shared their knowledge and technique to overcome them.

Throughout the latter part of the 1930s, the powerful combination of readily available small balloons and rapidly improving radiotelemetry technology bode well for the future of unmanned balloonsondes. By 1938, Johnson considered his radiosondes "now well in hand [and] able to obtain reproducible results." His lightest sondes could routinely reach altitudes in excess of 15 kilometers and stay there for over 10 hours, relaying data to the ground. He was confident enough in the basic technique to consider broadening the application of radiosondes to geophysical measurements that might reveal, for example, the influence of solar radiation on the upper atmosphere, time-varying geomagnetic influences on cosmic-ray flux, particle flux from solar flares, and the correlation between radio fadeouts and solar phenomena.[16]

A Survey of Practitioners

In fact, radiotelemetry was developing so fast and the use of balloonsondes with it, that by 1938 lawyers for the Bartol Foundation worried that, with their scientists flying so many sondes, the Institute might be liable if someone was injured by material falling out of the sky. This prompted Swann to carry out a survey of colleagues conducting balloon flights and to report the results to his board so they could determine if there were real insurance risks.

Swann wrote to 10 groups engaged in ballooning, and from their responses found that some 1300 flights had been made in the past five years. No accident claims were ever reported, and none of the groups had liability insurance.[17] Compton knew of no problems, nor did Millikan, who in 1937 alone was in charge of some 250 flights in collaboration with the Weather Bureau (most were radiometeorographs, but some performed cosmic-ray measurements). Millikan's cosmic-ray group, which by then included William Pickering in addition to Victor Neher, flew electroscopes and ranks of Geiger counters and enjoyed the use of the Weather Bureau's meteorological instruments, manpower, and hydrogen.[18] Compton's group at Chicago, which included war refugees Bruno Rossi and Marcel Schein, flew Dewey and Almy multiballoon chains from Stagg Field on the university campus in crowded Chicago. The group's efforts gained momentum after J. Martin Benade, a visiting professor of physics from Punjab University who had led one of Compton's "world survey" teams,[19] outlined a general design for a reliable cosmic-ray radiosonde that used a photoelectric circuit. Under R. L. Doan and with Compton's supervision, members of the Ryerson laboratory staff built ever-more refined radiosondes capable of transmitting data from argon-filled ionization chambers that weighed as much as 18 kilograms.[20]

Brian O'Brien reported to Swann that he had conducted some 60 flights, 25 of them carrying "considerable weight," and all were retrieved safely and without incident.[21] After his Explorer flights, O'Brien still was not comfortable with the wide discrepancies in the data from all quarters on the distribution of ozone in the atmosphere. In 1934, Götz, Meetham, Dobson, and the Regeners found maximum concentration at some 25 kilometers, which was considerably lower than previous estimates.[22] But Brian O'Brien's 1934 results favored neither center. Nor did he find an appreciable concentration of ozone at lower levels, below the 18.5-kilometer maximum detected from *Explorer*. O'Brien concluded that too many variables were at hand to come to any conclusion. Seasonal effects could be at the root of the discordant results. He argued that "only direct

FIGURE 8.2. Brian O'Brien holding his "Icaroscope" circa 1946, with Gordon G. Milne. The Icaroscope was used with high-speed motion picture cameras to automatically limit light transmission during atomic tests in the Pacific. AIP Niels Bohr Library.

[and frequent] observation with instruments carried into the stratosphere can give a detailed answer as to the precise form of the [ozone] distribution curve."[23] He also recognized that manned balloons could not reach high enough to make definitive observations. As we saw in Chapter 7, O'Brien's graduate students developed simple radio transmitters that could relay broad-band spectroscopic data from his tiny photoelectric sensors. In all of his flights from the Rochester area, he never encountered any liability problems.

Serge Korff, a student of J. Q. Stewart at Princeton who at first collaborated with Millikan and then moved to the Carnegie Institution of Washington and then to Bartol to work with Johnson, had conducted some 40 flights, each with from 4 to 10 balloons. He and Johnson built a wide variety of cosmic-ray radiosondes starting in 1936, for both improving their technique and gathering useful data. Assisted by Korff and others, Johnson flew pressurized 6-kilogram payloads and 4-kilogram unpressurized payloads containing two small coincidence counters to an atmospheric pressure level of 3.5 centimeters of mercury using trios of 73-cubic-meter cellophane balloons as well as larger chains of latex balloons.[24] Korff told Swann that a few balloons always remained as the sonde descended to the ground, so the downward velocity was small and the apparatus quite conspicuous. Therefore, the chance of such an instrument striking a person, Korff felt, was one in two million: "It is at least one hundred times more probable that that same person will be killed in a traffic accident."[25]

Responses such as these were tabulated by Swann, and helped to reduce Bartol's premium on liability insurance from $25 to $6 per flight; it was hoped that they would be further reduced to $3 per flight once the number of flights increased. Thus, by the end of the 1930s, scientific studies with unmanned sondes had become moderately routine for established scientific institutions in the United States. Specialized knowledge and shop technique were required, to be sure, along with funding for travel, logistics, radio, and balloon gear. But it was funding that was well within the normal scope of doing cosmic-ray physics or even geophysics and atmospheric physics. Although most of the balloons would burst at altitude, their payloads often could be reused, since the remaining balloons from the multiballoon system would gently return the payloads to earth, as Korff noted.

Finally, the number of flights tabulated by Swann were insignificant compared with the overall Weather Bureau balloonsonde program. By the end of the decade, the Weather Bureau and Navy were consuming thousands of radiometeorographs annually.[26] This rate of consumption increased the availability and lowered the unit costs of both the basic radiometeorographs and registration balloons. By 1940, standard radiometeorographs based on Diamond's design for the measurement of temperature, pressure and humidity weighed less than 1 kilogram, including batteries, transmitter, pressure unit and insulating box, and cost $25 per unit, which was well within the Navy's original specifications.[27] The Dewey and Almy balloons cost between $2.00 and $10.00 each.

Low cost, lack of danger, and a constantly improving technology brought new players into the radiosonde game, and also led to regulations. By the end of the decade, the Federal Communications Commission began to regulate the operational frequencies of radiosondes. Operators had to be licensed to specific frequencies. The general trend, guided by the National

Bureau of Standards, was to shorter and shorter wavelengths; as vacuum tube technology improved, the assigned carrier frequencies rose from the 100-meter range into the (then) short-wave region, commonly between 3 and 8 meters (39 megacycles). Short-wave transmitters thus became physically smaller and the transmissions were less susceptible to spurious reflections. By 1941, typical oscillator tubes cost only $0.30 and ground receivers on the market were "cheap and reliable."[28]

Even as the world went to war, some groups still flew a few balloonsondes for cosmic-ray studies. In 1940, T. H. Johnson found good evidence for protons in the cosmic-ray spectrum, which was established beyond doubt by Marcel Schein and his colleagues in 1941, in agreement with the requirement from the observed east–west effect that the primary flux was positively charged.[29] Soon after, Korff confirmed the existence of neutrons in the cosmic-ray spectrum using sondes.

Scientists and Sondes

If one were only to look at the "discoveries" made with balloonsondes, compared with those made from manned flights, the balance would be tipped toward the former. From the discovery of the stratosphere itself by De Bort at the turn of the century to the unambiguous establishment of the solar constant by Abbot in 1914 to the elucidation of the nature of the intensity-altitude characteristics and the high-altitude spectrum of cosmic rays, balloonsondes had demonstrated their promise and superiority. But a chronicle of "discoveries" is a poor indicator of what the scientific groups engaged in upper atmospheric research actually believed to be an appropriate research mode at the time. The best indicator is the behavior of the scientists themselves: What did they choose to do? What were their actions?

Swann actively supported research at Bartol to develop radiosondes, lightweight battery packs, and miniature Geiger counter circuits. He also supported cloud chamber development and laboratory high-voltage generators. Millikan and Compton did much the same, although they concentrated on the counters and radio devices. Benade and Doan at Chicago, and Johnson at Bartol, along with the others, identified their radiosonde work as developmental rather than operational in the mid-1930s, but those who continued with it to the end of the decade boasted that balloonsondes were now operational. Everyone acknowledged that radiotelemetry made flights feasible in equatorial or polar regions, areas in which physical retrieval was not feasible; Johnson added that cosmic-ray radiosondes would make it possible to conduct systematic world surveys that would shed light on the many idiosyncrasies in the intensity-versus-altitude curve, and that converting coincidence pulses to telemetry was an "almost ideal type of signal from the standpoint of both the transmitter and receiver."[30]

None of the American centers were as catholic in their use of sondes as was Erich Regener's Stuttgart group. Before he was purged from Stuttgart by the Nazis in late 1938, Regener's examination of the high atmosphere was part of a long-standing program of meteorological, atmospheric, and cosmic-ray research. In addition to working with his son on the ozone problem, he sent ionization chambers, Geiger counters, and electrometers into the atmosphere in research conducted in conjunction with Georg Pfotzer, Reinhold Auer, and many others at Regener's Physikalisches Institut der Technischen Hochschule. As we have seen, Regener never courted manned ascents. But unlike Millikan or Compton, Regener did not seem to object to manned ballooning. In fact, Auguste Piccard and Max Cosyns thanked the Stuttgart professor for providing some special ionization chambers for their flights in the *FNRS*; Regener characteristically maintained many collaborative research efforts with likeminded scientists of his day.[31] Although he was happy to lend advice and material assistance to Auguste Piccard and Cosyns, Regener confined his own attention to unmanned sondes, and at one point even seemed to harbor a sense of competition to maintain his preferred mode of research when Auguste Piccard entered the stratosphere for the first time.

American scientists who had preexisting interests in cosmic rays or ultraviolet solar spectroscopy—such as Swann, Millikan, Compton and O'Brien—exploited manned ballooning platforms if they were made available. But all of them turned just as quickly to unmanned sondes as a more natural mode of doing research when manned expeditions ceased. Even Max Cosyns, who flew again under the old FNRS balloon in 1934, designed radiosondes for cosmic-ray research in the late 1930s.[32] And Jean Piccard, with John D. Akerman, worked on radiotelemetry techniques from balloons in the late 1930s and at one point considered flying a cosmic-ray experiment. But no scientists other than the Piccard brothers pursued manned ballooning. To most scientists, such as Serge Korff and E. T. Clarke of MIT, the manned method was "so prohibitively expensive that only a few such ascents have ever been made."[33]

Indeed, there were acknowledged advantages to manned missions, such as the positive role afforded by multi-institution collaborations and the higher precision possible. But most of all, it was the payload capacity that advocates touted. Cellophane had been introduced by Regener, Johnson, and Piccard not only to overcome problems of ozone susceptibility, but to try to improve payload capacity as well as altitude control in sondes to levels possible in manned flights. But cellophane left much to be desired, and the enormously greater payload of manned flights made it possible to fly large laboratory devices that were familiar to the uninitiated experimenters. As a result, scientists could be introduced to ballooning without having to build a new form of detector. And in a few cases, when these instruments were flown, many of them with heavy shielding, the

FIGURE 8.3. Georg Pfotzer's triple-coincidence Geiger counter array, flown in July 1935. Photograph courtesy the Max Planck Institute for Aeronomy, Lindau.

result was a refined picture of the cosmic-ray spectrum and its directional characteristics.

Although manned ascents could not achieve the altitudes unmanned sondes commonly attained and could add comparatively little to *in situ* measurements of composition, wind, pressure, temperature, cosmic-ray counts and solar spectra, such flights achieved wide publicity and financial support as they were considered heroic episodes that extended the human experience to near space. They made the public aware of the existence of the upper atmosphere as a finite entity, which might someday become a new realm for high-speed commercial travel.

Without doubt, when scientists were left to their own devices, and when they required high altitude data, they turned to unmanned sondes. Even when they had a choice, some, such as Millikan, realized that the presence of man, and the complex environment required to sustain him in the stratosphere, was detrimental to the conduct of scientific experimentation when very high precision was required. Then as now, vibration, outgassing, and thermal variations caused by the human presence often could mask the very same subtleties sought by the scientific investigations that had been used as the rationale to fly in the first place.

Notes

1. For a detailed review of radiosonde technology, see Clarke and Korff (1941).
2. Swann (1936a), p. 680.
3. Bowen, Millikan, and Neher (1937), pp. 80–88.
4. Pfotzer (1985), pp. 85–87; Carmichael (1985), p. 104; Pfotzer (1974).
5. Rossi (1981), p. 36.
6. Pfotzer (1936), p. 23. Cited in Winckler and Hoffman ("Resource Letter CR-1 on Cosmic Rays,"), p. 4. See also Pfotzer (1974).
7. Weisz (1941), p. 18.
8. Compton (1933), p. 388.
9. Rossi (1985), p. 56.
10. Cochrane (1966), p. 99.
11. Stair and Coblentz (1938), p. 191.
12. Coblentz and Stair (1939), p. 573.
13. Cochrane (1966), p. 353.
14. Curtiss and Astin (1935), pp. 35–39. Curtiss and Astin's telemeteorograph (or radiometeorograph), consisted of three revolving electrical contacts, two that moved according to "some meteorological indicating element" and one, "the sweep hand," that kept constant time calibration driven by clockwork. When the sweep hand contact hit a set of two fixed contacts, the telegraph "key" was closed, a double signal was transmitted to a ground station and was recorded as two tics on a strip of moving paper tape. When the sweep hand passed either of the two instrument-monitoring contacts, a single signal would be sent. The positions of the resulting single tics on the tape, calibrated by the positions of the time marks, constituted the readout of the meteorological instruments in the sonde.
15. Curtiss and Astin (1936, 1938a, 1938b).
16. T. H. Johnson to Vannevar Bush, 3 August 1938, Swann Papers, APS.
17. Swann to S. A. Korff, 25 October 1938, Swann Papers, APS.
18. W. H. Pickering oral history interview, 1981, California Institute of Technology Institute Archives, pp. 13–25; pp. 24–25. Interview taken 1978. See also W. H. Pickering, oral history interviews, 1982, 1983, SAOHP/NASM. See also Neher (1985).
19. Compton (1967), p. 162.
20. Keen (1937), 355–373.
21. B. O'Brien to W. F. G. Swann, 27 September 1938, Swann Papers, APS.
22. Götz, Meetham, and Dobson (1934), pp. 416–446; Regener and Regener (1934a), pp. 788–793; Chapman (1943), p. 93.
23. O'Brien (1936), p. 70; see also O'Brien oral history interview, 9 March 1987, SAOHP/NASM.
24. Johnson (1937), pp. 339–354; Clarke and Korff (1941).
25. Korff to Swann, 26 September 1938, Swann Papers, APS.
26. Estimates range from 15,000 meteorographs per year (Snyder and Bragaw (1986), p. 128), to 30,000 (Cochrane (1966), p. 354, n. 160).
27. Snyder and Bragaw (1986), p. 129.
28. Clarke and Korff (1941), pp. 230–231.
29. See Simpson (1985), p. 389.
30. Johnson (1937), p. 340.

31. See Cosyns, Kipfer, and Piccard (1933).
32. Pfotzer (1974), p. 224; Cosyns (1939), p. 73.
33. Clarke and Korff (1941), p. 218.

CHAPTER 9

Project Helios

FIGURE 9.1. Jean and Jeannette Piccard and their conception of *Pleiades II*. NASM.

The Post-War Context

During World War II, Swann's Bartol Foundation, O'Brien's Institute of Optics, and virtually all of the military and civilian institutions identified thus far were totally immersed in war. Jean Piccard and the Aeronautical Department at Minnesota worked on numerous projects for the Navy and for the Army Air Corps, but tended mainly to the accelerated training of engineers and technicians, while many of their colleagues in physics left to join central laboratories for war research such as the Manhattan Project.[1]

Although a vast amount of weather data was gathered during the war using ground stations, aircraft, and balloonsondes, apparently not much research and development took place in balloons or balloonsonde technology itself. Aside from improvements in balloon tracking methods using radio, the powerful war-born technique of radar, and the development of aerodynamically shaped kite-balloons ("Kytoons"), the fundamental characteristics of balloonsondes remained largely unchanged. What was different was the military's awareness that it needed better long-range weather forecasting, better knowledge of upper-air wind patterns, and the radio propagation characteristics of the atmosphere.[2] The Navy and, especially, the Army Air Forces also recognized that these needs would have to be met by extending civilian research within the military and the civilian agencies developed as part of the war effort. But in the last years of the war, no consensus emerged over how such research would be controlled. To retain wartime scientific and technical engineering expertise, the government established a "Research Board for National Security" (RBNS). Planned as a peacetime extension of the wartime Office of Scientific Research and Development, the RBNS was to have the authority to contract with universities and industry to "explore the long range, fundamental scientific aspects of defense technology."[3]

In March 1945, the Navy suggested that one of the RBNS's possible projects might be to improve high-altitude meteorological forecasting to aid the "transportation of carriers such as balloons over enemy territory. These carriers might themselves send back meteorological intelligence or might transport propaganda or destructive agents."[4] Both the Army Air Forces and Army Ordnance wanted the RBNS to initiate studies that would improve the radio tracking of ballistic projectiles. This required "research in connection with propagation of radio waves in the ionosphere,"[5] which, although well beyond balloon height capabilities, could still be served by radio propagation experiments on balloons. When the RBNS did not survive the first year of postwar reorganization, the military contracted directly with civilian groups to improve knowledge of the upper atmosphere. Thus the Navy's Office of Research and Inventions (ORI), later known at the Office of Naval Research (ONR), and the Air Force's

Air Materiel Command, became the major patrons of stratospheric
ballooning.

Wartime advances in radio and plastics production would eventually
revolutionize ballooning techniques, but it was the new awareness of the
military importance of upper atmosphere research that gave ballooning
an immediate boost. University centers of wartime research, such as the
University of Minnesota, New York University (NYU), and Tufts were
drawn into ballooning projects after the war for many different reasons.
The Balloon Group Research Division within the College of Engineering
at NYU developed controlled-altitude free-balloon systems, with support
from the Watson Laboratories of the Air Materiel Command.[6] And a
group of physicists at the University of Minnesota formed two linked
groups to study ultra-high-altitude ballooning technology for the Navy,
which they then used for the study of cosmic-ray phenomena. The Minnesota group, in particular, formed around newly established and some
pre-existing interests in the department, but it was stimulated as well by
its collective wartime experience and the proximity of Jean Piccard. Piccard's entrepreneurship and undiminished drive to reenter the stratosphere got the ball rolling in the Minnesota area while the physics department was rebuilding.

Jean Piccard's *Pleiades II*

Jean Piccard remained in Minnesota throughout the war, teaching and
conducting aeronautical studies, and from time to time tried to offer his
ballooning expertise to the Reserve Officer's Training Corps. He continued to plan and dream, of course, and emerged from his wartime activities
still hoping to reach the stratosphere once again. In the last years of the
war, while Jeannette Piccard continued to search for better positions for
herself and her husband (Jean was an untenured professor at the time,
and did not gain tenure until 1946), Jean dreamed about a 30,000-meter
flight. By February 1945, he had once again attracted the attention of
aviation writers and editors who eagerly took his stories as they looked
to the end of the war.[7]

Jean based his plan for a *Pleiades II* on the multiple balloon concept
he had used in the *Pleiades* before the war. The design of the balloon
cluster was his uppermost priority, and it had to be completely new, both
in geometry and construction. In the spring of 1945, he contacted the
Celanese Corporation of America, the Bakelite Corporation, Du Pont,
and General Electric's Plastics Division searching for suitable balloon
material and possible financial support. He estimated that he would need
100,000 square meters of thin, strong plastic film for his balloons, but to
his frustration, none of these sources could supply this material. Jean was
convinced that plastic film had to be used, but what type and where he

would get it, how the balloons would be fabricated and then clustered, and the design of the cluster—all remained unknown.

Jean described his plans to Lt. Cmdr. D. F. Rex of the Navy's Office of Research and Inventions (ORI) in April 1945 as well as to his old friend and patron W. F. G. Swann. Still searching for funding in a prewar mode, Piccard hoped that a combination of private donations and Navy support could be found to build another sealed gondola and a vast cluster of plastic balloons.[8] Rex, soon to head the geophysics section of the Program Branch of the Planning Division of ORI, did not reply immediately; at the time, ORI was still one month away from being recognized officially as the umbrella under which the wartime Office of the Coordinator of Research and Development, the Bureau of Aeronautics' Special Devices Division, the Bureau of Ships' Naval Research Laboratory, and the Navy Patent Office were to be consolidated into a research coordinating unit for the Navy.[9] Rex and his military colleagues were planning intensely for the postwar world, but were not entertaining any new projects save for those that could help to end the war. Rex was therefore in no position to pursue any outside proposals, although he later played a central role in supporting Piccard's efforts and made sure that ORI was involved in them.

In a letter designed to supplement Piccard's proposals, Swann enthusiastically endorsed a return to the stratosphere feeling that in the past 10 years nothing had surpassed the observations taken with his equipment on the Piccard and Explorer flights: "I think the value of those observations is only just beginning to be recognized. We were about ten years ahead of our time when they were made."[10] Swann, still confident of the value of manned ballooning, was willing to cooperate and endorse Piccard's efforts, but did not limit his own cosmic-ray studies to manned platforms. He was preoccupied with reconstituting the research agenda of Bartol and wanted to preserve his enlarged staff, as well as keep his best staff, and so had to find interesting things for them to do. His right-hand man, T. H. Johnson, did not return to Bartol, preferring the larger sphere of the Aberdeen Proving Grounds. Swann knew that he had to move quickly to keep Bartol competitive with respect to staff and funds.[11] To be sure, Bartol was now active in laboratory nuclear physics and high-energy physics, but cosmic rays were still Swann's personal interest, and he took every opportunity to observe them from aircraft and balloons. By the fall, Swann was discussing his plans for flying heavy equipment aboard large bombers with Lyman Briggs, hoping that through Briggs he might obtain access to aircraft and even support from the National Geographic Society, where Briggs was still chairman of the Committee on Research. To Briggs, Swann repeated his feeling that they had found the cosmic-ray maximum just after Regener and Pfotzer and had detected a large horizontal flux, which led Swann to conclude that the electron component of cosmic rays was, in fact, atmospheric. Swann believed this conclusion was gradually gaining favor so he was anxious to verify it, and hoped that Briggs would approach Gen. Hap Arnold on his behalf.[12] But a manned balloon flight was also very much in Swann's mind, since high-flying aircraft were not yet available.

Piccard, meanwhile, was searching for support and refining plans for his 100-balloon *Pleiades II*. It was a carbon copy of the prewar agenda: from measuring temperature, pressure, humidity, composition and degree of ionization, to cosmic-ray, solar and ozone studies, to "Any other important observation which any of my friends in physics suggest."[13] Piccard himself became interested in measuring the temperature of the balloon and parts of the gondola with thin-wire thermometers, rapid action devices that might solve hysteresis problems that had plagued earlier measurements. All declared interests now were within the realm of aeronautical diagnostics; his instruments were diagnostic sensors of the craft itself, not of the environment that the craft found itself in.

As with the original *Pleiades*, Piccard believed that a balloon cluster of 100 balloons, each with a 2800-cubic-meter capacity, provided higher efficiency in both weight and cost than a single huge gas bag. There would be no rip cord problems and no valves. This time, his gondola would be made of soft aluminum, but, as before, it would be a 2.2-meter sphere. Dowmetal would still be used for interior fittings, but Jean now wished to avoid a Dowmetal skin, fearing that the growing number of power lines spanning the Great Plains might spell disaster for a magnesium gondola. However, this concern was more an outgrowth of the old feeling that Dow had not been a sufficiently sympathetic supporter, nor had it shown any promise of being one now.[14]

Collaboration in the Proposal Process:
Piccard Meets O. C. Winzen

In early November 1945 Jean Piccard met Otto C. Winzen, an immigrant engineer/entrepreneur who was chief engineer for Minnesota Tool and Manufacturing Company. Winzen had approached the university asking for advice on how best to design a fast-acting altimeter for dive bombing, and was led to Piccard among others. Through Winzen, Piccard met F. D. Price, president of the company, to whom he divulged his real dreams. Piccard told Price of the wonder of cosmic rays and of the many other scientific motives for his planned flight, but admitted that simply making the flight was an end in itself. He equated scientific discoveries to geographic discoveries. The romance of exploration was always a good selling point: "Going with open eyes into new territory is always an interesting and useful thing to do." Piccard envisioned other possible payoffs. The windows designed by Dow for his 1934 gondola had to be made frost-free, and Piccard claimed that this design was later adopted by Boeing

FIGURE 9.2. Otto Winzen. Winzen Collection, UM.

for its Flying Fortress. Speaking to an industrialist, Piccard emphasized, "Science always pays."[15]

Piccard and Winzen collaborated on the altimeter problem, but also drew together in planning for the balloon flight. Preliminary proposals were drafted once again for Rex of ORI, and for Col. H. M. McCoy of the Army Air Corps. By January 1946, they decided that the proposal should come from Winzen at Minnesota Tool, and that Piccard should be the central figure and scientific director. Anticipating funding from the Navy, the Army Air Corps, and the Manhattan Engineer District, Winzen collected the names of those Jean had contacted, and drafted a letter of intent to the Assistant Secretary of the Navy. Although some who were named sponsors (including Rear Adm. C. E. Rosendahl of Lakehurst's Lighter Than Air (LTA) Division) stiffly responded that no official decision had yet been made, all were interested in the flight.[16]

Rosendahl's reserved though friendly acknowledgment of Winzen's presumptive letter was enough to lead Winzen and Piccard to again approach ORI in February. Both ORI and Rosendahl's Office of the Chief of Naval Airship Training and Experimentation became more interested in Piccard's proposal when, in March, they obtained preliminary endorsement from the Office of the Assistant Secretary of the Navy for Air. In effect, Winzen and Piccard played different parts of the Navy off against one another, in a gradually escalating series of endorsements that led to actual support. Rex of ORI, in particular, as the new head of the geophysics section of the Program Branch, did not want to see any university-based Navy-supported research emerge from anywhere else but ORI.[17] The Program Branch was distinct from the Scientific Branch of the Planning Division in that, although the division itself was charged with developing and maintaining research contracts with civilian scientific groups and making sure that the contracts reflected the interests of bureaus within the Navy, the military officers making up the Program Branch were concerned with the latter activity, whereas the Scientific Branch tended to be interested in the former. Both, however, had a hand in "matters of policy which interrelate the scientific fraternity of the nation with the Navy," according to Capt. R. D. Conrad later in the year.[18] Even so, there was no traditional base of cooperation and coexistence to draw upon in a peacetime world, and as a result of very different missions, not only within different branches of the services, but within each division of any one branch, the Navy was far from being a single-minded patron. In the short run this made the Navy ripe for entrepreneurs like Winzen, but in the long run lay trouble.

Piccard lacked Winzen's appreciation of the complexity of the Navy, and as well seemed distant from the scientific agenda for *Pleiades II*. Although Swann became his source for scientific justifications, Piccard still harbored doubts about solar spectroscopy or ozone studies; those he thought unnecessary he believed would be uninteresting to the Navy as well. But Swann, and Minnesota astronomer Willem J. Luyten, argued that solar spectroscopy was important, if only to reach the magnesium doublet at 280 nanometers to confirm theoretical predictions. Uppermost in Piccard's mind, however, was how to convince the Navy to participate. He believed the Navy would be more interested in the new logistics he had in mind, which included the idea, now standard practice, of launching the balloon gondola from a truck, moving downwind.[19]

A Corporate Site for Research: General Mills

Although the preliminary proposals were made by Winzen's company, the formal proposal submitted in April 1946 was under Piccard's name alone because Winzen's Minnesota Tool ceased to exist by the early spring.[20] Winzen and Piccard looked for a new institutional sponsor, possibly the university or General Mills, Inc., of Minneapolis. Since General Mills maintained many informal connections within the University of Minnesota faculty, and was a dominant industry in the Minneapolis area, it was an appropriate place for Piccard and Winzen to approach concerning their ballooning initiative, and indeed, the company greeted their proposal with enthusiasm. The fast-growing giant grain conglomerate had a reputation for promoting self-sufficiency in industrial research and development. Its founder, James Ford Bell, believed in diversification and worked to make General Mills capable of manufacturing anything from "flour to locomotives." He achieved his greatest success before the war in improving cereal puffing guns and packaging equipment.[21]

Before the war, General Mills was only known for its ability to produce cereals and chemicals; thus it had some trouble obtaining wartime contracts beyond the highly lucrative ones for foodstuffs. The war gave General Mills an ideal opportunity to diversify. In response it created a small engineering research and development section to explore how to turn its experience in ruggedized precision tooling into implements for war. It soon found itself designing and producing fire control devices, radio transmitters and receivers, gun sights, submarine antennae, and servomechanisms for the Army and Navy and the National Defense Research Committee.[22] Its early successes in supplying the British with roll correctors for naval shipboard fire control were followed by a number of U.S. military contracts for gun sights and torpedo directors.[23]

At war's end the research division contained 46 chemists and 8 physicists with advanced degrees; and 34 electrical, mechanical and chemical engineers, metallurgists, and administrative staff with expertise in the development of precision optical and mechanical devices for fire control and aeronautics. The mechanical division that produced the devices had over two thousand employees. Arthur D. Hyde, vice-president and director of research at General Mills, was faced with the problem of keeping this expert group together after the war. The most obvious way was to continue to be of use to the military.[24] As part of the company's general policy to diversify, Hyde created an Aeronautical Research Laboratory in the Research Division and looked for useful projects to work on.[25] The last years of the war had been the company's best years since its founding in 1929, but the phenomenal growth of the company during the war challenged company founder James Ford Bell and his chief aides to find more uses not only for their core grain products but also for their technical staffs in the chemical and mechanical research fields.[26] Thus Hyde's moves were well within company policy and tradition, but this did not include philanthropic support; in the spirit of the postwar world, initiatives such as Piccard's *Pleiades* could well lead to new and highly lucrative military contracts. Piccard and Winzen were, in addition, desirable additions to the staff of Hyde's Research Division: they brought with them at least one highly likely ONR project.[27]

By May, the Special Devices Division of ORI approved the concept of the Piccard–Winzen proposal, subject to contract negotiation. Piccard and Winzen, along with their new General Mills colleagues in the newly formed Aeronautical Research Laboratory, met with staff officers from Special Devices, the Navy's Lighter than Air Division at Lakehurst, and the Planning Division of ORI in order to come to terms over mission goals, design, governance, cost, and each group's role in the project, especially Piccard's.[28] Even though both General Mills and the Navy were uncertain about Piccard's duties, they accepted his *Pleiades* concept for a two-man, 10-hour flight. The 2,000-kilogram payload and gondola seemed to present no problems for them; it was based directly on Auguste Piccard's original design, and could "be made in any welding shop." The balloon cluster was the problem to beat.[29]

The design and configuration of the balloons and the cluster took center stage, along with planned testing procedures for the balloon clusters. At the time that Piccard and Winzen were just formulating their proposal, before their association with General Mills, they were still not sure what substance to use for the balloon skin. They thought about using 30-meter gores of Fortisan balloon fabric, available from Goodyear, but preferred some form of plastic, arguing that ordinary balloon fabric was inefficient because it was a composite of woven cloth and a sealant, either rubber or dried linseed oil. Although this cloth provided strength, the sealant was a drawback: "Half the weight is dead weight for one of the two purposes."[30] Plastic film, on the other hand, was more efficient because "every molecule . . . serves the two purposes simultaneously." They knew that they wanted a cluster of plastic balloons, but still had to find the right type of plastic. Ultimately, with Winzen's urging, Piccard accepted Goodyear's prewar rubber-derived product Pliofilm. Piccard had tried to avoid the product because he wanted to avoid Goodyear.

Swann's Scientific Agenda

Piccard and Winzen's proposal did not describe possible scientific experiments, nor the need for them. Piccard made little pretense of being interested in the scientific agenda, leaving its description to the supporting letter Swann had written in March which discussed potential contributions to cosmic-ray physics. Swann advocated a manned return to the stratosphere: "The kind of apparatus which can be taken in a gondola permits accuracy of observations far beyond that attainable in pilot [unmanned] balloon flights." Swann added that, in his opinion, no better data had been taken since the Explorer flights: "The data obtained on these flights served a fundamental purpose in establishing the key phenomena which have been largely responsible for our present understanding of the meaning of the primary cosmic radiation."[31] Swann's rhetoric was biased, to be sure, but was perfect for the proposal. Indeed, Swann was a long-time master of "sales methods in promotion" and many of his now-prominent students, such as E. O. Lawrence, learned at his side.[32]

Swann's arguments were based on a concept of a scientific technique that was perceived to be no longer in its initial stages, but at a stage of refinement where "higher accuracy becomes of predominant importance." He hoped to vindicate his observations of the horizontal cosmic-ray flux from the Explorer flights, feeling that they had been ignored by cosmic-ray physicists. Swann also wanted to reobserve the variation of cosmic-ray intensity with zenith angle, which was also of interest to other groups, such as Millikan's postwar team at Caltech, headed by H. V. Neher and W. H. Pickering, who were then planning to fly unmanned sondes carrying complex triple-coincidence counters.[33] Swann also wanted to fly heavily shielded counters again to try to distinguish electron from meson showers, and thereby confirm the existence of mesons in the upper atmosphere. He also hoped to obtain primary particle tracks using photographic plate stacks and cloud chambers. He noted in passing that the manned effort might also be valuable for ultraviolet spectroscopic observations, atmospheric conductivity, and electrical measurements. In total, it was very much a pre-war scientific agenda.

Justifying Navy Patronage

Justifying the manned effort in the name of pure science was not an ORI concern. The Navy was interested in the military application of what might be learned about the upper atmosphere, and wanted to know whether it could be a site for naval operations. Throughout 1946, the Planning Division and the Special Devices Center of the Office of Naval Research (newly formed out of ORI) haggled with the Office of the Chief of Naval Operations, the Bureau of Aeronautics, and the Office of the Chief of Naval Airship Training and Experimentation over their respective roles in the stratosphere project. Meanwhile, Adm. Harold Bowen, chief of naval research, signed a long memorandum to the chief of naval operations prepared by D. F. Rex of ONR that established ONR's interests and priority. ONR had three major programs underway dealing with . . .

upper atmosphere research intended to provide fundamental information, not already available, concerning the physics and chemistry of the atmosphere at great heights from the earth's surface. Current interest in reaction propelled missiles and extreme altitude flight make high altitude studies of cosmic-ray phenomena, atmospheric composition, high altitude meteorology, electromagnetic effects, and the absorption and transmission of radiant energy both timely and necessary.[34]

The three programs ONR intended to support were the Army/Navy B-29 program to perform a wide variety of scientific studies at 12,000 meters, a rocketsonde program using captured German V-2 rockets and new American-made Viking and Aerobee rockets, and the Piccard–Winzen manned stratospheric balloon program, which had several advantages over the first two:

Certain technical advantages pertain to such an ascent all of which cannot be realized by any other known means. These include the availability of an observer aloft throughout the flight, a measuring platform almost stationary with respect to its environment, and ample time available at any level for taking observations.[35]

In its own official histories of its Office of Naval Research, the Navy points out that it was driven to support science because it needed to study "its operational environment in very special ways."[36] Long-range communications, ballistic missile behavior, global reconnaissance—all fell within the purview of the new Navy. As a result, Piccard could not be happier with the Navy's enlightened attitude toward ballooning. To Piccard, the Office of Naval Research represented a new breed of military, whose motives were not different from his own; or so he thought.

Navy Planning, Patronage and Control: The Birth of *Helios*

Jean Piccard's increasing desire to obtain military patronage emanated from his acquired distaste for the strings that came with publicity-conscious commercial and corporate support. His dislike for commercialism now outweighed his distrust of the military. Although he at first entertained thoughts of obtaining a mixture of the two, he was so warmly received by Rex and the Navy that he had to rethink his strategy. Winzen was a strong advocate of military support, and the two agreed that Winzen should seek military patronage with Piccard's proposal before Jean approached commercial sources.[37] Both wanted one major donor so that there would be no question of allegiance or conflict of interest. Although they had high hopes that the Navy contract would be approved, until it was reality Piccard was unwilling to reveal its details, even to Swann. Piccard reflected:

We expect that this contract will relieve us of all necessity for publicity and in fact will require us to shun publicity even to the extent of not telling our friends what we are doing. Since the contract will not be drawn with a private corporation we shall not be obliged to chew a specific gum over the radio while we are in the stratosphere, nor smoke a specific tobacco, nor drink a specific drink. We hope that the advantages will offset the disadvantages.[38]

The disadvantages remained unstated, but may have been the inevitable classification the Navy would slap on the project. But Jean could not contain the old Piccard relish for publicity, notwithstanding his words to Swann above. During this time, Jeannette primarily, but with Jean's approval, approached news services about signing exclusive contracts to sell their story.[39]

Jean had no concept of the complexities of the postwar Navy. He was no longer dealing with individuals, but with separate branches and offices, each having very different agendas. Planning took on a life of its own, and can best be seen by reviewing policy making for the scientific portion of the mission. All concerned parties worried more about governance over the mission than its purpose. When discussion turned to explicit payload experiments some seven months after Piccard's initial proposals were submitted to the Navy and Army, various branches of the Navy were still maneuvering to determine who would control what was to become Project Helios. A newly formed Technical Committee for *Helios* and members of ONR's Upper Atmosphere Panel met on December 19, 1946, to discuss the matter.[40]

The joint meeting of military officers and civilian scientists was supervised by Lt. Comdr. George W. Hoover of the Special Devices Center, ONR. Special Devices's interests were still strongly aligned with those of its original parent body, the Bureau of Aeronautics. First and foremost, Hoover wanted to see the "establishment of a platform from which to drop missiles in order to obtain aerodynamic characteristics from the physiological aspect for supersonic flight." *Helios*, to Hoover, would collect "data for furthering the study of the construction of cockpits for protection of the human element." Hoover's wishes were endorsed by the Technical Committee, for all parties wanted the Navy to gain experience in various operational aspects of flying, launching, and retrieving such devices. But, speaking for the Nuclear Physics Branch of the Planning Division, on the ONR Upper Atmosphere Panel side, Urner Liddell urged traditional balloon applications for science: determining the concentration of mesons and neutrons as a function of altitude, and studying the solar spectrum to determine the composition of the atmosphere. The Navy was not of one mind.

Liddell, at the moment head of both the physics and nuclear physics sections of the Scientific Branch of ONR's Planning Division, collaborated with Swann to establish design and weight requirements for cosmic-ray instruments, but the ONR Upper Atmosphere Panel had other interests as well, since it consisted of both military and civilian scientists from a broad range of agencies. They knew that the Weather Bureau wanted to send aloft barographs, humidity devices, and air samplers, and undertake studies of airborne particulates; the National Bureau of Standards was interested in studying the propagation of infrared radiation for its signalling potential; the University of New Mexico wanted to reexamine the ozone content of the upper atmosphere; and the Departments of Agriculture and Public Health wished to send organic samples aloft. Other groups wanted to make sound propagation experiments, study genetic reactions to electromagnetic radiation, and conduct general physiological testing.[41]

With so much interest in using *Helios*, Liddell and others convinced Hoover that a larger payload was required. More meetings were called to establish payload priorities and weight and space requirements, but in the interim, planners began to realize that the potentially complex array of experiments for *Helios*, even if there were to be multiple flights, could only be done with automatic scientific instruments requiring only the minimum of attention from the aeronauts. History was repeating itself, almost. Although planners envisioned a series of launches, they still thought in terms of a single "all-up" flight that would do everything.

By the time of the second conference in mid-January, the Navy had strengthened its resolve to exploit *Helios* for testing supersonic flight components. Especially complex and demanding was the missile drop program. Batteries of small drop missiles 18 centimeters in diameter and 2 meters long, weighing 90 kilograms and characteristic of missiles used in the Bureau of Ordnance Project Bumblebee, were to be modified by a Cornell University contractor and by General Mills, both of whom were interested in recovery techniques. Hoover argued, however, that:

We have reached a stage in aviation where cockpits must be jettisonable and reactions to and effects of supersonic speed are of paramount importance. The original purpose of the project [Helios] was to establish a platform from which to test physiological factors. Such tests would provide mere aerodynamic information but would [not] be designed to determine potential reactions upon the individual at supersonic speeds. New planes now under design may force man into a position beyond his own limitations physically, and unless adequate data is secured at this time deaths may ensue later.[42]

The Navy retraced many steps taken by the Army Air Corps in planning for *Explorer* before the war, but there were major changes in the postwar environment of military research. Now the Navy designed the scientific agenda and was specifically interested in testing flight materiel in the stratosphere, which was no longer the realm of possible future activities, but of current concern. Instead of thinking only of a single flight, the Navy looked toward a series of flights that would lead to an ongoing capability, but competing interests within the Navy rushed to be first on the flight roster. As yet there was no clear line of governance or decision making.

Hoover claimed primacy for the supersonic drop tests. He stressed the importance of these tests because he felt that the results from Project Bumblebee were unsatisfactory. A manned platform would reduce the number of uncontrolled variables. But virtually all of the above experiments, even Swann's, as the Navy envisioned them, could meet the needs of the Navy. In fact, the Navy was becoming particularly adept at interpreting all upper-air research as military-related.

Alan T. Waterman, ONR's deputy chief and first chief scientist, and others in the Scientific and Planning Divisions, remained concerned about image. Wishing to attract the best civilian talent, he argued that the ONR should do nothing to hinder the interests of the best civilian scientists it could find to participate. Riding on *Helios*, if it was planned right, should be a plum for a good scientist, one that ONR would like to use to attract such people to its focused interests. But *Helios* had to make sense. With Capt. H. B. Hutchinson, ranking officer on the Navy's Upper Atmosphere Panel as an ally, Waterman reached a compromise with Hoover, who agreed to invite 18 scientists to propose "basic research" experiments. Even though the invitation was given out on very short notice in mid-January, 1947, within a few weeks 17 groups had responded positively.[43]

Waterman wanted *Helios* for pure science, at the very least to keep valued scientists connected with the military during peacetime. In the young ONR, his well-connected staff already knew ". . . outstanding men in whose specialties we were interested . . ." through its collective wartime experience.[44] The rapid and enthusiastic response he received to the idea of Helios participation demonstrated the value of his contacts in the scientific community and strengthened his argument that to attract good scientists ONR had to offer good projects. Waterman looked beyond a single flight to establish a programmatic capability to service a wide range of scientists. Hoover, more involved with the military mission for *Helios*, deferred to Waterman on scientific matters. Waterman in turn dictated the scientific agenda, immediately overriding Piccard's vague objections to spectroscopic instrumentation.[45]

Jean Piccard's Role as General Mills Contractor

What follows should by now be a familiar story. Jean Piccard was gradually excluded not only from policymaking and technical execution of the flight, but from the flight itself. Still, the official Navy news release announcing *Helios* highlighted Jean Piccard's services. The 61-year-old professor still possessed the image that the Navy wished the public to see, which would reflect its support for exploration and adventure in the name of science.[46]

Beyond being responsible for the balloons and gondola design at first, Jean was little more than a titular coordinator in his position as a contractor for General Mills, and got into trouble there too. Winzen was responsible for gondola construction, and all technical services, and the Navy took care of the scientific payload. Jean did not sense this until it was too late, and eventually saw Winzen as a conspirator and enemy. But Winzen held all the cards. He was now the head of the General Mills Aeronautical Research Laboratories, and directly under him were two parallel lines of activity; Jean Piccard represented one line as scientific supervisor of a division consisting only of himself, Jeannette as consultant, the veteran balloonist and designer R. H. Upson from New York University as part-time consultant (who in fact reported to Winzen and was based at the University of Minnesota, in Akerman's aeronautical engineering department) and a one-man stratosphere laboratory. Upson was a renowned balloon designer, hailed by Minnesota colleagues as the father of the natural shape balloon. The other line was headed by J. J. Ryan, as technical supervisor, and it consisted of the bulk of the Laboratory with seven sections, five laboratories, and a model shop.[47] Although the public image of the laboratory operation was aided by Piccard's presence, the operation itself did not depend on it. As a result, Piccard had little control over policy, planning, or procedures in the laboratory. He later blamed the failure of *Helios* on this setup.

Jean began as a summer consultant for General Mills Aeronautical Research Laboratories; under the ONR contract, both he and Jeannette were retained through the winter and spring of 1947. John D. Akerman reduced Piccard's teaching load and was pleased to see that Jean had at last succeeded in his life-long ambition, which "also justifies our long range assumptions ... at the time of Dr. Piccard's affiliation with the University of Minnesota."[48] Akerman's feelings for Piccard were not shared by General Mills colleagues or by the Navy. Although Piccard was responsible for the procurement and testing of suitable balloon film materials, as well as the design of the gondola, he felt that he was treated more as a lower-level employee because he was constantly subject to direction and review, such as the weekly status meetings that began in January 1947, at the direction of Winzen.[49] Piccard was not comfortable working under what he felt were onerous constraints beneath his station. He did not like working under Winzen, did not appreciate being told to conduct small technical projects that were unrelated to *Helios*, and especially did not want to hear that General Mills, as a government contractor, would retain rights to patents. His dissatisfaction increased during the spring and became a very sore point between Winzen and Piccard. Meanwhile, other aspects of *Helios* drew quick fire from Jean Piccard.

Who Flies?

As the Navy continued to negotiate within itself to establish priorities, and continued to monitor Winzen, Piccard, and General Mills, and did its best to maintain the interest of outside scientific groups, it set the flight date for *Helios* as June 21, 1947, the summer solstice. More bureaus of the Navy entered the program, and a far-flung organization took shape: the ONR contract was handled in Sands Point, New York, coordination was in Washington, design and construction in Minneapolis, and flight operations in Lakehurst, New Jersey.

By early 1947, the original A Century of Progress pilot, and now Rear Adm. T. G. W. Settle, became the coordinator of technical operations personnel and assumed the responsibility of establishing procedures for handling and tracking the Helios flights, as well as recovering the balloons and gondola. Settle, based at Lakehurst as airship adviser to Adm. C. E.

Rosendahl, the chief of naval airship training and experimentation, wished to play a leading part in *Helios*. As early as September 1946, he called for Navy-wide collaboration, "utilizing the specialized experience of Mr. Jean Piccard," but when the time came to propose flight crews, Settle did not honor what everyone knew to be Piccard's wish.[50] Instead of nominating Piccard and his wife as crew, Settle chose proven Navy fliers such as Harris F. Smith and Henry C. Spicer, and made his old friend C. L. Fordney the scientific observer.[51]

Piccard acted predictably. He felt that once again, his name was being used. Even when other forces within the Navy that were sensitive to Piccard's wishes capitulated later in January, designating Piccard as the scientific observer, Piccard still objected to the wording of the agreement since it reduced his chances for a new world's altitude record.[52] At first, General Mills backed Piccard and refused to sign the contract that included Settle's all-Navy flight roster. Adm. P. F. Lee changed the wording of the contract to indicate that Jean Piccard would be an acceptable copilot to a pilot designated by the Navy Department. But this was not acceptable to Piccard, who was able to hold up General Mills compliance for a short time by insisting that the contract be further altered to say that he would be in command and would choose his companions.[53] As inventor and designer of the new aerostat—the radical multi-balloon system based upon his Pleiades design—Jean felt it was unfair to put him under "the command of a man who is not at all a stratosphere pilot."[54] He was unwilling to admit that Settle's candidates were indeed experienced in the stratosphere, especially Fordney.

All that Jean had at hand was his name. In his own mind, as the daimyo of manned ballooning, he remained central to the effort, and did enjoy the title of project scientist that General Mills bestowed on him. But this title was never acknowledged by the Navy, which wished his role to remain undefined. While haggling continued, General Mills staff began to take the Navy's point of view, and Jean's position weakened. Thus as work on the gondola and balloon cluster continued, an old pattern reemerged: Jean Piccard wanted control and engaged in verbal puffery; after miscalculating the degree to which he would be tolerated, he tried to back down. But even when Piccard announced that Henry Spicer, of Settle's staff, would be acceptable as pilot if he were copilot, the Navy did not agree, and this drew from Jeannette a spate of criticism about the Navy's apparent duplicity which further damaged relations. According to Jeannette, both Piccards had to be allowed to fly, or they would break off all relations with General Mills and the Navy.[55]

Jeannette's protestations indeed led to a break between the Piccards and General Mills, as well as with the Navy. In fact, Settle eventually chose Spicer as pilot, while Jean Piccard continued to be listed as "scientific observer" through April. But for the time being, the question of crew smoldered above an even hotter problem: the organization of the program itself.

Who's In Charge?

As annoying as the Piccards were, the real problem for *Helios* was to get the administrative details straight and establish a clear line of authority for the program. The Navy was certainly in charge, and the utility of *Helios* was clear to factions within the Navy. But each faction had its own idea of how *Helios* should be used, and who within the Navy was in charge. Settle's Helios organization charts, prepared between February and April 1947, were confusing structures that kept the Offices of the Chief of Naval Operations and the Chief of Naval Research separate but equal; although Settle's own position as chief of operations for *Helios* took center stage in the chart, there was no clear line of authority.[56]

Hoover, Hutchinson, Rex, Winzen, Liddell, Settle, and others representing factions within ONR and the Bureau of Aeronautics Lighter Than Air Division struggled with the confusing administrative details of who would be responsible for what through January and February, while a mock-up gondola was constructed at General Mills under Jean's direction and the roster of scientific experiments was developed through endless modifications. The missile-dropping experiment continued to be important to the mission, but Hoover, faced with pressure from Waterman, became willing to reconsider if it jeopardized the general Helios program. The missile-drop remained intact through February when all scientific investigators and ONR staff visited Minneapolis in February to examine the gondola and perform integration tests. In addition to Swann, those present and involved in the inspection at the General Mills plant were T. H. Johnson, formerly of Bartol and now of Aberdeen Proving Grounds; Marcel Schein, Chicago; C. G. Montgomery, now at Yale; E. O. Salant, NYU; J. A. Wheeler, Princeton; and Brian O'Brien, Rochester. They were enthusiastic about ONR's Helios plans and worked hard to provide cloud chambers, multiple-coincidence counters, cosmic-ray emulsion stacks, ozone spectrometers, sky radiation spectrometers, and a host of meteorological items.[57]

Bolstered by Alan T. Waterman and his ONR colleagues, *Helios* became an exciting prospect to those seeing it as a mode of conveying heavy experiments to high altitudes for long periods of time. Most of the experiments were labeled "nonexpendable" and "automatic." Only O'Brien's proposals were "restricted," and that was because his university reserved the right to release all press announcements. In fact, O'Brien was given priority for these flights as he was considered to be the scientist with the most experience in balloon-borne spectroscopic experimentation, from his Explorer days.[58] Piccard's experiment was minor and pedestrian: a

FIGURE 9.3. Project Helios staff from General Mills meet with the Navy in early 1947. Third and fourth from right are the Piccards, fourth from left is Otto Winzen. Winzen's hand is resting on Commander George W. Hoover's shoulder. Winzen Collection, UM.

simple shielded resistance thermometer to measure gondola skin temperatures.

Although the scientists were enthusiastic about *Helios*, none of them restricted their research to the planned flight. Each was well established and was quickly reestablishing prewar momentum. *Helios* was to be a relatively unobtrusive endeavor, requiring little more than a temporary loan of an automatic instrument. But the Navy remained unable to co-ordinate the scientific agenda with military interests and logistics. Throughout the late winter and spring of 1947, meetings were held in Washington and Minneapolis between ONR, Lakehurst officials, General Mills personnel, and participating scientists. On at least three occasions, urgent requests went out to the scientists for ever-more detailed infor-mation on their experiments (size, weight, exact dimensions, power re-quirements, etc.) as the gondola was now under construction and could still be modified. New requests went out to the scientists to reduce di-mensions when the military found fresh uses for *Helios* payload space. Specifically, sound propagation experiments were added in late February. The Signal Corps, one of many military groups interested in monitoring nuclear explosions, believed that *Helios* would be useful for testing ways of detecting and distinguishing various atmospheric and artificial noise sources such as air bursts, aircraft engines, rocket motors, and large bombs.[59] And agencies interested in ballistic missile guidance research wished to strengthen Helios's planned sky brightness and infrared horizon visibility studies.

Clearly, although most of the basic science experiments and scientific goals for *Helios* were carryovers from the 1930s, the military-related experiments quickly became more sophisticated and quite distinct from their prewar predecessors. The requirements of high-speed and high-altitude flight, or long range reconnaissance, radio communications, and nuclear test detection had become paramount during and after the war; accordingly, military patronage for the conduct of aeronautical research was vastly increased in scale and oversight. Hoover therefore classified the substance of the project as "restricted," with the result that Navy public affairs officers, submerging *Helios* to Neptune's depths, called it the "Piccard Flight" in their attempt to foster an image of heroic stratospheric exploration in the name of science, which seemed a very acceptable face for military research and development.[60]

The Navy public announcements also glossed over doubts about the design of the balloon cluster itself and the confused and bewildering bureaucratic maze created by the multiple interests engaged in the project. Winzen and on-site ONR officers knew that Piccard's dream of 100 balloons working smoothly together from the time of inflation to the launch needed detailed testing and evaluation. One of the most persistent problems they perceived was how to avoid interference between adjacent balloons during ascent, when the partly inflated Pliofilm gas bags would swell and change shape radically. Single-strand chains would not create such a problem, but would be so long that the rope connecting system would be extremely heavy. Jean had in mind an approximately spherical or hemispherical system, but somehow the balloons had to be kept untangled. Winzen himself came to see *Helios* as overly complicated; at best it was a prototype for future missions using fewer balloons; smaller clusters or even single balloons, unmanned, which could rise to the same altitude.[61] Winzen made this point time and again to scientists who he hoped would remain advocates of ballooning after *Helios*. The cluster design was at best a compromise at a time when huge monolithic plastic cells were not even thinkable.

The ultimate practical application was to use single plastic cells. The military could think big, of course, whereas most scientists tended toward more modest solutions. John Winckler, a young physicist at Princeton who was soon to move to Minneapolis, was given a tour of the General Mills facilities by Jean Piccard, and thanked his guide for the introduction:

I'm sure I need not emphasize the importance of this work for cosmic-ray investigations. We hope the development work on the balloons will proceed well, not only for the manned flight, but also for the benefit of those of us who want to make single balloon flights at various parts of the earth.[62]

However, scientists within or outside ONR were not in charge of *Helios*. Waterman was the closest to having policy authority, but he was tied to the program. Scientists's continued strong interest in flying single cells or small clusters, from a wide range of geographical climes much in the manner argued by Millikan a decade before, had to wait. The immediate and far more complex demands of military reconnaissance and aviation required the huge cluster as envisioned by Piccard as an expedient. But even though the single-cell idea became stronger as *Helios* stumbled through the testing phase, no one in the Navy was in charge to face it. Somehow, the two-headed administrative dragon had to be conquered.

Technical and Scientific Coordination Problems

In March 1947, as other high-altitude initiatives such as the V-2 missile testing program at White Sands were beginning to yield exciting new glimpses into the ultraviolet spectrum of the sun and the cosmic-ray intensity profile at great height, Urner Liddell worried that the scientific component of *Helios* was not being coordinated properly. There was no "scientific officer" within ONR, and one was needed in the Planning Division. Waterman took on this task, but quickly delegated it.[63] Throughout March, General Mills personnel, including Piccard, worried about environmental controls and weight limits, looking for lightweight batteries and other weight reducers. There was also the overriding problem of designing a safe system for launching the monster balloon cluster safely. These problems caused Winzen to ask Hoover for a two-month launch delay in April.[64]

Administrative problems reached critical proportions in late April. H. F. Smith, stationed at General Mills, argued that *Helios* was "long on administrative and short on actual engineering assistance . . . we are rapidly reaching an impasse which will prevent any collaborator from installing his gear . . . It is impossible to build a research tool in Minnesota using blueprints still in Washington." He called for a streamlined procedure, with coordination in Minneapolis rather than Washington and New York, to mount "a belated effort to put a dowdering Juggernaut back on the tracks before it overwhelms us."[65] But Settle still controlled the operation of the mission from Lakehurst. His interests and procedures conflicted with those at General Mills and within the ONR Special Devices Center and Planning Division. There were also constant procurement problems with the Navy's Bureaus of Ordnance and Aeronautics, which claimed oversight authority as well.

Beneath all this, but festering in the minds of those closest to the project, was the knowledge that the balloon tests were not going well in Minneapolis. In fact, the Pliofilm cells, as each balloon was called, were failing inflation and launch tests. Even single balloons were not proving reliable, since Pliofilm was found to be very sensitive to sunlight. Pliofilm, a prewar Goodyear rubber-derived product, was initially thought to be an excellent balloon material; as already mentioned, Akerman and Piccard experimented with it as a possible alternative to cellophane during their first foray with constant-level balloon sondes. Before the war, Pliofilm output at Goodyear was only a few pounds per day, but its great commercial appeal as a transparent food-packaging material, or in heavier gauge as cartons, umbrellas, raincoats and shower curtains, helped production jump to tons per day within a very short time.[66] Both war production and lack of competition made Pliofilm abundant and cost-effective as a food packaging material at war's end.

Jean Piccard did not want to use Pliofilm at first, hoping that there might be a commercial alternative to a Goodyear product. But Pliofilm at first seemed to be an ideal balloon material; it was as amenable to shaping and sealing as anything yet on the market. As with all new balloon materials, however, development, handling, and testing procedures all had to be worked out. Piccard differed with Winzen and Upson on how this should be done, and the Navy people wanted their own way as well. Three distinct groups at General Mills thus argued over the design of the test procedures, and at one point Jean Piccard sided with six members of General Mills research staff in an effort to oust Winzen. When General Mills vice-president in charge of research Arthur D. Hyde demanded, in front of Henry Spicer and an ONR official, that "all of them . . . draw their pay as of that minute, including Jean . . . it was amusing to see Jean define his position of loyalty."[67] Jean Piccard had miscalculated again, but it should have left no doubt in his mind who was in control.

After the other engineers quit, as part of a general shake-up at the General Mills Aeronautical Research Laboratory, Hyde gave those loyal to Winzen one month's pay bonus in an attempt to restore morale and unity. A momentarily chastened Jean Piccard held on by the skin of his teeth, determined to reap his aerial reward. But Winzen's triumph was brief; the Pliofilm cells failed to perform properly in the flight tests he designed. In February and March 1947, Pliofilm cells continued to rip and deform. Drawn in thin sheets, the substance leaked helium faster than desired and deteriorated with age and exposure. Thus, after the shake-up, Winzen was happy to have Piccard oversee the reevaluation of Pliofilm, for Piccard would thereby assume responsibility for its choice.

Jean Piccard had already turned to Brian O'Brien of Rochester for quantitative tests to determine how ozone and ultraviolet radiation could weaken Pliofilm. O'Brien's laboratory staff at the Institute of Optics were expert in evaluating the physiological and chemical effects of ultraviolet radiation and ozone. They irradiated three samples of 0.75 mil Pliofilm, 1 mil Nylon, and 1.2 mil polythene (polyethylene) with an open carbon arc source that reproduced ultraviolet solar radiation equivalent to that at an altitude of 34,000 meters for 12 hours under high ozone concentrations. The Pliofilm sample degraded from a "yellowish soft pliable sheet to a stiffer, very fragile sheet which I was unable to bend without breaking," noted Fred W. Paul, an O'Brien associate.[68]

O'Brien's tests continued throughout April, but they did not please Piccard, who wanted more tests. O'Brien and Paul had found that the new fully synthetic substance polythene was far more resistant to great temperature ranges, and remained more pliable after prolonged contact with high ozone concentration and ultraviolet radiation than either Nylon or Pliofilm. But polythene was permeable to helium and not available in quantity as thin sheeting, and so there was no immediate substitute. Piccard, unhappy with the results and fearful of anything that might deny him a flight to the stratosphere, did not relay the negative results on Pliofilm to Winzen at General Mills. By mid-June, a suspicious O'Brien asked Piccard if he had tried to confirm the laboratory studies by analyzing recovered Pliofilm samples after flight tests. To be sure that his message was getting through, O'Brien sent ONR's Urner Liddell copies of his letters and test results.[69]

The Demise of *Helios*: How Not to Run a Program

Rips in Pliofilm were matched by fissures in the communication and coordination structure of the project itself. Not only had Jean Piccard kept O'Brien's test results from General Mills, but conflict reigned between the various concerned Navy agencies. With each subsequent test failure of the Pliofilm balloons, Spicer, Settle's on-site program officer at General Mills, strengthened his argument that *Helios* had to be appreciated as a test program and not an operational capability. He told both Settle and H. B. Hutchinson that "I can not think of a technical branch in the Navy that will not benefit greatly from the work now in progress on this project and the experience being gained by all of us."[70]

But Capt. Hutchinson, head of the Program Branch in the Office of the Chief of Naval Research, as well as ranking officer on the Navy's Upper Atmosphere Panel in ONR's Planning Division, could only see that little progress was being made; he wanted an operational balloon system that would have direct scientific and military payoff, and all he saw was competition and confusion. Hutchinson now decided that he had to take a direct role in determining governance. He called for a general meeting to straighten out all procedures and to streamline all phases of coordination of engineering, operations, and scientific experimentation. The meeting was delayed several times until May 19, 1947 amidst growing rumors that *Helios* might well be canceled.

Hutchinson was more concerned with organization than with the technical problems of Pliofilm, although it was the latter problem that was creating much of the anxiety among those closest to the project in Minneapolis. He felt that solving the organizational problems would help solve the technical problems. By now he knew that there were many competing interests, and, through Smith's and Spicer's reports, that coordination at a distance was cumbersome and ineffectual in managing those interests. Hutchinson had already transferred the responsibility for the scientific coordination of *Helios* from Washington to H. F. Smith in Minneapolis, but it was too little too late, and did not solve the problem of how to deal with responsibilities for operations still based at Lakehurst under Settle.

The May 19 reorganizational meeting in Washington between officials from ONR, the Navy's Bureau of Aeronautics (BuAer), the Office of the Chief of Naval Operations (CNO) and General Mills revealed deep problems in coordinating all aspects of the program. Some of these had to do with complaints that Hutchinson received from Adm. Nicholsen of the Design Division in BuAer about poorly developed and conflicting technical requirements for the manned flight, complaints from Smith about the integration of scientific instruments, and the fact that to date no launching of a single Pliofilm balloon had been fully successful.[71] Inflation (both of the balloon and the costs involved) and launching headaches were spelled out by General Mills engineers, who asked for an additional $72,000 to continue tests to determine why none had yet succeeded. In response, the "captains and kings there assembled" started asking why the initial outlay of some $140,000 had resulted in seven failures. Captain Bolster, acting chief of naval research, later confided to H. F. Smith that the Navy would look into the expenses of General Mills.[72]

Those at the meeting were divided into four factions: advocates of manned flight from Lakehurst under Settle; advocates of scientific ballooning from ONR under Hutchinson and Waterman; advocates of a manned aerial platform for military reconnaissance and testing from ONR's Special Devices Center, CNO, and BuAer; and General Mills, which advocated developing the technology itself. After considerable haggling about the lack of progress and the purpose of the project itself, most of the representatives admitted that the structure of Project Helios presented deep, possibly insoluble problems. Eventually, it was suggested that *Helios* be abandoned for a simplified program of upper atmosphere research and balloon development.

Some of the 37 representatives in attendance, mainly from the ONR Planning Division, pointed out the defects of the manned Helios mission compared to sending each experiment aloft in unmanned gondolas under single balloon cells. Single unmanned experiments under single balloons were vastly simpler, both technically and administratively, than orchestrating a complex battery of integrated experiments under a 100-balloon cluster. As the single balloons were developed, they could be tested by sending single scientific packages aloft. All agreed that Piccard's cluster design, in particular, was "still too formidable to attempt."

Advocates for splitting up *Helios* argued that General Mills had learned much about building and launching the balloons themselves, and most of the scientific instruments were designed to be automatic anyway. In his new capacity as scientific coordinator, Smith publicly advocated an ONR program in unmanned scientific launches, to be coordinated solely by ONR with General Mills as prime contractor for operations.[73] Hoover, speaking for the Special Devices Center, endorsed continued development of balloons for "future possibilities," but admitted that his immediate needs could very well be met by unmanned flight. Hoover indicated that General Mills had recently submitted a new proposal for balloon development. T. R. James of General Mills elaborated that, with sufficient funding, Pliofilm could be replaced by the new polythene, which was far more resistant to tearing and to ultraviolet radiation. But as yet not much of the new substance was available; consequently, it was very expensive to produce in sheets, and samples were still not strong enough to hold helium. James, supporting the growing consensus for unmanned flights, pointed out that not enough polythene, "the preferred material," would be available for at least a year for the manned flight, but there was enough around to begin constructing single cells for flying small packages.

Polythene resin, or polyethylene as it is now called in the United States, was first synthesized in Britain in March 1933 at the Alkali Division of Imperial Chemical Industries, Ltd., at Winnington, Cheshire. Produced through a chemical reaction between ethylene and benzaldehyde at 2,000 atmospheres pressure at 170 degrees centigrade, this high-molecular weight polymer was far from easy to synthesize and therefore stimulated no immediate interest among the giants such as Du Pont, which was fully engaged in exploiting Nylon, even though by 1936 polythene was found to have excellent film-forming properties.[74] The conversion of high-pressure polythene from a "laboratory curiosity to a commercial product" came only when it was found to have excellent electrical insulation characteristics.[75] Just before the war, polythene was first used to insulate submarine electrical cable, and during the war it saw wider use in radar fabrication in Britain. It was also used as a waterproofing material by the U. S. Navy.[76] Wartime needs helped to refine the high-pressure methods of producing polythene, but completely restricted its commercial application. Thus, although there was far more tonnage available both in Britain and in the United States after 1943, when both the Bakelite Division of the Union Carbide and Carbon Corporation and the Du Pont Company obtained production licenses, there was as yet little or no exploitation of its film-forming properties.[77] With the end of the war, Union Carbide and Du Pont moved rapidly to adapt polyethylene for commercial use; its fabrication into large sheeting was first attempted by the Plastics Division of the Visking Company, which was soon absorbed as a division of Union Carbide.[78]

In the United States, polyethylene rapidly proved to be an excellent low-temperature packaging substance. It remained flexible over a wide temperature range, was permeable to oxygen and carbon dioxide while remaining waterproof, and was chemically inert. To be useful as a balloon skin, however, its permeability had to be reduced, and methods for fabrication also had to be developed. These problems delayed its introduction during 1946 and 1947, the critical years in which Project Helios was struggling to make Pliofilm work.

Waterman's representatives at the May 19 meeting called by Hutchinson, including Urner Liddell, T. J. Killian, and Roger Revelle, all realized that Pliofilm was unsuitable, that polyethylene showed promise, that a progressive series of testing ever larger clusters would push the manned program back indefinitely, at least to 1949, and therefore that in the interim they should consider unmanned tests. A single cell of the size General Mills had been building could carry 27 kilograms to 31,000 meters. Thus, on technical points alone, Hutchinson was able to recommend to the Office of the Chief of Naval Research that *Helios* be completely reevaluated and that a new proposal by General Mills for unmanned single flights should be entertained by ONR as quickly as possible. Although the meeting participants agreed on the reevaluation, Waterman wished to point out that the original rationale for *Helios* was still justified and had a place in ONR's scientific mission:

A high altitude flight was viewed with increasing importance because only through such a program could a stable platform be maintained at various heights in the atmosphere from which many very important special studies could be made. These investigations could probably not be made in any other manner, and it was this fact that had justified the project originally.[79]

Waterman added that the proposed development program should include not only a manned flight as a goal, but also the development of a heavy lift capability for unmanned payloads. He saw the unmanned flights as critical steps en route to an eventual manned flight; in this manner Waterman deftly achieved a compromise, giving manned enthusiasts such as Spicer, who had been designated the pilot for *Helios*, some hope that he would eventually fly, while developing a real capability for flying heavy scientific payloads comparable to those presently being flown into space for a few moments aboard V-2 rockets from White Sands, which delighted many cosmic-ray physicists but frustrated them as well because the rocket spent so little time in space.

Even though both Lakehurst and ONR wished to remain involved in manned ballooning, some of those who attended later technical meetings at General Mills were far from comfortable with the prospect. H. A. Tanner, a senior chemist at the Naval Research Laboratory, felt that the General Mills balloons were not fully proven for manned flight because they had not demonstrated resistance to temperature extremes, and as yet there was no safe and "less abusive" method of launching. Nor had they licked the nasty problem of the behavior of slack balloons in a cluster. Tanner agreed that balloons now available could be used for unmanned flights, singly and in small clusters, to both explore the upper atmosphere and to study the behavior of the balloons themselves, and that only then might manned flight be attempted. Until then "a manned flight would be a stunt comparable to going over Niagara Falls in a barrel."[80]

Hutchinson concluded the May 19 meeting by saying that if some form of development program were to lead to a manned capability, BuAer had to be involved, but for the time being, *Helios* would become a general balloon development program. In tune with Waterman's arguments, Hutchinson reorchestrated Helios's priorities. He had Hoover's compliance that Special Devices interests could for the time be served by unmanned flights, and so convinced his CNO colleagues that an unmanned program would satisfy immediate Navy needs. Because CNO controlled Lakehurst operations, Settle was effectively outflanked. Spicer, his representative and advocate, had to admit that a manned flight was still a long way off. This virtually eliminated Settle's central role. Thus Hutchinson unraveled the administrative jumble by removing Settle from the middle of the dual-keel Helios organization chart, eventually replacing it with a linear ONR operative design.[81]

Within four days, *Helios* was under "complete reappraisal" by ONR. Before the minutes of the May 19 meeting were even distributed, BuAer and NRL moved to suspend all work, and a general "roll-up" of the operational teams had started in Settle's office: The operational staff was disbanded, and plans for a piloted flight were canceled. The Navy still wished to perform missile drops from unmanned clusters, so to this end—and only because of this end—*Helios* remained an active project, but one retrofitted to planning status.[82] Coincidentally, BuAer had been conducting its own tests of modified Dewey and Almy Darex J-2000 and Darex J-1100 balloon production samples since February and stepped them up in June with the cancellation of *Helios*. But the Darex designs were equally unreliable for ultra-high-altitude sustained flight.[83] Thus BuAer elected to wait and see what would reemerge from the shards of *Helios* in the alliance of General Mills and ONR.

Jean Piccard's Reaction to the Cancellation of *Helios*

Jean Piccard had no appreciation for Navy politics, nor did he have any vision of ballooning other than a manned flight. He did not attend the May 19 meeting, and reacted predictably when the minutes of the meeting were distributed. H. F. Smith, who publicly advocated the new unmanned program, privately told Jean Piccard that he still hoped that "means may yet be devised to lift Drs. J. P. and J. P. to 100,000 feet over the sea—

preferably the land."[84] Smith's silent advocacy was not necessary to keep Jean Piccard's dreams alive.

Jean and Jeannette Piccard had been conducting their own tests of Pliofilm and polyethylene, and had convinced themselves that success was at hand. Thus they were deeply disappointed and embittered that *Helios* was delayed indefinitely when the gondola was nearly ready and the scientific apparatus had been pledged. Piccard distributed a long and pointed letter to all Helios participants, military and civilian, blaming both Spicer and Winzen for the failure of the flight tests. His own indoor tests of single balloons had been very successful, he argued, and all he needed was more time and money to work up to the originally planned cluster, using polyethylene.[85]

At first T. R. James of General Mills sympathized with Piccard in case his pleas might extend the contract. James agreed with Piccard that they were close to an operational prototype once sufficient polyethylene was available.[86] Piccard still hoped the Navy would reconsider; in his first distributed letter early in June he concluded: "I cannot imagine that the project will be stopped at this stage when so much successful work has already been done by you [the scientific collaborators] and by us and when the developmental experiments on ground handling and launching have just reached a most promising stage."[87] Piccard assumed that the participating scientists would support him in advocating a manned flight. In fact, no record exists to indicate that any of them did so in writing. Swann and O'Brien moved quickly to use unmanned ballooning and B-29 aircraft flights, as we shall see.

With T. R. James's approval, and again independently of Winzen, Jean performed additional tests of the balloons and the cluster concept. This emboldened him to tell all scientific collaborators in another long epistle in late June that he was looking for new sponsorship and that his direct involvement was essential because, "regardless of previous poor results obtained by operators inexperienced with these balloons . . . [i]t may be superfluous to remind my scientific collaborators that almost every apparatus works better in the hands of its inventor."[88] Jean objected to Winzen's testing procedures, and since his relations with Winzen were on the rocks, he believed he could only gain by separating himself from Winzen. But General Mills thought otherwise, seeing Jean's comments as an attack on their good name. To make matters worse, in this very sensitive time, Jean also chose to fight patent applications that Winzen and General Mills had taken out on plastic balloons. Justified or not, it was not a good move, and proved to be Piccard's fatal blunder.[89]

James was particularly displeased with Piccard's criticism of General Mills research and testing staff and informed Hutchinson that both Jean and Jeannette Piccard were to be terminated as of July 1, and that they had been notified on May 29. Piccard's demands and disregard for General Mills as a competent ONR contractor were too much for James to

FIGURE 9.4. Scale testing of the behavior of balloons in clusters in the Field House on the University of Minnesota campus. Jeanette Piccard is crouching at far right. Winzen Collection, UM.

bear. The Piccards claimed that their employment at General Mills guaranteed them a return trip to the stratosphere, but James countered to Hutchinson that this was an unrealistic expectation for a 61-year-old physics professor. In any case, "Under present conditions, we cannot promise this trip, and, even if we could, to employ any person for a consideration of this type might seriously handicap the project."[90] General Mills was not the National Geographic Society, its cereal and chemical products and Navy contracts had little to do with romance and exploration. James performed damage control by firing the Piccards, but still had to convince ONR that General Mills was in control of balloon testing. He outlined the "facts" as they stood, in contrast to Piccard's allegations, and did all he could to assure Hutchinson in June 1947 that the new polyethylene balloons being developed at General Mills for exploratory testing, would soon prove worthy for unmanned ascents.

Jean Piccard had lost everything. His only recourse now was to retreat to the campus. There he licked his wounds and quickly resumed his search for support, convinced that he was right and he and Jeannette would eventually return to the stratosphere. He felt that the project had been mismanaged by the Navy and by General Mills, and that it had been canceled only because they ran out of money. To Auguste he confided that, on the bright side, his double salary from General Mills and from the university meant that for once, they had been able to put some money in the bank.[91] He spent the summer and fall snapping at Winzen's patent applications and pleading with those he knew in the Navy and in General Mills to reconsider. Both Jean and Jeannette did what they could to discredit Winzen and others, such as Spicer, who Jeannette claimed was the culprit who had demanded fatal outdoor tests. In his anger, Jean even blamed Winzen for the idea of Navy patronage and for being limited to Pliofilm owing to the Navy's demand for secrecy. However, Jean himself had approached Rex of ORI long before he met Winzen, welcomed the blanket of secrecy as a haven from having to provide continual commercial endorsements and was only willing to consider polyethylene when Pliofilm failed. Jeannette, equally bitter, was convinced that Winzen had conspired to rid the project of the Piccards. But it is clear that the distributed letters Jean and Jeannette wrote snuffed out any hope of being reconsidered by the Navy or General Mills.[92]

To save his dream, Jean next went beyond the Navy and tried unsuccessfully to gain an audience with Jimmy Doolittle. He found even the Institute of the Aeronautical Sciences in New York aloof. Thus, he and his wife turned to grass roots fund-raising, as they had in Detroit. James and Settle did agree to release movies of the inflation tests for the Helios balloons along with unclassified promotional materials so that the Piccards might use them for their lecture tours, but James and Arthur D. Hyde firmly stated that General Mills no longer had an association with Jean or Jeannette Piccard, and Settle made no public endorsements of their efforts.[93] The Piccards used the movies to try to drum up excitement and support for what was still "so vital a part of our lives."[94] But all responses were the same: sympathetic but firm refusals: No one was interested or able to support their return to the stratosphere. The National Advisory Committee for Aeronautics was not interested in 1948. Even though it was directly involved in establishing meteorological standards of reference for the upper atmosphere, manned ballooning did not fit into its agenda or small budget, since other organizations were already providing data from rockets, acoustic soundings, and meteor observations.[95] Bradley Dewey of Dewey and Almy Chemical Company probably said it best when he told Jean that he did not want to be responsible for their deaths if something went wrong. However, he added:

Despite my feeling, I recognize the fact that if you have got the bit in your teeth, you will probably flounder around until you find some way of getting in trouble, and if you are destined to do this, I would like to be sure that you at least have the best equipment possible. Put another way, I won't spend a cent on the venture, I won't be known as the sponsor of it, and I won't let my company take either position, but I will co-operate with you if the Navy wants to finance the co-operation in a modest way.[96]

But the Navy had learned its lesson: Manned ballooning was as yet too cumbersome a mode to gain quick access to the upper atmosphere, and the Piccards were too unrealistic to entrust with such a venture. What was needed was a long-term balloon development program, out of which might come a manned flight capability. Without Navy support and with General Mills hardened against them, the Piccards were left behind.

Notes

1. For background on the reorganization of science during World War II, see, for instance, Kevles (1978), chaps. 20, 21; Stewart (1948); Dupree (1957), Chap. 19. From his experience with the behavior of the liquid oxygen respiration system during his Detroit flight, according to Donald Piccard, Jean Piccard later patented a liquid oxygen converter that was built by the Pioneer Central Division of Bendix, under military contract during the war. Don Piccard to the author, 18 July 1988.
2. The military also realized that the technical infrastructure of the aerological services had to change to accommodate these new needs. See Bates and Fuller (1986), pp. 57–61.
3. Kevles (1975), p. 27.
4. See "Memorandum for the Coordinator of Research and Development 'Suggested Problems for the Research Board for National Security'," from R. D. Conrad, OCR&D, 6 March 1945, p. 15, Records of the Navy Office of the Coordinator of Research and Development, NARA, RG 298.
5. Commanding General, Army Air Forces (Director, Research and Development Division) to RBNS, 8 February 1945, Records of the Navy Office of the Coordinator of Research and Development, NARA, RG 298.
6. See Moore (1952), pp. 395–404; Moore, Smith, Murray, et al., (1950); and Spilhaus, Schneider, and Moore (1948), pp. 130–137. On the general activities of Spilhaus and the development of meteorological equipment at Fort Monmouth, see Bates and Fuller (1986), pp. 57–59.
7. S. Ralph Cohen (Editor, *National Aeronautics*) to Jean Piccard, 17 February 1945; Jean Piccard to Editor, 7 February 1945, PFP/LC. On Jeannette's search for a suitable position, see W. F. G. Swann to Jeannette Piccard, 21 May 1945, Swann Papers, APS; Wallace E. Conkling (The Bishop of Chicago) to Benjamin T. Kemerer, 27 April 1945, PFP/LC.
8. Jean Piccard to D. F. Rex, 28 April 1945, PFP/LC.
9. See Hagan (1986), pp. 26–27.
10. Swann to Jeannette Piccard, 21 May 1945, Swann Papers, APS; also in PFP/LC.
11. T. H. Johnson to W. F. G. Swann, 27 October 1945; Johnson to Swann, 27 June 1946, Swann Papers, APS.
12. Swann to Lyman Briggs, 29 October 1945, Swann Papers, APS.
13. Piccard to Swann, 12 October 1945, Swann Papers, APS; also in PFP/LC.
14. Jean Piccard to Willard Dow, January 1946, rough draft not sent. Modified letter sent 11 February 1946, PFP/LC. Willard Dow replied in February that the Dow Chemical Company was not able to participate in another flight, and he questioned the scientific need of Piccard's plan. See Willard Dow to Jean Piccard, 14 February 1946, PFP/LC.

15. Jean Piccard to F. D. Price, 14 November 1945, PFP/LC. Don Piccard has noted that the windows were used on the B26 Liberator, as the B17 Flying Fortress had electrically heated windows. Don Piccard to the author, 18 July 1988, p. 16.
16. C. E. Rosendahl to O. C. Winzen, 25 January 1946, PFP/LC. Assistant Secretary Sullivan took no direct interest in the matter.
17. D. F. Rex to Commander Burwell, 15 February 1946, Project Helios File, ONR.
18. Captain R. D. Conrad, quoted from minutes of first Naval Research Advisory Committee meeting, 14 October 1946, Quoted in Hagan (1986), p. 41.
19. Piccard to Swann, 7 March 1946; Swann to Piccard, 8 April 1946, Swann Papers, APS.
20. Winzen states that the company went out of business in March 1946. See "Otto C. Winzen–Balloon Data," 1 September 1976, OW/UM. Donald Piccard recalls that the Navy did not find Minnesota Tool an appropriate contractor for the balloon proposal. Donald Piccard to the author, 18 July 1988.
21. Gray (1954), p. 230.
22. Arthur D. Hyde to ONR, 24 July 1946, Project Helios File, ONR. See also Gray (1954), p. 230; Crouch (1983), p. 640.
23. One curious project was their pigeon-controlled guided missile control system for an aerial torpedo. See Capshew (1986), and Don Piccard to the author, 16 August 1988.
24. Hyde to ONR, 24 July 1944, Project Helios File, ONR.
25. On General Mills planning for the postwar era, see Gray (1954), p. 234–235; 254.
26. Ibid., pp. 238–244.
27. Winzen claims that he had personal assurances that Navy contracts would be his "if I joined a substantial organization which would serve as the financial background for my work." See "Otto C. Winzen–Balloon Data," 1 September 1976, OW/UM.
28. See Memorandum, H. F. Smith to D. F. Rex, 24 May 1946, Project Helios File, ONR.
29. Jean Piccard to D. F. Rex, 11 April 1946, "Proposition for Pleiades II Flight," p. 5; see also "The Gondola, Inside Arrangement," no date, Project Helios File, ONR.
30. Ibid., "Proposition," Appendix 5.
31. Swann to Jean Piccard, 5 March 1946, appended to Jean Piccard to D. F. Rex, 11 April 1946, "Proposition for Pleiades II Flight," Project Helios File, ONR.
32. E. O. Lawrence to Merle A. Tuve, 8 April 1924. Merle Tuve Papers, Library of Congress, quoted in Cornell (1986), p. 111.
33. Biehl, et al. (1948), pp. 353–359.
34. Chief of Naval Research to Chief of Naval Operations, "Request for Naval Cooperation with Stratosphere Research," 16 October 1946, Project Helios File, ONR.
35. Ibid.
36. Hagan (1986), p. 4.
37. Jean Piccard to C. E. Rosendahl, 4 February 1946, PFP/LC.
38. Jean Piccard to Swann, 8 July 1946, Swann Papers, APS.

39. John Wheeler (North American Newspaper Alliance, Inc.) to Mrs. Jean Piccard, 13 February 1946, PFP/LC.
40. "Joint Meeting of Technical Committee (Project HELIOS) and Members of Upper Atmosphere Panel, ONR 19 December 1946." This record constitutes the minutes of the first joint meeting between the ONR Panel and the Technical Committee. Project Helios File, ONR.
41. Ibid.
42. "Minutes of Helios Technical Committee Meeting 17 January 1947," (H. F. Smith to E. C. Settle—Special Devices Center, 21 January 1947, p. 3), Project Helios File, ONR.
43. Ibid. The letters were sent out to a wide variety of scientists, including H. G. Houghton, head, Department of Meteorology, MIT; J. Kaplan, Institute of Geophysics, UCLA; Victor H. Regener, Department of Physics, U. New Mexico; E. J. Workman, New Mexico School of Mines, Research and Development Division; Horace Byers, Department of Meteorology, University of Chicago.
44. Alan T. Waterman comments, in Minutes, 14 October 1946, NRAC. Quoted in Hagan (1986), p. 42.
45. On Piccard's early opinion, see Piccard to Swann, 8 July 1946, Swann Papers, APS.
46. "Navy Contracts for Dr. Piccard's Services for Balloon Ascent to Stratosphere," Navy Department. Release date 21 December 1946, PFP/LC.
47. "Organization Chart—Aeronautical Research Laboratories," attached to "Roster of Navy Officers and Scientists Attending Scientific Conference—Project Helios," 1947 Helios Folder, PFP/LC. In fact, Upson is credited with designing the "natural shape" or zero pressure balloon concept, in collaboration with other General Mills engineers. See Dwyer (1979), pp. 10–11.
48. John D. Akerman to S. C. Lind, 18 January 1947, PFP/LC.
49. G. O. Haglund to Jean Piccard, et al., 8 January 1947, Ballooning 1947 Flight Folder, PFP/LC.
50. T. G. W. Settle to Chief of Naval Operations, 17 September 1946, Project Helios File, ONR.
51. T. G. W. Settle to Special Devices Center Director, 7 January 1947, Project Helios File, ONR.
52. In November, Jean sought out NAA advice on barographs for establishing altitude records. See Jean Piccard to Lowell Swenson, 6 November 1946, PFP/LC.
53. The Navy could submit a list of candidates to Piccard, but it had to include names "congenial" to him, such as his son Donald Piccard or his wife Jeannette. See P. F. Lee to General Mills, 24 January 1947; Jean Piccard to O. C. Winzen, 31 January 1947, Ballooning 1947 Flight Folder, PFP/LC.
54. Jean Piccard "Memorandum," 2 February 1947, Ballooning 1947 Flight Folder, PFP/LC.
55. Jeannette Piccard to O. C. Winzen, 26 February 1947, 1947 Ballooning Flight Folder, PFP/LC.
56. See Enclosure A to T. G. W. Settle to Chief of Naval Research, 28 February 1947, Project Helios File, ONR.
57. See "Research Program, Project Helios," 3 February 1947, Project Helios File, ONR.

58. E. C. Settle "Memorandum for Distribution," 21 February 1947, Project Helios File, ONR.
59. Ralph Cole (Engineering Division, Watson Laboratories) to ONR, 13 February 1947, Project Helios File, ONR. On the monitoring of nuclear explosions, see Ziegler (1988).
60. Helios was at first an internal code name. See Public Affairs Memorandum, 16 March 1947. For codes of all upper atmosphere flights, see 9 December 1946 memorandum. Project Helios File, ONR.
61. O. C. Winzen to W. G. Stroud (Princeton Physics), 25 February 1947, Project .Helios File, ONR.
62. John Winckler to Jean Piccard, 9 June 1946, PFP/LC.
63. Urner Liddell memorandum, 28 March 1947, Project Helios File, ONR.
64. Winzen to George Hoover, Special Devices Center, 7 April 1947, Project Helios File, ONR.
65. H. F. Smith to Bert H. Hickman (ONR), 18 April 1947, Project Helios File, ONR.
66. Allen (1949), pp. 259–260.
67. Henry Spicer to "Dear Admiral," 26 April 1947; Jean Piccard to "All Scientific Collaborators," 10 June 1947. Project Helios File, ONR.
68. Fred W. Paul to Jean Piccard, 10 March 1947, PFP/LC.
69. Brian O'Brien to Jean Piccard, 16 June 1947, PFP/LC.
70. Henry Spicer to "Dear Admiral," 26 April 1947, Project Helios File, ONR.
71. "Meeting to Confer on Coordination of Engineering and Scientific Phases of Project HELIOS," 19 May 1947, Project Helios File, ONR.
72. Fred Smith to Jean and Jeannette Piccard, 1 June 1947, PFP/LC.
73. Ibid.
74. Allen (1968), pp. 8–9; 21–24.
75. See Seymour and Cheng (1986), p. viii; pp. 2–3.
76. Ibid., pp. 27–28; Raff and Allison (1956), p. 16; ONR Historical Files, "Skyhook '60" Folder.
77. Allen (1968), p. 45.
78. Ibid.; Kresser (1957), pp. 10–11; ONR Historical Files, "Skyhook '60" Folder.
79. "Meeting to Confer on Coordination of Engineering and Scientific Phases of Project HELIOS," 19 May 1947, p. 4, Project Helios File, ONR.
80. H. A. Tanner, "Conferences at General Mills 17, 18 June, Report of," NRL RSRS Files, RG 11704, box 99 S78-1(119), Folder 7, 30 June 1947.
81. On the original organization chart, developed by Settle, see T. G. Settle to Chief of Naval Research, 28 February 1947, revised April 1947, Project Helios File, ONR.
82. C. M. Bolster, Acting Chief of Naval Research to Chief of Naval Operations, 23 May 1947. See also circular letters, 2 June 1947. Project Helios File, ONR.
83. Chief, BuAer to Chief, Naval Research, "Report of Tests on Performance of High Altitude Neoprene Balloons," 29 July 1947, Project Helios File, ONR.
84. Fred Smith to Jean and Jeannette Piccard, 1 June 1947, PFP/LC.
85. Jean Piccard to "All Scientific Collaborators," 10 June 1947, Project Helios File, ONR.
86. Letters, Jean Piccard and T. R. James to Hutchinson, 6 June 1947, Project Helios File, ONR.

87. Jean Piccard to "All Scientific Collaborators," 10 June 1947, Project Helios File, ONR.
88. Ibid.
89. Jean Piccard to George Shepard, 4 June 1947; William C. Babcock to Jean Piccard, 9 August 1950, PFP/LC.
90. T. R. James to H. B. Hutchinson, 30 June 1947, Project Helios File, ONR.
91. Jean to Auguste Piccard, 4 June 1947, PFP/LC.
92. Jeannette Piccard to Harris Smith, 6 June 1947; Jeannette Piccard to Henry Spicer, 25 June 1947; Jean Piccard to R. E. Harris, 26 June 1947. PFP/LC.
93. T. G. W. Settle to Chief of Naval Research, Director of Public Information, 7 July 1947, Project Helios File, ONR.
94. Jean and Jeannette Piccard to John Nicholas Brown, 2 December 1947, PFP/LC.
95. R. G. Robinson to Jean and Jeannette Piccard, 28 April 1948, PFP/LC.
96. Bradley Dewey to Jean Piccard, 19 October 1948, PFP/LC.

CHAPTER 10

The Legacy of Helios:
Skyhook, Stratolab, and Modern
Scientific Ballooning

FIGURE 10.1. 280,000-cubic-meter balloon launch from the carrier *Valley Forge*, as part of Project "Skyhook '60." Navy photograph.

Project Skyhook

Within one month of Project Helios's demise, and with Piccard and Settle out of the way, ONR reactivated the General Mills contract to fabricate balloons for ONR contractors "requiring high altitude transportation facilities for equipment."[1] Following discussions between Arthur D. Hyde and the branch office of ONR in Minneapolis, it was agreed that the Helios program would continue at a reduced level ($50,000 per year) and would provide design, construction, and test-flight services for single-cell plastic balloons of many types, primarily the open-neck, constant-altitude plastic cell. What had started out to be another hasty stunt in the stratosphere became a deliberate program to develop and market a real capability for stratospheric flight.

Project Helios was officially reborn as Project Skyhook in September and October 1947.[2] The Skyhook program revolved around General Mills, as it was able to produce the single polyethylene cells, a secure launch site—Camp Ripley, Minnesota—containing complete facilities for tracking and reasonable recovery, ground crews, and logistical support. Hyde did not wish to "sell" balloons to other ONR research contractors; General Mills would rather establish a balloon-launching facility to provide omnibus services "on order." ONR agreed to this, and had several ONR contractors as possible users on tap, including John T. Tate of the University of Minnesota, Marcel Schein of Chicago, and Robert D. Sard of Washington University. Hoover had been in touch with Schein and Tate, and each wanted to secure several 31,000-meter flights for cosmic-ray research, but required flight services.

The first Skyhook balloon flew on September 25, 1947 from the airport at St. Cloud, Minnesota, and carried a 29-kilogram payload to 31,000 meters under a 5,600-cubic-meter plastic cell.[3] Throughout the rest of 1947 many of the original Helios scientific contractors were drawn into the revised program of launching single balloons and single payloads. Brian O'Brien flew his Helios spectrograph under a Skyhook balloon, and Swann's Bartol associate Martin Pomerantz was soon building radiosondes for flights near the geomagnetic equator.[4] Similar groups at Chicago, New York University, and the Naval Research Laboratory joined a multitude of newer groups during the late 1940s and 1950s to send packages (mostly cosmic-ray telescopes and plate stacks) aloft, as did similar groups formed in the United Kingdom and around the world.

General Mills required the specialized services of numerous contractors and subcontractors for *Skyhook*, and gained them through ONR's network of Navy technical facilities and civilian contractors. The Naval Research Laboratory provided telemetry and radio control mechanisms,[5] and in several years the University of Minnesota Physics Department conducted applied research on natural balloon shapes and launch procedures.

University of Minnesota Physics Department Balloon Group

The Minnesota Physics Department became the focus of cosmic-ray ballooning when, soon after the initiation of its ONR-supported flights, they discovered the heavy component in the primary cosmic-ray flux. In a series of Skyhook balloon flights in the spring of 1948, conducted in conjunction with physicists from the University of Rochester, the group flew stacks of newly developed nuclear emulsions along with a small cloud chamber, all enclosed in a pressurized 0.8-meter aluminum sphere, to altitudes as high as 29,000 meters. Phyllis Freier of Minnesota along with Bernard Peters and Helmut Bradt of Rochester concentrated on nuclear emulsions, while Ed Lofgren, Frank Oppenheimer, and E. P. Ney of Minnesota developed the cloud chamber.

As noted in previous chapters, photographic plates had been used since the 1930s to record the passage of charged particles in the reduced silver grains in their emulsions. Ordinary photographic plates, however, consist of a thin emulsion deposited on a comparatively thick glass substrate. Stacks of ordinary photographic plates, therefore, could only record paths, interactions, and trajectories in a discontinuous manner. In the 1940s, Cecil F. Powell and his Bristol University cosmic-ray physics colleagues worked with the Ilford Company to design a more suitable photographic medium and created the "nuclear research" emulsion: thick (on the order of 1 millimeter) self-supporting emulsion sheets that when stacked provide a continuous record in three dimensions. These emulsions became available immediately after World War II, and Powell's group put them to good use, discovering the pi-meson (or pion) in 1947 in the first continuously detailed records of nuclear disintegration processes in the atmosphere.[6]

The Minnesota discovery flight carried two sets of nuclear emulsion stacks above and below the cloud chamber, whose lead plates acted as absorbers. The balloon spent three hours above 28,000 meters and four hours above 20,000 meters, recording heavy primaries in its upper stack of Ilford C2 photographic plates.[7] Repeated flights through the summer established that the primary cosmic-ray flux included significant amounts of helium nuclei and some nuclei as heavy as iron. The heavier nuclei, although only 1 percent of the primary flux, accounted for a significant fraction of the total mass and energy of the flux.[8] These classic papers came from the first sustained series of unmanned cosmic-ray balloon flights in the United States with nuclear emulsions and the new plastic balloon technology provided under the embryonic Skyhook program. This powerful combination, funded by a joint program of the ONR and Atomic Energy Commission (AEC), as well as considerable local support from General Mills, 3M, and other Twin Cities civic interests, helped to establish the new unmanned balloon program and the reputation of the Minnesota group.

Al Nier, then a senior physicist in the department who had recently returned from the Manhattan Project, recalls that the success of the group was the result of a number of interrelated factors. First, the most prestigious member of the department, John T. Tate, was able to attract first-rate experimentalists eager to try something new. Second, Tate was an effective entrepreneur who believed that the department had to establish an engineering capability independent of any particular project. And third, the department enjoyed a close association with the very interests providing the new balloons.[9] In the 1930s, Tate, as editor of both the *Review of Modern Physics* and of *The Physical Review*, both watched and at times mediated in the debate between Millikan and Compton. Aware of the need to send instruments as high as possible, he became much impressed with the potential of the new balloons, which could travel higher in the atmosphere and remain there for longer times than any technology had been able to achieve in the past.

During the war, Tate headed up major divisions for the wartime Office of Scientific Research and Development (OSRD), and with his colleague A. O. Nier and other members of the Minnesota Physics Department, made important contributions to war research coming away with invaluable new contacts and a new view of how the military might aid research in the future.[10] During the war, Tate was one among two dozen prominent scientists who had urged the Navy to develop a centralized office for research. And in the postwar era, Tate and Nier retained close contacts with the Office of the Chief of Naval Operations and the Office of Naval Research. Tate also was befriended by James Ford Bell, the president of General Mills, and enjoyed considerable local respect among Minneapolis business people as his contacts with Washington during the war had brought in many lucrative contracts. As Nier recalls, Bell was able to raise $250,000 for the rebuilding of the Minnesota Physics Department after the war, gathering funds from General Mills, Minneapolis Honeywell, the *Minneapolis Star and Tribune*, Minnesota Mining, and Northern States Power. This funding allowed the department to build a large and well-equipped shop, as well as attract new faculty.[11]

These contacts served Tate well. He felt responsible for the rebuilding, care, and feeding of a department which had a number of young physicists looking for new research programs after the war. He also knew that the department would be well served if it moved in directions of interest to the military, and so watched the development of *Helios*, noting that the balloons left behind could certainly be used to study cosmic-ray phenomena. A number of physicists in the department had already distinguished themselves in atomic and nuclear physics, and J. W. Buchta, the department chairman, was about to bring in Frank Oppenheimer and Ed Lofgren and also recruited a number of younger staff and students for cosmic-ray studies. Tate and Buchta were certain that both ONR and the new Atomic Energy Commission would fund their proposals because they were interested in any activity that addressed problems in nuclear physics and the structure of matter.[12]

Tate approached ONR and the AEC to establish two interlocking research groups for balloon development and for cosmic-ray physics using the balloons. The cosmic-ray physics group was formed first, and there was considerable overlap in senior personnel. The physicists were all quite aware of Piccard's presence, but they knew that Piccard wanted to fly with the equipment, whereas the physicists preferred to stay on the ground.[13]

After their successful series of flights during the first season, the study of the heavy component of primary cosmic rays became the "bread and butter" research activity of the group for many years. Beyond producing a long series of publications, their work fostered several generations of Ph.D. theses on high-energy cosmic-ray phenomena, looking for mesons, strange particles, solar high-energy particles, and the highest energy interactions they could find in a period when accelerators were beginning to offer an alternative source for studying high energy interactions.[14]

Balloon Development at the University of Minnesota

The Minnesota group started its own balloon development studies in 1951, supported by ONR. Otto Winzen had left General Mills to form his own company in 1948, taking Skyhook with him, and other members of the original General Mills group also left to begin small ventures that depended on government contracting. Even though a number of commercial services were available, including General Mills, ONR still desired the services of the university for applied research into the theory of balloon design, the refinement of launch procedures, and the development of long-duration high-flying aerostats, in support of the U. S. Air Force's *Moby Dick* reconnaissance and weather balloon program.[15] ONR, like its Air Force counterpart, the Air Force Cambridge Research Laboratory, readily entertained similar projects in areas of critical technological development at New York University and later at Tufts University. As in wartime, it was good practice to ask a number of contractors to address the same problems. Multiple contracting not only curried favor with commercial and academic groups, which helped to build a broad, loyal political constituency, but provided a greater chance that the assigned problems would be solved. When the academics involved had similar interests in building new instruments for observing natural and artificial phenomena, the resulting combination was often exploited by military public affairs offices as an example of an enlightened military ready to support civilian enterprises that might benefit all humanity.[16]

The Minnesota High Altitude Balloon Project included three dozen scientists and engineers and was headed by Charles L. Critchfield, professor of physics, who divided his time between the balloon project and Los Alamos; Edward P. Ney, associate professor of physics and alternate scientific director; John R. Winckler, associate professor of physics who had come to Minnesota from the Princeton cosmic-ray group; Homer T. Mantis from the Engineering School who was trained in meteorology; and Gilbert Perlow, recently recruited from NRL, where he had conducted cosmic-ray observations with V-2 rockets. This group examined the operational characteristics of balloon systems, including methods of launching. They also examined factors determining balloon shape such as load, amount of initial inflation, variation with altitude, and meteorological conditions. Many of the flights were made in connection with other ONR projects granted to the physics department in support of its research in cosmic-ray physics, and much of their work was conducted jointly as operational and experimental flights with Winzen Research (Project Skyhook) and with General Mills. In their first decade, the physics department group refined flight operations, operational flight control equipment, diagnostic testing, inflation tests, telemetering, and communications, and emerged as a leading university group in the development of the field.

The Growth of Skyhook

Soon after the appearance of the first discovery paper from Minnesota in July 1948, cosmic-ray physicists from military laboratories such as NRL, and many from universities looked to *Skyhook* as a new and valuable mode of conducting research.[17] Skyhook balloons provided longer look times than could sounding rockets and were far more economical, costing less than $2,000 per flight. Over the many years of the Skyhook program, most major balloon cosmic-ray groups in the United States used its facilities.

Skyhook did far more than support cosmic-ray studies in its early years. Under the influence of the military it became a "research tool" for obtaining supersonic flight environmental data.[18] Brian O'Brien also used *Skyhook* to continue the ozone studies he had signed up to do on *Helios*. Although primarily concerned with the management of the Institute of Optics, O'Brien found time to prepare his Helios spectrograph, which was the old hanging sun spectrograph on both Explorer flights.[19] The venerable device itself required virtually no alterations, save for the addition of a Kapok cushion that would protect it on landing and also insulate it from water damage. The photographic recording system was started by an automatic timer initiated by a recording barograph provided by the University of Minnesota.

The June 16, 1949 flight began at Camp Ripley and rose to an altitude of 30.5 kilometers before landing in a small lake in the northern part of the state. O'Brien's team followed the balloon by automobile, retrieved the film with help of local fishermen, and then processed it in a local laboratory. Characteristically, O'Brien never got around to reducing his results, but he did report to ONR that the operational portion of the project was a success. A graduate student completed the analysis in 1952 and found that the vertical ozone distribution was the same as that measured in 1934 and 1935 during the Explorer flights, even though the total amount of ozone above the spectrograph was greater in the summer of 1949.[20]

The scientific and balloon development program under *Skyhook* prospered through the 1950s and 1960s, supported largely by ONR's Nuclear Physics Division and by the AEC. NRL continued to send up nuclear emulsions, and numerous cosmic-ray groups hitchhiked on University of Minnesota flights in much the same manner that Rochester did in 1948. Virtually all major groups—including Rochester, Chicago, and the Canadian Research Council—found rides available in the early 1950s.[21] Even with the opening of the Berkeley Bevatron in the mid-1950s and its attainment of high-energy particle fluxes that could match energies found in cosmic rays for the study of particle physics, the balloon work continued to prosper. Cosmic rays themselves were interesting because of their astrophysical and geophysical implications. And still, no artificial source could match the energies found in the most energetic rays.

The diversity of the Skyhook program can be seen not only in the types of packages flown, their weight, sophistication, the number of outside groups entertained, and the sites for launching, but also in the role of the balloons themselves. In one hybrid rocket program, balloons became a launch platform for small rockets in packages called "Rockoons."[22] James Van Allen, on returning to his alma mater, the University of Iowa, to continue his upper-atmosphere rocket program initiated at the Applied Physics Laboratory (APL), realized that Skyhook-type balloons could loft small rockets in excess of 16 kilometers, and then the rocket could be fired through the flimsy balloon to an altitude of over 80 kilometers. Rockoons used cheap ($1,500) Deacon and later Loki and other small solid-fueled rockets and carried a wide array of cosmic-ray instruments into the ionosphere on budgets that university physics departments could handle. Van Allen, as well as Herbert Friedman of NRL, flew Geiger counters, ionization chambers, and other radiation detectors on Rockoons to gather data on the distribution of ionizing radiation at different altitudes as a function of latitude and season, and on the X-ray radiation from solar flares during 1956, which was designated as the pre-IGY (International Geophysical Year) year. Among the many things accomplished with Rockoons, the Iowa group refined their understanding of the existence of ionospheric currents in the equatorial and auroral geomagnetic zones, and NRL refined knowledge of the relation between X-ray emission during solar flares and radio fadeouts on earth.[23]

The tiny Rockoons were cost-cutting expedients in an era when few could afford the luxury of larger sounding rockets, but they also were ideally suited to shipboard launching around the world. Since the Navy had specific reasons for developing a global launch capability, it willingly supported Van Allen's geophysical studies. Van Allen's wartime services to the Navy, in APL's proximity fuse project, included a naval commission and duty in the Pacific training handlers and evaluating the effectiveness of the fuses. He was on solid ground on shipboard.

By November 1951, 160 Skyhook flights had taken place and in each following year hundreds of flights were conducted by dozens of groups. And by 1957, over 2,500 flights had been conducted under General Mills auspices alone.[24] The majority of these were devoted to air sampling in the Air Force's nuclear test detection programs.[25] As the scientific program continued, the balloons and facilities became larger, and were able to accommodate heavier and more sophisticated experiments. Balloon volumes increased from the thousands of cubic meters into the tens of thousands of cubic meters. Balloon skin substances became stronger and lighter, methods of sealing the gores and the designs for gores changed continuously, and the basic design of the open natural-shaped cell was continually refined. Multiple launch sites in Minnesota, Texas, Alaska, California, and on shipboard provided the all-important latitude data required by cosmic-ray physicists who turned their attention more and more to geophysical problems even though many, such as Marcel Schein, were still after the highest-energy particles. Meteorology and biology were served as well; their practitioners and patrons still sought to improve understanding of the large-scale flow of upper-atmosphere jet streams for global reconnaissance programs and for the survivability of high-altitude manned flight.

The vast majority of Skyhook flights focused on gathering cosmic-ray data; although the Navy presented *Skyhook* to the public as a program of basic research in nuclear physics, a large part of it was also involved with the growing nuclear test detection program.[26] The National Science Foundation (NSF) soon joined the ONR and AEC to support *Skyhook*, and together they provided a wide base of support for many different interests. As a result, the majority of Skyhook flights were dedicated single-interest ventures that kept logistics and coordination requirements to a minimum. Many small launches took place worldwide, although the Navy sponsored some gargantuan efforts. One of the largest was known as Project ICEF (International Cooperative Emulsion Flights) which was initiated in cooperation with NSF in 1959 and 1960 and involved some 25 universities and research centers throughout the world. Its star attraction was a 280,000-cubic-meter balloon that was used to loft a 1,000-kilogram gondola carrying Marcel Schein's cosmic-ray emulsion stack payload to 37,000 meters. The Navy became involved on a grand scale. The old Essex class carrier *Valley Forge* was assisted by six destroyers and four Super Constellations in the Caribbean Sea, all under the direction of a rear admiral. Schein's University of Chicago Public Relations Office proudly hailed it as a "new role for the 'Valley Forge.' Only weeks before she had been on antisubmarine patrol of the North Atlantic; now she sheltered the mobile base of a peacetime scientific expedition with international ramifications." University officials, inheritors of the legacy left by "A Century of Progress" a quarter of a century earlier, were delighted with their continuing good relations with an enlightened Navy: "Professor Schein's experiments converted the massive 880-foot flight deck of the 'Valley Forge' from a military landing field to a laboratory workbench for the biggest cosmic-ray hunt ever attempted at high altitudes."[27] But the average Skyhook project payloads were 300- to 500-kilogram gondolas carrying smaller emulsion stacks developed by many university and government groups worldwide. Marcel Schein had been at the center of this effort, along with E. P. Ney; both had been involved with Navy ballooning efforts since the late 1940s and *Helios*.[28]

In addition to carrying out cosmic-ray and geophysical studies, constant-altitude Skyhook flights were able to act as stabilized observing platforms by the late 1950s. Both the Navy and Air Force sought stabilized platforms for balloon reconnaissance; by June 1954, ONR was delighted to report that it had succeeded in photographing a total solar eclipse from a Skyhook balloon at 14,000 meters using four cameras mounted on "a stable gyromagnetic oriented platform."[29] An expression of the continuing desire to improve stabilization as well as pointing accuracy can be found in ONR's initial support for Princeton astronomer Martin Schwarzschild's Stratoscope. The first Stratoscope built in 1957 was a stabilized 30-centimeter reflecting telescope capable of recording high-resolution images of the sun's photosphere. The second was a giant 0.9-meter infrared telescope that was originally designed in the early 1960s for direct imaging but was first applied to infrared spectroscopy of planets and galaxies. Schwarzschild insisted from the outset that his needs be met by an unmanned automatic laboratory, and that his efforts lead to expertise in unmanned robotic craft, such as NASA's Orbiting Astronomical Observatory program.[30]

In contrast to Schwarzschild's sophisticated unmanned Stratoscope, High Altitude Observatory astronomer Gordon Newkirk's Coronascope project began as a manned ONR program because he believed that he lacked the stature and ability to attract enough funding to develop stabilized unmanned platforms. But when Schwarzschild offered his 30-centimeter Stratoscope gondola to Newkirk after it had completed two successful seasons, Newkirk happily turned to unmanned flights of ever-improving coronagraphs to study atmospheric scattering and the nature of the outer atmosphere of the sun. Newkirk's coronagraphs eventually found a berth on the manned Apollo Telescope Mount flown in 1973.

In fact, astronomers were early users of manned balloon platforms because their requirements for stabilization and pointing accuracy (for long-exposure non-solar studies as well as high resolution-narrow band pass coronal studies) were thought to be beyond what technology could provide in the 1950s. Accordingly, physicist John Donovan Strong of The Johns Hopkins University turned to manned ballooning in the Stratolab program to conduct infrared studies of the planets with his "satellite-substitute" vehicles. Although we cannot cover in any adequate manner the many complementary and contrasting approaches to scientific ballooning that existed in the late 1950s, nor the extent of the continuing relationship between a reborn manned ballooning program and scientific needs, we can mention that a familiar pattern re-emerges where both Strong and Newkirk were converted to manned ballooning through Navy advocates but quickly reverted to unmanned sondes when automation and stabilization became adequate. These episodes barely touch on the many pre-Sputnik attempts at scientific ballooning that survived into the NASA era, but they give an idea of the types of studies by outside scientists whom the Navy and Air Force wished to support. Those that started as manned projects were without exception a result of the efforts of promoters of manned ballooning and spaceflight within the military.

After *Helios*: Promoting Manned Ballooning

As with the *FNRS*, *A Century of Progress*, *Explorer*, and *Helios*, the Stratolab manned balloon series had its ardent advocates. Learning the lesson of *Helios*, the military advocates of advancing balloon technology in the postwar era developed their art and science using unmanned sondes, but their goal was manned flight. Charles B. Moore, while a graduate student in chemical engineering at New York University, had become fascinated with the possibilities that polyethylene had opened up for ballooning. During World War II, he worked for the U.S. Army Air Corps at Fort Monmouth developing electronic devices for weather operations research. After the war, he and his NYU professors and colleagues formed a Balloon Group Research Division in the College of Engineering to develop controlled-altitude free balloon systems. Their work was supported by the Watson Laboratories of the Air Materiel Command.[31] Moore moved from NYU to General Mills to further the application of polyethylene to ballooning, and in 1949 replaced Otto Winzen, who had left the previous year. With the endorsement of General Mills but not its support, Moore became the first to fly under a plastic balloon on November 3, 1949. He collaborated on this effort with General Mills' Mechanical Division personnel and others sympathetic to manned flight.

Moore's flight under a plastic balloon came to the attention of the Navy through two close contacts: ONR's M. Lee Lewis, who had been attached to the General Mills balloon project and then in 1951 became the ONR balloon project officer; and Lt. Malcolm Ross, who replaced Lewis first at General Mills in 1951, and then became balloon project officer at ONR in its Air Branch in 1953. Ross had training in physics and meteorology and during World War II he served in Fleet Weather Central, Pearl Harbor, and was aerological officer aboard the U.S.S. *Saratoga*.[32] Both Ross and Lewis joined with Moore and two of the leading physicists from the University of Minnesota, Ed Ney and John Winckler, to investigate how a manned program might be established. By the early 1950s, Moore was engineer in charge of balloon operations for the General Mills Aeronautical Research Laboratories, which provided him with sufficient backing to form, with Ross and Lewis, a powerful advocacy group that gained approval within the Navy for detailed technical feasibility studies.[33]

Moore and Ross never forgot the lure of a new world's record, but they pursued it quietly in the context of developing the capability for manned spaceflight. They predicted that the 1935 record of 22.1 kilometers set by Stevens and Anderson, which Moore regarded as the "swan song" of the classic era of rubberized cloth balloons, could now be broken by at least 9.1 kilometers with the new polyethylene balloon cells. It was therefore possible to expose manned systems to the "outer aeropause" and "it would not be at all impractical to recommend that the first experiments relevant to a manned satellite be conducted on a high-level balloon."[34] Manned ballooning, to Moore, was not an end in itself. It could best be justified only as a "tool and a bridge" to manned spaceflight: "It is believed, therefore, that polyethylene balloons will afford the scientists pursuing studies in space medicine a tool and a bridge with which to study, inexpensively and immediately, problems which must be solved before man goes into space."[35]

Moore's plan for a manned program was rejected initially by the chief of naval research, but in a few years, with persistence, he eventually found a patron in the new ONR head, Admiral Frederick Furth. In 1954 Furth approved a research program within the ONR Air Branch.[36]

Stratolab

Jean Piccard's old Helios gondola, in storage at Lakehurst, was shipped first to Winzen Research in Minneapolis for refitting and then to the Naval Ordnance Test Station at Inyokern (China Lake, California) for integration in the new Stratolab project. The Navy and Air Force wanted to test the use of pressure suits under near-space conditions, and so the Air Force Aeromedical Laboratory at Wright Patterson Air Force base provided pressure suits similar to those worn by test pilots in rocket-powered aircraft.

Ross, who by December 1955 had returned to civilian life but continued on at ONR as a physicist in its Air Branch, and Lewis tested the suits and the balloon systems in two flights in an unpressurized rectangular gondola. The first took Ross and Lewis to 12,000 meters on August 10, 1956 and the second took H. Froehlich and K. Lang to 13,000 meters on September 24, 1956. Both flights included some geophysical studies in addition to physiological testing and training.[37] For the first sealed-cabin Stratolab flight in the refurbished Helios gondola, Ross and Lewis won out as pilot and observer with Moore as backup. As with *Explorer*, the site for the launch was the fabled Stratobowl. The first flight on November 8, 1956, reached 23,500 meters, setting a new unofficial world's altitude record that broke the 21-year-old record of *Explorer II*. Within a year, *Stratolab* reached 27,000 meters, but in the same year, two flights of the Air Force's similar Manhigh gondola reached 29,700 and 31,300 meters. *Manhigh*, initiated in 1955 and managed at Holloman Air Force Base in its Aeromedical Field Laboratory, was a solo flight program designed for physiological studies alone, the highlights of which were the planned parachute bailouts of J. Kittinger from altitudes in excess of 31,000 meters.[38]

Stratolab was a means of testing procedures for high-altitude manned flight, improving pressure suits and pressurized cabins, and addressing general problems in human physiology experienced under the stressful conditions of high-flight. At the center of the program was Capt. N. L. Barr, of the Navy's Bureau of Medicine and Surgery, who was particularly interested in the ability of two-man crews to perform complex activities under "space-equivalent" conditions.[39] But in the tradition of *Helios* and *Explorer*, Stratolab also represented an effort to court the interests of scientists in need of a high platform for sophisticated studies of the atmosphere and celestial objects.[40] Beyond making the program marketable, Barr wanted his flight crews to have realistic and sufficiently complex tasks to perform. Meaningful experiments provided by participating scientists were the best solution.

Early Stratolab flight crews performed numerous small tasks and experiments, none of which required highly detailed planning but did provide some data on the high atmosphere and near space as well as an evaluation of the work abilities of the crew. The flight of *Stratolab 2* under a 57,000-cubic-meter balloon on October 18, 1957, carried, for example, heart and respiratory gear to monitor physiological fitness and make evaluations of human fatigue and mental performance. Ross and Barr quickly found that the tasks they were required to do, which included moving around the cabin, created almost intolerable distress and fatigue. Wearing Air Force MC-3 partial pressure suits under Navy cold weather suits, and carrying a back-pack parachute system with emergency bailout seat pans greatly restricted motion. Ross argued that this attire and protective equipment "were not fully compatible with the concept of Strato-Lab as a research 'laboratory'."[41] If they did not have to move around, they would have been better off, but as their purpose was to carry the laboratory into the stratosphere and beyond, the only option left was to improve the flight gear.

Geophysical and astrophysical studies that kept the crew occupied (which they claimed helped to reduce the psychological impact of "earth separation") ranged from managing air sampling bottles flown for Ruven Smith of the Naval Ordnance Test Station (NOTS) to sky brightness measurement devices provided by Richard Tousey of NRL. Tousey also provided a set of interference filters that were to be used visually by the crew and that would record sky brightness at various elevations. E. P. Ney and his student Robert Danielson provided small gamma-ray detectors, and the Naval Observatory provided small telescopes for lunar observation. The telescopes were not used; they had been intended for a study of stellar scintillation in a program developed by John Hall and A. H. Mikesell. In fact, a good number of smaller projects had to be canceled for one reason or another: The scintillation experiment was dropped because the flight plan was delayed, a radiometric experiment was canceled because the equipment failed just before launch, and hoped-for photography of *Sputnik I* was not carried out. Much of the data that was gathered was either negative or equivocal. Yet, Ross and Lewis argued that the experience gained in the flight would lead to "definitive" experiments in astronomy and other scientific areas.[42] This feeling was bolstered by a record of the stability and damping characteristics of the gondola during flight. A spot of sunlight directed onto a panel and photographed by an automatic time-lapse camera demonstrated very little gondola rotation, and horizontal stability to within one degree, except for one period when the crew deliberately "rocked the boat." Ross was happy to report that the stability of the system seemed to meet the requirements of one of their biggest planned astronomical projects, John Strong's 40-centimeter infrared telescope, which "will allow the first significant evaluation of this high-altitude capability from a manned plastic balloon."[43]

In 1958, Ross invited astronomers and physicists as well as physiologists to propose projects to ONR that required flights into the stratosphere to observe the heavens unobstructed, or to study how the human body reacted to near-space conditions: "The visionaries dream of satellites and space ships for this application. Some day their dreams will materialize. But why wait?" He saw ballooning as "the logical step toward space astronomy," and argued that a combination of Stratoscope-type unmanned systems and Stratolab manned systems were capable of addressing "virtually all the major problems in astronomy . . . without the usual atmospheric disturbance."[44]

FIGURE 10.2. Cdr. Malcolm Ross and model of *Stratolab 5*, John Strong's 40-centimeter infrared telescope system. Navy photograph.

Continued overtures by General Mills, Raven Industries, Otto Winzen, and Ross convinced Strong and Newkirk to participate, as well as Harvard astronomers Hector Ingrao and Donald Menzel, who discussed plans for a 61-centimeter Cassegrain reflector operated by a two-man crew carried aloft by a 85,000-cubic-meter balloon to observe the planets.[45] Both Ross and Moore's rhetoric for manned balloon flight is reminiscent of earlier Explorer-era entrepreneur/pilots such as Piccard, Settle, and Stevens: They were all eager to return to the mysterious stratosphere; they all felt that their flights demonstrated the feasibility of safe and economical travel in the stratosphere; and they all hoped to go to higher altitudes, stimulate the use of astronomical telescopes to observe planetary surface detail, travel into storm clouds to study meteorological conditions, and even examine the electrical conditions and growth of cloud droplets in strong updrafts.

Ross saw unmanned ballooning as a prerequisite for manned flight and thus looked for any scientific pursuit that might take advantage of a balloon platform in order to promote the advance of balloon technology. As he pushed for the continued testing of larger and more capable manned systems he advocated ever-larger balloons that could carry greater payload weights with increased reliability. Fiercely developmental in his rhetoric, he argued that before any of these balloons could be used for manned flight, they had to be tested with unmanned payloads. Thus Marcel Schein's ICEF payload was, for Ross, a test of the balloon design for *Stratolab 5*, which carried Ross and Victor A. Prather USN (flight surgeon/observer) to some 35,000 meters in an unpressurized gondola. Wearing Mercury astronaut-type suits equipped with full body detectors to determine the incident cosmic-ray particle flux, Ross and Prather on May 4, 1961 left the carrier *Antietam* under a 280,000-cubic-meter balloon. They accompanied a large battery of automated instruments into the upper stratosphere that measured temperature, pressure, wind patterns, and water vapor content. They also tested reconnaissance systems and infrared horizon and sky spectrometers "using satellite type equipment and techniques." Ross claimed that these devices, "when coupled with camera and atmospheric information, [were] expected to be quite valuable as a test of equipment and techniques under development for satellite measurements."[46] This spectacular ascent established a new world's record, but ended in tragedy when Prather drowned in his pressure suit after a planned landing at sea.[47]

Stepping Stones to Space

The Navy's Stratolab, along with the Air Force's Manhigh, evolved out of a desire to explore near space once again in the spirit of Fordney, Settle, Kepner, Stevens, and the Piccards. But in the space age, balloon advocates argued, practical questions had to be addressed as well: How would we learn to live and work in space? Some were convinced that the best way to learn would be to first try out the systems under a stratospheric balloon. The high frontier was no longer the stratosphere; it lay in space itself. Thus to justify the developmental character of both Manhigh and Stratolab, it was argued that balloon systems could be improved through the testing of instrumented systems, and that manned programs were the preliminary stages leading to spaceflight.

By the end of 1958, Winzen Industries had built six sealed cabins for Manhigh and Stratolab which logged some 95 hours of flight under "space-equivalent" conditions between 25,000 and 30,000 meters. This experience led Winzen to promote what he called "Satelorb," which was a large sealed cabin capable of sustaining a crew of two for a week or more at 30,000 to 45,000 meters, or at an altitude where physiological conditions other than weightlessness (such as radiation levels) began to approximate spaceflight.[48] Winzen also envisioned huge floating laboratories, complete with up to three observation decks, that would be capable of supporting research during indefinite stays in the stratosphere.[49]

In January 1959, Vitro Laboratories of Silver Spring, Maryland, sponsored a two-day workshop to review the use of manned balloons in the space age. Everett E. Beson, chief project engineer, General Mills, Inc., argued that a balloon-launched manned rocket would be a safe and reliable method by which to simulate spaceflight in suborbital ballistic flights.[50] The ultimate expression of Van Allen's economical Rockoon, more reminiscent of Kurt Stehling's 1950s vision of a balloon-launched moon probe, and based upon Project Farside, was Beson's proposed Skyhook-type balloon which would carry its payload to 30,000 meters. The solid fueled rocket would then fire sending the capsule 30 to 60 kilometers higher. This scenario was thought to be both cost-effective and reliable, in contrast to the far greater dangers and expense of launching from the ground, which required large liquid-fueled multistage rockets.

Beson told his audience that "man must crawl before he walks"; he must go through a series of tests before he is qualified for spaceflight. Much of the work to date had only been done in ground simulators or during high-speed flight; Beson therefore argued that "There still is too large a gap between the capability of existing ground training devices and what is required to fully simulate a true space journey." Thus "no one can deny that if a trainer incorporating the best features of the several types of ground and air-borne equipment could be elevated to operate in a space-equivalent altitude, that this would provide the most adequate means of space flight preparation."[51]

The technical concept reports of Beson, Ross and Moore, and Winzen were read by middle managers in NASA. But upper management in NASA's manned spaceflight program did not take them seriously, nor did they approve any major feasibility or requirements studies. Robert Gilruth, then chief of NASA's Project Mercury, expressed the collective opinion: There were simply too many critical elements in the design of manned spaceflight systems (ballistic reentry characteristics of the capsule, life-support systems design for high-g and high-temperature conditions, communications above the earth's ionospheric layers) that a balloon gondola could not simulate.[52]

Ross and Moore's arguments for using manned balloon flights as training for spaceflight were not heeded by the military or NASA, but unmanned ballooning techniques did indeed prepare many scientists for satellite-borne experimentation. Some early exponents of X-ray astronomy entered the field as a result of the experience they had gained in cosmic-ray physics through ballooning.[53] Similarly, Newkirk's Coronascope and Schwarzschild's Stratoscope taught their instrumentation teams many lessons that were put to good use in both the Orbiting Solar Observatory and Orbiting Astronomical Observatory satellite series, as well as in Skylab.

In the space age, manned stratospheric ballooning lost visible purpose. Astronomers had been the only scientific group willing to become involved in manned ascents in the mid- to late 1950s, and by the early 1960s they all bailed out, or were pushed out. The influential astronomer Donald Menzel, a 1960s equivalent to Millikan in some respects, agreed with Beson, Ross, and the other advocates of manned ballooning that it had some merit. But Menzel was not ready to commit himself to manned balloons unequivocally; at best they were only a temporary necessity. The balloonist, argued Menzel, should not be a pilot but a "scientist who goes along as a non-expendable (we hope) servo-mechanism, to perform certain duties which it would be difficult to relegate to some kind of automatic computing equipment."[54] Menzel added: "We shall have to ask ourselves if man is really necessary in the balloon. If we can replace him with some automatic robot, ballooning becomes simpler and less expensive." He acknowledged, however, that beyond the wide array of scientific problems that could be addressed from properly stabilized balloons, such craft also had a role to play, manned or unmanned, "for testing equipment that later may be used in a satellite. The relative ease of recovery of equipment from a balloon will facilitate thorough testing before we employ it in a satellite."

The few astronomers who did participate in manned ventures were no longer enthusiastic after the first flights. Their experiences and the trade-offs they faced are beyond the scope of our present study, but indeed they were reminiscent of the experiences of their counterparts in the 1930s. Without exception, when sufficiently stable automatic platforms became available along with sufficient funding, all turned to unmanned sondes, satellite berths, or completely different modes of observation after they tasted manned ballooning.

Today, as in recent years, NSF, NASA, The University Consortium for Atmospheric Research (UCAR) and other institutions have sponsored robotic infrared, near ultraviolet, X-ray, and high-energy astrophysical, geophysical, and atmospheric projects with detectors, collectors, and sensors flown routinely under balloons launched from the National Scientific Balloon Facility near Palestine, Texas. Many groups at institutions such as NRL, MIT, Chicago, Johns Hopkins, Goddard Space Flight Center, Minnesota, and University College London have lofted all sorts of telescopes and detectors to study a wide range of phenomena. Plate stacks, interferometers, mass-spectrometers, X-ray detectors of varying design, meteorite samplers, and an ever-widening array of exotic devices have been flown and continue to fly.[55] The legacy inherited from Skyhook can be traced back to the lessons learned on Helios and the experiences of several generations of scientists, who, requiring data on conditions in the highest reaches of the earth's atmosphere, entered the very different worlds of manned and unmanned ballooning.

Conclusions

The legacies of the Piccards and *Explorer* came together in the immediate postwar era in the Navy's Project Helios, which itself evolved into the two major scientific ballooning programs of the modern era, Skyhook and Stratolab. Although initiated by Jean Piccard as his *Pleiades II*, *Helios* soon became a military program because the Navy sensed an immediate need to gain access to the upper atmosphere as well as to the scientific manpower of the nation.

The role of the Navy in *Helios* was very different from the military experience in the 1930s in manned ballooning. Although Naval aeronauts certainly took ultimate control of the *A Century of Progress*, and the Army Air Corps was definitely behind *Explorer*, neither directed the scientific agenda of the flights, nor demanded that they meet explicit military needs such as aerial reconnaissance, aerodynamic drop testing, and nuclear test detection. *Helios* reflected a new pattern of large-scale military funding for scientific studies at a time when the various reasons for establishing the pattern were far from reconciled. Factions within ONR represented both the old and the new Navy, and their attitude toward *Helios* reflected their interpretation of the ultimate purpose of the new Office of Naval Research. In such a new agency, which was trying to learn how to exist in peacetime when the only models to follow were developed during an all-consuming war, it is not surprising to find differences of opinion. As ONR was struggling to establish itself as a central agent for basic research within the framework of the Navy, and at the same time was still struggling to determine how it would operate, it often found itself facing internal discord or disagreements with other bureaus in the Navy that were not yet comfortable with its existence. In effect, the details of support and management associated with handling a complex manned balloon flight reflected the internal conflicts ONR was experiencing in its first two years.

The demise of *Helios* was the result of the poor management of an untested and radical technique for balloon flight, poor coordination of a complex research program, and conflicting agendas within the Navy. The Navy followed a prewar agenda at first by trying to do too much in an "all-up" flight of a bewildering array of experiments, managed by too many people from too great a distance, and carried by a cluster of balloons that had never been proven. It made little sense to continue pursuing a manned ballooning program when the balloon skins were still unreliable and untested and when their collective behavior in an untested cluster was completely unknown. Added to this, the Navy then piled on applied projects that further complicated the flight profile, and cast into serious doubt just who was in charge.

Piccard's 100-balloon cluster never reappeared in Navy or Air Force programs—there were simply better ways to fly once huge single cells became available—and constitutes the primary technical obstacle that *Helios* was unable to overcome.[56] But save for the balloon cluster and the added military agenda, the balance of *Helios* itself was identical to prewar missions. This was Piccard's model, carried over from the 1930s, not only with respect to patronage but also the service it would render to science. The Navy, too, wished to create an image of serving science; its Office of Naval Research was to be its primary mechanism for keeping alive the networks of scientists created during wartime. It was therefore in its own interest to cut *Helios* short when the science was found to be suffering. The Navy soon learned that *Helios* was not going to meet the diverse needs of each scientific investigator; coordination of the scientific mission was completely at an impasse. Unlike Army Ordnance, which created a civilian panel to coordinate the science done with V-2 missiles fired from White Sands after the war, the Navy tried to do this itself, and when it failed, it realized that flights of single experiments made more sense.[57] ONR realized this first, and through an alliance struck between Hutchinson and Waterman was able to convince BuAer and CNO that a manned program was premature. Thus the Navy decided in favor of a carefully planned progressive program of developing unmanned sondes capable of carrying larger and larger payloads under monolithic polyethylene gas bags.

Even though *Helios* never flew, it provided an important nucleus of scientific and technical talent sought out by the Navy to exploit new advances in plastic film technology, and revealed how the Navy could patronize scientists if it wished to direct them toward research programs that complemented its own needs: pure science informed by institutional needs, or what some Navy planners then called "basic" research. Bureaucracy had overpowered entrepreneurship in the stillborn Project Helios; the ubiquitous fascination with sportsmanlike exploration that marked the prewar manned efforts reemerged as the military's need to know how to make both machines and man live and work in the stratosphere. Although they had perceived a need, the Navy had not yet decided the best way to meet it.

Only when ONR's basic science advocates gained control over the general program, by separating the manned and unmanned efforts, and established a simple and direct means of giving scientists access to the balloons, did the capability gain a wide constituency representative of both the military's and the scientific community's needs. With *Helios* simplified into separate Skyhook and Stratolab components, balloon development, scientific missions, and military interests were better reconciled. Scientific payloads could be flown as simplified unmanned projects unless both principal investigator and sponsor agreed that man should be present.

True to its plan, Skyhook helped to develop the technology needed to make Stratolab feasible. But as Ross and Moore pushed to establish Stratolab, courting scientists and military largess, its original mission disappeared. No longer capable of extending man's reach, it was sold first as a shortcut for astronomers, then as a step into space, and finally as a way to test instrument systems planned for spaceflight. Ross was an advocate and explorer much in the tradition of the Piccards, intent upon improving the technology of manned spaceflight. Therefore it is not surprising that Ross acknowledged Jean Piccard, "whose influence has been profound . . ."[58] Piccard, on the other hand, did not have the same influence on the Minnesota physicists, for even though he brought ballooning to Minnesota, he never was able to see ballooning in the way they did. Yet they appreciated his legacy, which allowed them to establish the general tenor of modern balloon studies. They did so largely because a man like Piccard ended up, for better or worse, on their doorstep in the 1930s.

Notes

1. R. C. Young (ONR Branch Office) to Chief of Naval Research 7 October 1947 "Subj: Reactivation of General Mills Contract N6onr-252, T.O.I." Project Helios File ONR.
2. See, for instance: Chief of Naval Research to Director, NRL: "Stratosphere Balloon Project–Project SKYHOOK" Report on General Mills contract dated 10 October 1947. NRL RSRS files RG 11704 Box 99 S78-1(119), Folder 10: 13 November 1948.
3. [Technical Information Office, ONR], "Project Skyhook," November 1961 press release. Skyhook '60—Press Releases folder, ONR historical files. See also "Otto C. Winzen—Balloon Data," 1 September 1976, OW/UM.
4. The Bartol group initially turned to Dewey and Almy Darex balloons because they were less expensive. Swann and his Bartol team also chose to use their Helios cosmic-ray instrumentation aboard ocean-going ships around the world and on B-29 flights, which ONR was also supporting jointly with the Army Air Force. A few B-29 bombers specially adapted for flight at 12,000 meters were the heart of the Joint Army Air Forces/Navy Upper Atmosphere Research Program recently initiated at the Naval Ordnance Test Station at Inyokern, California, sponsored by ONR and the Army Air Corps. Swann, with additional support from the National Geographic Society, along with many other scientists, participated in this program in the late 1940s. When Swann found the B-29 program frustrating due to flight schedule changes caused by constant maintenance problems with the aircraft, his Bartol group returned to ballooning. On the "Joint Army Air Forces/Navy Upper Atmosphere Research Program," see: D. F. Rex "Agenda for Conference" with enclosures, in Menzel Papers, B-29 folder. Harvard University Archives. See also: Melvin Payne to W. F. G. Swann, 9 May 1946; W. F. G. Swann to W. F. Swann, 10 July 1947; W. F. G. Swann to J. Piccard, 14 February 1948; 2 March 1949, Swann Papers APS. See also: "Research & Development Project card (JRDB),"

N6ori-144, 3 February 1947. Attached to T. H. Johnson to W. F. G. Swann, 23 January 1947, Swann Papers APS.

5. T. R. James to NRL: 4/26/48. NRL RSRS files RG 11704 S78-1(119), Box 100 Folder 13; routing slip with request for instrumentation for Skyhook: radio control mechanisms, radio to aircraft communications, air pressure telemetering.

6. Sandström (1965), pp. 11–17; Janossy (1948), pp. 402–404; Simpson (1986), p. 7; See also: Sekido and Elliot (1985), p. 127, 210. Powell's Bristol group built their own plastic balloons during this time. The formation of the Bristol group warrants historical attention.

7. Freier, Lofgren, Ney, Oppenheimer Bradt and Peters (1948a), pp 213–217.

8. Freier, Lofgren, Ney, and Oppenheimer (1948b), pp. 1818–1827; Bradt and Peters (1948), pp. 1828–1837.

9. Al Nier oral history interview, SAOHP/NASM, pp. 72–81rd.

10. At different times during the war, John Tate headed up Division 6, "subsurface warfare" which developed antisubmarine weapons. including radio telemetry for marine convoy buoy monitors and aircraft monitoring of enemy submarines. Tate also headed Division 3 (Rocket Division) during its early period after the NDRC reorganized in late 1942, and Section C-1 of Division C (Communications). See Stewart (1948), pp. 11–12; 286; Baxter (1946), pp. 32; 423–424.

11. A. O. Nier to the author, October 8, 1988. See also: [The Bird Dogs] (August 1961), p. 96; and A. O. Nier oral history interview, 1 March 1984, NASM/SAOHP, p. 72rd.

12. See: E. P. Ney oral history interview, 29 February 1984, SAOHP/NASM, p. 5rd.; A. O. Nier oral history interview, NASM/SAOHP, pp. 73–75rd. On ONR's interest in cosmic-ray physics, see: "Joint Meeting of Technical Committee (Project HELIOS) and Members of Upper Atmosphere Panel, ONR 19 December 1946" Project Helios File ONR. On Tate's early involvement with the on-going debate between Compton and Millikan in the 1930s, and with the progress of cosmic-ray physics, see: De Maria and Russo (1987).

13. See Nier oral history interview, p. 73rd; Ney oral history interview, p. 5rd.

14. John Naugle oral history interview, 20 August 1980, SHMA/AIP, pp. 21–34; p. 33rd.

15. On Moby Dick, see: Crouch (1983), pp. 644–648. On Minnesota's involvement, see: University of Minnesota Physics Department, "Progress Report on High Altitude Plastic Balloons" Contract NONR–710 (01), folder: June 15, 1952–December 22, 1952 (ONR), Department of Physics, University of Minnesota. Co-sponsored by the Army, Navy and Air Force. Volume V "Confidential Security Information."

16. On multiple contracting during World War II, see: DeVorkin (1980), pp. 595–623.

17. See K. W. Patrick to Chief of Naval Research "Proposed Measurement of Cosmic Radiation, Project Skyhook," 9 August 1948, Box 100 Folder 15; William A. Baum to W. J. Miller, "Cosmic Ray balloon experiments to be conducted by the Rocket Sonde Research Section," 28 July 1947, Box 99 Folder 8; W. J. Wehmeyer to "Skyhook" Project Liaison Officer, 17 June 1949, Box 100 Folder 22, NRL RSRS files RG 11704 S78-1(119).

18. See W. A. Gorry, Officer in Charge, U. S. Naval Unit, White Sands Proving Grounds to Chief, Bureau of Ordnance, "Project 9-U-J-1 (Skyhook), report of" 23 May 1949, NRL RSRS files RG 11704 S78-1(119), Box 100 Folder 21, p. 1. The Helios missile drop program was revived by the Bureau of Ordnance's Special Devices Center to try to obtain data useful in the design of cockpits that might be ejected from high-speed aircraft in an emergency. Three Skyhook balloons raised 110-kilogram missiles to 31,000 meters, the missiles were released and allowed to fall freely to 6,000 meters, at which point a parachute would open to break the descent. The parachute, operated by ram pressure, was a critical factor in the design of the cockpits. By late 1949, the missile drop program was declared unsuccessful and was canceled.
19. Krolak (1952), p. 4. See also: Brian O'Brien oral history interview, NASM/SAOHP, 9 March 1987.
20. Krolak (1952), p. 36.
21. University of Minnesota Physics Department, "Progress Report on Research and Development in the Field of High Altitude Plastic Balloons" Contract NONR-710(01), NR 211 002 with ONR (and Army and Air Force), 20 January 1953 to 4 February 1953. ONR Historical Files.
22. Van Allen (1983), Chapter 3, see p. 21.
23. Van Allen (1983), Chapter 3; Friedman (1981), pp. 40–41; see also H. Friedman and James Van Allen oral history interviews, SHMA/AIP and SAOHP/NASM.
24. Moore (1952), pp. 395–404; see pp. 397, 403.
25. See "The Story Behind Plastic Balloons," (General Mills, Inc., Department of Public Relations, September 1957), p. 9.
26. See DoD Press Release #884-55 (16 September 1955), ONR Historical Files, "Skyhook '60" folder; "The Story Behind Plastic Balloons" (General Mills, Inc., Department of Public Relations, September 1957).
27. Garber and Holton (1960), in ONR Historical Files folder #43.
28. See "News Release" DoD Public Affairs "Seven Ship Navy Task Group to Participate in Cosmic Ray Research Program This Month," 17 January 1960, No. 43–60; "Giant Navy Balloons to Investigate Cosmic Rays," 20 December 1959, No. 1470–59; "Skyhook Balloon Launched from Valley Forge," *Naval Research Review* (1960.3), pp. 24–25. Winzen Research provided launch crew and facilities aboard the U.S.S. Valley Forge.
29. DoD Press Release, "Navy Balloon in Stratosphere Photographs Solar Eclipse," 30 June 1954. #641-54 ONR Historical Files, "Skyhook '60" folder.
30. Martin Schwarzschild oral history interviews 1979 SHMA/AIP, 1983 SAOHP/NASM.
31. Moore (1952), pp. 395–404; Moore, Smith, Murray, et al. (1950); and Spilhaus, Schneider and Moore (1948), pp. 130–137. On the general activities of Spilhaus and the development of meteorological equipment at Fort Monmouth, see: Bates and Fuller (1986), pp. 57–59.
32. "Malcolm D. Ross (Commander, USNR)," Skyhook files, Folder 22, ONR Historical Files.
33. See: "Project Stratolab" fragment from an Arthur D. Little corporate publication, NASM Technical Files LTA files, and Moore (1952), pp. 395–404; Crouch (1983), pp. 651–652.
34. Moore (1952), p. 403.

35. Moore (1952), pp. 403–404.
36. Crouch (1983), p. 652; Ross and Lewis (August 1958), pp. 45–52; p. 47.
37. Ross and Lewis (1958), p. 47.
38. Crouch (1983), pp. 653–656.
39. Barr (1957) cited in Ross and Lewis (1958), p. 46.
40. Charles B. Moore, "Project Stratolab: ADL Staff Member has Key Role In Record-Setting Balloon Flight," pp. 4–5; 11, in Arthur D. Little corporate publication. Fragment in NASM Technical file.
41. Ross and Lewis (1958), p. 48.
42. Ross and Lewis (1958), p. 51.
43. Ibid.
44. Ross (1958), p. 5.
45. See, for instance Ingrao and Menzel (1964), pp. 207–229, based upon remarks at a November 1960 conference; and Menzel (1959), pp. 1–6.
46. "Stratolab High #5 Background Information" Technical Information Office ONR, ONR #18-61. Fragment in NASM Technical File.
47. Crouch (1983), pp. 655–656; Stehling and Beller (1962), pp. 252–254.
48. Winzen (1959), pp. 562–569.
49. Winzen (1958), pp. 436–459.
50. Beson (1959), pp. 125–129.
51. Beson (1959), pp. 126–127.
52. Robert Gilruth oral history interview #6, 2 March 1987, GWS/NASM.
53. Hirsh (1983), p. 62.
54. Menzel (1959), p. 2.
55. See section no. 4 in Riedler (1979).
56. Jean Piccard's son Donald Piccard followed his father in ballooning, concentrating on sport balloons. In the late 1950s, Donald developed a balloon cluster and flew it on 18 September 1957. It was a low altitude demonstration in an open basket under a dozen cylindrical cells. Large balloon chains were, however, used by a number of people, notably the French astronomer/balloonist Audouin Dollfus in the 1950s. See Dollfus (1950). I am indebted to Ron Doel for making this source available.
57. On the Army's plan, see DeVorkin (1987).
58. Ross (1958), p. 9.

CHAPTER 11

Anticipating Apollo

FIGURE 11.1. "A Step Up" editorial cartoon for the *Columbus Dispatch* by Eugene Craig. Photograph appeared in John E. Moreau, *First Men on the Moon* (1976). Copy courtesy Mr. Moreau. Reprinted with permission from *The Columbus Dispatch*.

From the aëronautic standpoint, we faced the problem of constructing a craft in
which a pilot and his assistant and many instruments could be lifted ten miles
into the sky and be permitted to work there . . . Our problem, then, was to find
conditions that would permit two men to live up there in more or less normal
working order, and a means of getting them to the desired height.

Auguste Piccard, 1933[1]

In essence, the success of our effort in manned exploration of the new medium
of space depends upon our ability to carry the essential elements of a workable
human environment, tightly sealed in spacecraft, to far-distant reaches.

James E. Webb, 1964[2]

A Link with the Past

Although NASA paid little heed to Malcolm Ross, Charles Moore and
Otto Winzen and their pleas for a continuing program of manned ballooning as a stepping stone to space, Robert Gilruth did invite Jeannette
Piccard to the Marshall Space Flight Center as a special consultant in
the mid-1960s. The lone survivor of a distant age of heroic stratospheric
flight, she remained an ardent spokesperson for the romance of manned
flight at high altitudes. Gilruth brought her to Houston to "just get the
word [out] about what it is we're doing." He thought her qualifications
were perfect, as he recalls telling her: "You're famous, the people still
remember what you did, and it can help our program."[3] At one point
during an interview for *Ballooning*, in which she left little doubt about
the difficulties she and Jean Piccard had with Settle and others running
"A Century of Progress," Jeannette wished to preserve the memory of
her husband as a man dedicated to the pursuit of science very much in
the spirit of the wonders Apollo would bring to mankind.[4]

There is, indeed, a remarkable similarity between how science was used
to justify manned ballooning in the 1930s and how NASA and its allies
in government and science promoted Project Apollo as a scientific mission. In both eras, the technical goal was the same; as the quotations
from Auguste Piccard and James Webb attest, each was an enterprise to
carry man into a new and distant realm. And in each, there was much
rhetoric promoting manned ventures, which ultimately produced similar
reactions in the world of science.

The Rhetoric

John F. Kennedy's "Second State of the Union" address on May 25, 1961
was the culmination of a lengthy process whereby NASA was transformed
from an agency with developmental goals in space engineering, technology
and science, into one with a single mission: "of landing a man on the
moon and returning him safely to earth."[5] John Logsdon, Walter

FIGURE 11.2. Jeannette Piccard on tour for NASA, in front of a mock-up Apollo Command Module at North American Aviation, March 22, 1965. NASM.

McDougall, and others have shown clearly that Kennedy's was a political decision; his rhetoric, reasoning, and decision making were conditioned by Cold War anxieties.[6] Kennedy had been advised by James E. Webb, his NASA Administrator, and knew that the goal he set could be accomplished "before this decade is out." It would be a spectacular achievement that would get America going again, reestablish its world prestige, and help it feel better about itself and its leaders. But a manned mission to the moon required extraordinary focus and marshalling of forces. More than money would be needed, Kennedy pointed out to Congress and the nation. Success was possible only if "every scientist, every engineer, every technician, contractor and civil servant gives his personal pledge that this nation will move forward, with the full speed of freedom, in the exciting adventure of space."[7]

FIGURE 11.3. NASA Administrator James E. Webb in a Gemini trainer at Marshall Space Flight Center, August 7, 1965. NASA.

The scientist was first "on tap" in Kennedy's rhetoric among those who had to be mobilized, in what was a surrogate for war, to take the nation to the moon. Kennedy was no more specific about the scientist's role in this process, of course, than he was specific in his January 1961 inaugural speech when he argued that cooperation in science and space might promote peaceful relations with the Soviet Union: "Let both sides seek to invoke the wonders of science instead of its terrors. Together let us explore the stars." If Kennedy's rhetoric represented his thinking, or even the advice he was receiving, space, science and the scientist became symbols of a surrogate army, and that army became NASA. In the few short months between his two inaugurals, Kennedy's rhetoric went from diffuse and reflective, to concrete and effective; NASA was expected to do the same.

NASA was in its third year when Kennedy presented his dramatic challenge to the world. In its first years, as Allan Needell has pointed out, NASA enjoyed warm relations with the scientific community since the technology it fostered held out great promise for science. Access to completely new realms of natural phenomena, whether they lay in the stratosphere or in space, could not help but lead to new discoveries, so it was inevitable that scientists would be directly involved, especially those like James Van Allen or John Simpson who were already familiar with centralized government patronage for scientific rocketry and ballooning.[8]

NASA inherited scientific programs established during the International Geophysical Year, which by 1961, with access to space, returned much useful and revolutionary data. As in the 1930s, the question, however, was to what degree were scientists willing to buy into the sponsoring agency's stated mission, whether it be a spectacle in the stratosphere or landing a man on the moon?

To meet the demands of Kennedy's rhetoric, the popular image of Apollo became a carefully orchestrated media event; promoters established "science" in its broadest sense as a vaguely defined key word, synonymous with "space."[9] To some observers, NASA's rhetoric led it steadily away from reality to the point where, years after Apollo and in the midst of the *Challenger* tragedy, one *New York Times* reporter commented that "Some agencies have a public affairs office; NASA is a public affairs office that has an agency."[10]

Robert Seamans, NASA associate administrator under Webb, expanded upon the President's May 1961 speech the following August, commenting that "the United States must make this effort for urgent scientific, technological, political, and economic reasons."[11] He identified four major justifications for the national decision to land a man on the moon: "(1) the quest for scientific knowledge; (2) direct and immediate application of satellites into operational systems; (3) the risk of delay in our space competition with Communism; and (4) the technological advances and stimulus to our economy that will emerge from the space effort."[12] Throughout his widely reprinted essay, Seamans emphasized the role of science in the space program. His commentary strongly suggests that NASA by late 1961 wanted to have the needs of science appear to be the prime rationalization of JFK's accelerated program to land man on the moon. Seamans also cited an endorsement by Carl Sagan for the need of trained human observers and instrument manipulators on the moon; man would be superior to automatic instruments for studies in astronomy, mineralogy, and microbiology. Sagan's testimony gave Seamans powerful evidence that "manned spaceflight is essential" in scientific exploration.[13]

Hugh T. Odishaw, Executive Director of the Space Science Board (SSB) of the National Academy of Sciences, helped to design the board's public posture on the space program in the early 1960s. Odishaw saw NASA's manned programs (Mercury, Gemini and Apollo) as evolving toward a scientific payoff, even though the immediate goals were to establish a capability "tied to public interest in human exploits and to political interests in world leadership." To Odishaw, "the man-in-space program may, for our purposes, be considered a program of gross exploration;

'gross' only to distinguish its early phases from those that follow, which include quantitative scientific investigations."[14] Mercury and Gemini would lead to Apollo, which he viewed as the scientific payoff.

The relationship between the SSB and NASA was far from simple and harmonious. From 1958 though 1960, the board was concerned about NASA's "big engineering" proclivities, but in 1961, the pattern changed partly because the new NASA Administrator, James Webb, was an old friend and ally of SSB chairman Lloyd V. Berkner. Berkner, a Webb loyalist since the Truman era, overcame divisive elements on the board to endorse NASA's man in space program in February 1961.[15] Berkner was one of the most vocal and visible advocates of establishing postwar "big science" in America. The first full-time president of Associated Universities, Inc. (AUI), a naval aviator, a member of Richard E. Byrd's first Antarctic expedition, and a long-time student of the ionosphere who had been involved in Navy satellite proposals in 1944–1948, "Berkner was fascinated by what he perceived to be the transformation of science that occurred during his lifetime," as a result of World War II.[16] Typical of many scientists who had been at the centers of wartime research, Berkner perceived for science a central role in all aspects of modern society, and therefore considered it essential for scientists to involve themselves in large-scale efforts important to the nation, whether those efforts be national defense or landing a man on the moon. He carried this vision of the scientist/statesman into space as founding chairman of the SSB. Berkner approached space science just as he had faced the prospect of establishing a national facility for radio astronomy as AUI president: Science should organize in cooperation with government interests to engage in missions of national scope. But, far from an idealist, he also knew that real progress would come only when scientists united behind their spokesmen and institutions to lobby and educate potential government patrons.[17]

Odishaw, speaking for Berkner and the SSB, argued that, "from a scientific standpoint, there seems little room for dissent that man's participation in the exploration of the Moon and planets will be essential, if and when it becomes technologically feasible to include him."[18] Echoing a familiar argument used by both the Piccards and Swann some 30 years earlier, when automatic instruments were not yet reliable nor sophisticated enough to meet scientific needs, Odishaw asserted that the presence of man and his faculties would add the critical element of scientific judgment to the conduct of scientific research that could not be matched by automata. In the Space Science Board's statement, reported to Webb in March and released to the world in August, 1961, the scale of a manned scientific enterprise was "unlikely to be greatly different from that required to carry out the program by instruments alone." Admittedly, "the primary scientific goals of the program [were] immense . . ." but to Odishaw and the board, they were in keeping with national purpose. Historians Walter McDougall and R. Cargill Hall have both observed that the SSB's endorsement was "a mere genuflection to public sentiment," yet it gave Webb much of the ammunition he needed to convince Kennedy that his own science adviser (Jerome Wiesner) was unduly conservative about man in space and his opinion was not representative of the entire scientific community's attitude. After Yuri Gagarin flew on April 12, 1961, neither Webb nor Kennedy needed the SSB's endorsement, although it remained the most visible confirmation that man on the moon was good for science, an idea Webb was not shy about sharing.[19]

The SSB's endorsement may well have been a political nod, or even ammunition desired by Webb. But to some scientists far from Berkner's sympathies, it was a crushing sell-out. Merle A. Tuve, Director of the Carnegie Institution of Washington's Department of Terrestrial Magnetism, and no friend of Berkner, harbored "serious doubts" over the scientific value of the space program. When invited to attend the Iowa Summer Study in 1962, Tuve tried to alert Detlev Bronk, President of the National Academy of Sciences, to his deep concerns over the distortions inherent in such programs:

The distortions in the public presentation of these space programs amount almost to a deliberate misleading of the public, and the intimate marriage of pure science with military threats by the same rockets toward space constitutes a new danger added to the already threatening fact that the physical scientists are high priests of destruction.[20]

Tuve's fear of science becoming evermore intimately associated with the huge expenditures so familiar a part of military programs, and indeed, his fear that the two worlds were becoming one, was heightened by the SSB's endorsement of the NASA mission:

The fact that our most conspicuous leaders in scientific institutions and in the Academy have joined in these distortions, with the usual familiar remarks concerning military money which will be spent anyway, leaves me feeling personally out-of-tune with the times.[21]

Tuve saw the SSB's actions as a "desperate compromise with public honesty" that would lead only to a "verification that the surface of the moon is an exceedingly hostile environment for mammalian life."[22] He could not be more distant from the feelings of Berkner, Odishaw, and Seamans who believed that a manned program was needed for scientific reasons. All four stood outside NASA's walls looking in, all harbored personal demands for both science and NASA, but none fully appreciated how important it was to be sensitive to where the power center lay within NASA. This was the responsibility of NASA's highest ranking scientist, Homer E. Newell, who was faced with the survival of science within the space agency. Newell felt obliged to continually remind everyone that science was, indeed, a part of NASA from the start: "The mandate given to NASA in the National Aeronautics and Space Act of 1958 calls for a vigorous effort of science in aeronautics and space exploration."[23] Then

FIGURE 11.4. Homer E. Newell, who had been deputy director of NASA's Office of Space Flight Programs, participated in the 1962 Iowa Summer Study as the new director of NASA's Office of Space Sciences after Webb's first reorganization, and later became Associate Administrator for Space Science and Applications in a second reorganization late in 1963. NASA photograph circa August 1963.

deputy director of NASA's Office of Space Flight Programs, Newell was beginning the most difficult phase of his long career in space science, which began in 1946 when he became part of the Naval Research Laboratory's upper atmosphere rocketry program. When James Webb reorganized NASA in November 1961, Newell became director of the new Office of Space Sciences in direct competition with the now-separated and far larger Office of Manned Space Flight, created to focus effort on Mercury, Gemini and Apollo. Newell found himself caught in a reorganization that separated manned and unmanned constituencies and "fractured the space agency in a manner from which it has never recovered."[24] Although his new office gave science greater visibility, it also made it more vulnerable, as more and more of NASA's budget was directed to the manned space program. To keep space science alive even before the reorganization, Newell had little recourse but to link it to NASA's proposed Project Apollo, assuring the public whenever he could that out of NASA's "scientific research will come . . . knowledge about the universe and its laws; knowledge about the earth upon which we live; knowledge about physical life."[25] Newell's task was to highlight science within the constraints set by the agency. He could agree with his scientist colleagues and critics at the Iowa Summer Study in 1962 that the interests behind Apollo were not scientific, but all the while he had to make the situation look as good for science as possible in NASA, in order to convince scientists to sign on.[26]

Newell, an ardent representative of the interests of science within the agency, was constantly obliged by his NASA superiors and pressured by Congress to endorse manned spaceflight, or to explain what Kennedy's expanded program meant to the scientific interests of the nation.[27] An adept bureaucrat and experienced writer of technical reports and semi-popular reviews, Newell exploited his talents to produce a series of lengthy news releases, popular articles, and NASA pamphlets that put man in space. One pamphlet prepared in 1963 explained that although instrumented satellites—Explorers, Pioneers, Mariners, and Rangers—had already returned excellent data, the "necessity of man in space" could not be denied if "scientific exploration" was to go forward: "Whether systematized or not, whether planned or incidental, every look he takes, every glance will be exploration. And if he is an accurate observer, all of it will be science."[28] With Robert Jastrow, director of NASA's Goddard Institute for Space Studies in New York, Newell in the summer of 1963 prepared an essay for the *Atlantic* that NASA reprinted, entitled "Why Land on the Moon?" After identifying the "Soviet space challenge," they argued that the money spent on space would return dividends far greater than expenditures: "The science which we do in space provides the equivalent of the gold and spices recovered from earlier voyages of exploration." They admitted that "the driving force of the program is not in scientific research alone," but in the "urgencies of the response to the national challenge" of the Soviets. Yet the overall thrust of their apologia for the program was to show how man on the moon would solve some of the most fundamental questions about human existence.[29]

As spokesman for NASA's programs, none were more forceful or official than James E. Webb. Kennedy's choice as NASA administrator, Webb repaid Newell's public support for the "necessity of man in space" by publicly and privately endorsing the centrality of science to NASA's mission. When in February 1963 Webb heard that engineers at NASA's Manned Spaceflight Center (MSFC) in Houston told visiting Congressman Joe Karth (chairman of the Space Science and Applications Subcommittee of NASA's authorizing committee in the House) that "the space sciences program was making no contribution to the manned space flight program [and that it was] impractical and of a different order of value from that being conducted by the manned space flight office," Webb fired a long and stern letter to Robert R. Gilruth, director of the Houston center, to educate his hotheaded and impolitic engineers: "How can we get more of a sense of responsibility for people to weigh their words and use good judgement." Webb felt that if MSFC staff, especially Maxime Faget, really believed that science was of no value, then they have "missed one of the very important elements necessary to the program ... That all he [Faget] is now doing is based upon scientific work of the past, and that unless we continue scientific work in the present, we will be condemned to state-of-the-art vehicles and systems for the future."[30]

James Webb appreciated the power and generality of science more than his Houston subordinates. To Webb, scientific work was essential to determine "the precise knowledge of the environment," as this would help the engineer design the best vehicles for traveling in and through it. He wanted NASA to be a tight, productive ship: "The relationship between the scientist and the engineer is one we are trying to foster in every way we can, and in my opinion is just as important to the future space power of this nation as any specific hardware program we are carrying out today." Accordingly, Webb asked Gilruth to convey once again the science policy he established at Headquarters and elucidated in numerous press releases he and Newell had written. Sending copies of the releases to Gilruth, for wide distribution at MSFC, Webb concluded: "Our objective is to advance scientific knowledge and technological capability through important missions such as the manned lunar exploration. When this objective is not understood by those responsible for the conduct of the program, it is a matter of deep concern."[31]

Webb, as much as anyone, believed that harnessing scientific manpower was essential for the success of Apollo, and so construed Apollo as a scientific mission. In numerous press releases on NASA policy, and through his lectures and addresses, often reprinted by NASA Public Affairs as booklets and flyers, Webb used the term "science and technology" to describe a wide range of activities and expertise that would make the manned spaceflight program successful and worthwhile. Some of Webb's public statements during 1961 carefully separated science and the manned program, especially those made during his testimony at budget hearings, but the bulk of his rhetoric during the early 1960s highlighted science time and again.[32] He therefore helped promote the idea that the Apollo program was informed by scientific research and was supportive of further research. In November 1962, writing to President Kennedy recounting a recent meeting in which they discussed accelerating the manned lunar landing program, Webb enunciated his vision of the motive behind Apollo: "The objective of our national space program is to become pre-eminent in all important aspects of this endeavor and to conduct the program in such a manner that our emerging scientific, technological, and operational competence in space is clearly evident."[33] Preeminence meant landing on the moon first, as the culmination of a well-managed, orderly, and timely program that exploited the best American talent and demonstrated what it could do. Apollo would be, Webb hoped, confirmation of the Nation's scientific and technological credibility in the Cold War world. "To be preeminent in space," Webb advised Kennedy, "we must conduct scientific investigations on a broad front." This was the concept Webb promoted among scientists and scholars; deeply committed to it through action as well as rhetoric, Webb explained to a George Washington University Commencement audience in June 1961 that "in the minds of millions, dramatic space achievements have become today's symbol of tomorrow's scientific and technical supremacy."[34]

Webb's sincere rhetoric, born of a conviction that vigilance and preparedness in the Cold War world required a strong scientific arm, was translated into images, symbols and myth by his NASA minions: Commenting in 1966 on NASA as a myth maker typical of any institution interested in survival, H. L. Nieburg argued that NASA had a vested interest in continuing a perceived international science and technology race, as a replacement for the arms race, with the moon as a winnable goal. When describing the space race and domestic tensions between Air Force and civilian domination of space policy in this manner, Nieburg strengthened the association of space with science, and hence Apollo as a scientific program, or at its broadest, as a program in science and technology driven by the political forces of nations. Nieburg also warned that "The advocates of escalating the science-technology race also assert as an article of faith that science can solve any problem, provided only that a sufficient number of scientists be mobilized and given enough money."[35]

Vernon Van Dyke in 1964 envisioned national pride and Cold War pressures as the appropriate incentives behind Apollo.[36] John Logsdon, whose work on the national decision to send a man to the moon was stimulated in part by Van Dyke's early study, elaborated on the importance of pride, showing that it, along with superiority in science and technology, was at the very root of national security policy. Quoting Alton Frye, Logsdon argued that national security "derives . . . from its successes in science and technology, its economic and social strength, its prestige— even from its pride."[37] Whatever the mechanics of the decision to go to the moon, science became implicated not only as a promoter and beneficiary of manned flight, but also as a provider of the capability of spaceflight itself. One way or another, "science," broadly and vaguely construed, was to play a visible role in Apollo.

Kennedy, Webb, Nieburg and Van Dyke all suggested in their own way that science and scientists were players in the race to the moon. In similar fashion, the planners of *A Century of Progress* and *Explorer* viewed scientists as necessary participants in the assault on the stratosphere. Just as the National Academy of Sciences hoped to counter the 1930s "revolt against science" phenomenon by demonstrating through display that science, in cooperation with American industry, would bring the nation out of the Depression, Hugh L. Dryden, Webb's close NASA lieutenant, defended Kennedy's accelerated man-in-space program in June 1961 by predicting, in the words of a *Science* reporter, that "the space effort would yield new technological developments that would strengthen the nation's economy by providing new jobs and new industries."[38] Not only were scientists regarded as players in both eras, but they came to be considered as shapers of national recovery.[39]

Most certainly, "science" and "space" were synonymous in early NASA rhetoric. But that link was forged even before NASA existed. James R. Killian, President Eisenhower's science adviser, released a Science Advisory Committee Report, "Introduction to Space," in March 1958 that identified four salient justifications for supporting a space program: "(1) the compelling urge of man to explore and to discover, the thrust of curiosity that leads men to go where no one has gone before"; (2) "the defense objective"; (3) "national prestige"; and (4) "scientific observation and experiment which will add to our knowledge and understanding of the earth, the solar system and the universe." Michael Smith notes that these justifications defined the program over the next 10 years.[40] As John Logsdon and others have argued, the Cold War created pressures on the nation to preserve its national defense and prestige, which in turn prompted NASA's goal to land a man on the moon. Although NASA never tried to hide these motives from the public, it tied them and Apollo to the higher goal of scientific exploration: Killian's second and third justifications led to the Apollo decision, but his first made Apollo part of the human condition, and his fourth reconfirmed the legitimacy of Apollo as a scientific mission. Not only were "space" and "science" synonymous, but the needs of the latter justified the expense of conquering the former.

Killian's 1958 justifications found their roots in the fertile soil tilled for decades by popular and highly persuasive writers such as Arthur C. Clarke, G. Edward Pendray, Wernher von Braun, and Willy Ley. In the tradition of Hermann Oberth and Robert Goddard, these pioneer visionaries often spoke of the new worlds that rockets and space travel opened for scientific investigation. The brain trust of the space age prophesied that rockets would soon enter the realms of "civil and military aviation, of weapons development, and of fundamental scientific research."[41] But it was the latter that often dominated the pages of Clarke's and Ley's books; often these well-crafted volumes were indistinguishable from elementary texts on astronomy, with added chapters on how rockets work. Ley's works in the late 1950s conveyed the fascination of astronomy. He freely associated astronomy with space travel, mixing sentences about rocketry and the doing of astronomy: How and why rockets are built were explained alongside interesting astronomical facts.[42]

Acknowledging the many practical uses for rockets, and the popular enthusiasm for flights to the moon, G. Edward Pendray, a founder of the American Rocket Society, looked beyond them to a higher mission: "The true purpose and fulfillment of life is to know and understand; to see a fuller concept of the world and its place in the universe, and our own position in the cosmic scheme . . ." Pendray believed that rocket power would bring "new knowledge of the realms into which mankind has so far been unable to venture, and thus will stretch our mental horizons and enrich the fields of physics, meteorology, radiation and many another science.[43] Pendray's vision was repeated time and again in the 1950s and early 1960s, in its broadest sense by James R. Killian, but then in the context of manned flight by Wernher von Braun, Ernst Stuhlinger, and Willy Ley.

With von Braun, Willy Ley promoted the idea that astronomers stood ready to reap the benefits of landing on the moon.[44] Ley's easy association of astronomers' interests with space travel made the two likely partners in the public perception. To Ley, von Braun and Ernst Stuhlinger, the manned effort, either on the moon or in an orbiting space station, would stimulate interest in space science. As Stuhlinger later pointed out: "Perhaps the most profound significance of Project Apollo is its catalytic effect on the material support of large efforts with purely scientific goals."[45]

The Reaction

Apollo rankled some scientists from the very start. Various authors have examined the debate over the complexion of the U.S. space program—that is, whether it should emphasize manned or unmanned missions—and have shown that although NASA at first was constrained by the Eisenhower administration's policy, which favored a diverse program of unmanned scientific and commercial satellites, the space agency continued to openly and aggressively plan for a manned program.[46] Accordingly, when Kennedy gained office, he established an Ad Hoc Committee on Space, headed by Jerome Wiesner, that looked into NASA's activities and concluded that the space agency should reverse its internal priorities and reemphasize space science. Many scientists both within NASA and elsewhere "objected to engineering for man in space at the expense of quality space science."[47] R. Cargill Hall believes that the root of the tension between the scientific community and NASA engineers can be traced to administrator James Webb's reorganization and division of manned and unmanned spaceflight operations in 1961 to accommodate Apollo, and the subsequent policy shift and imbalance that this division created.[48] But Hall also notes that NASA was built by politicians and managed by engineers who either brought along or inherited the scientific programs and manned flight imperatives from the preexisting agencies and institutions absorbed by NASA. Thus, within NASA at least, the physical reorganization Webb instituted was also perceived as a move to put the engineers in charge to get a job done. The engineer's attitude toward space science during Apollo was hardly reassuring to the scientific community or to James Webb, as illustrated by Webb's adverse reaction to Max Faget's unfortunate conversation with Congressman Karth. Webb's admonishment hardly changed matters; in the recent words of a NASA engineer, Apollo represented an "engineering tour-de-force carrying a large science program on its broad coattails."[49] But from the scientific community's perspective, riding coattails meant in some cases perverting scientific missions into engineering stunts; for example, many perceived that Project Ranger was transformed from a scientific program into an engineering support project for Apollo.

When Project Apollo began officially in the summer of 1961, Project Ranger, an automatic lunar probe, was already deeply in trouble after a series of flight failures. Designed as scientific probes of the lunar surface, of the earth's geocoronal envelope, and of the space between the earth and moon, Ranger was transformed into an Apollo support program. The first Rangers had been at the center of a tug of war between scientists advocating a maximum return of scientific data, and engineers who wished to quickly prove the Ranger system itself by sending a probe to the moon before the Russians sent one. The engineers won, effectively "downgrading space science in favor of engineering."[50] And when future Ranger flights 6 through 9 were committed to the Apollo program, those scientists already involved in Project Ranger and other NASA space science projects complained bitterly to Homer Newell. This already tense situation was further inflamed when the press, trade journals, and NASA officials began to picture "all of NASA's unmanned lunar flights as engineering missions in support of the manned lunar landing."[51] The reporting was correct, of course; even before Apollo was funded, NASA administrator James Webb with his deputies Robert Seamans and Hugh Dryden all viewed the unmanned Ranger and Surveyor as developmental stages of Project Apollo.[52] The transformation of Ranger came quickly: Some instrument packages were canceled and landing sites were altered to fulfill expected Apollo needs.[53]

The fate of Ranger in 1961 alerted many scientists to the deep changes taking place in NASA. It led Vernon Van Dyke to ask in 1964 if scientists "regard the space program as a scientific program."[54] And did scientists feel comfortable with such an expensive program, which ultimately would be linked to, or would affect deeply, policy governing future support for science? Some scientists in touch with Congress in the early 1960s feared that their elected representatives were woefully ignorant of the needs and priorities of science. Of all the perceived motives behind Apollo—national prestige, military advantage and science—only science was identified with a special interest group. Thus part of the scientific community's anxiety that promoted its vocal criticism was its perception of being inextricably linked with the manned spaceflight effort. It was just at this time, in astronomy at least, when scientists realized that they had to educate their congressional patrons. In one non-space example, the Navy was then building an enormous radar dish near Sugar Grove, West Virginia, and asked several astronomers to advise them on what astronomical uses it could be put to as well. But when several congressmen heard that the construction of the dish was not succeeding, and was costing far more than predicted, they berated the project as an astronomical boondoggle. As a result, astronomical leaders realized that they had to organize to make known what the community's scientific priorities were, so that others would not make that decision for them.[55] In like manner, astronomers and other scientists, in the wake of the transformation of Ranger and Webb's summer 1961 reorganization, watched NASA's enormous funding growth and realized that the health of their profession might well depend on the health of NASA's man-in-space program. The scientific criticism, astronomer Leo Goldberg recently argued, was generally well founded, "motivated in no small part by a sincere conviction that the responsibility for spending large sums of money in the name of science could not be entrusted to NASA alone but had to be shared with the scientific community."[56]

There were two paths to take in the early 1960s: scientists could deny responsibility for the public funds to be spent on Apollo, or they could argue publicly that although Apollo could not be justified on the grounds of science alone, still it offered a chance to do some good science, and the best scientific program would come from the advice of the best scientists. As one NASA scientist put it, after designing the parameters for lunar sample collection, "the Apollo Project is primarily a glorious adventure . . . a magnificent feat, and a milestone in the history of the human race. No other purpose or justification is necessary."[57] Few doubted this, but organized science, as represented by Lloyd Berkner and his Space Science Board, decided that although Apollo could not be justified by science alone, good science gave it a higher and more significant justification. In the wake of Webb's 1961 reorganization and separation of the offices of Manned Space Flight and Space Sciences, which "now places science on a parallel footing with [NASA's] three major areas of technological activity, including man in space," Berkner advised his colleagues that it was time for science to organize its collective voice to be sure it would not "miss the boat."[58]

The Testimony of Scientists

Scientists expressed themselves about space research in many ways, from participating in workshops co-sponsored by NASA and the National Academy of Sciences, to testifying before Congress. Although since then some scientists have wondered how much the workshops or congressional testimony actually influenced policy, these forums helped crystallize the thinking of scientists at the time. The 250 scientists participating in the two-month (June 17–August 10) Space Science Summer Study in 1962 endorsed much of NASA's programming, but many remained uneasy about the place of science in the space program. In his memoir *Beyond the Atmosphere*, Homer Newell remembers Princeton astronomer Martin Schwarzschild confiding that astronomers "found it distasteful that NASA, not they, should be making the decisions." Newell also remembers that even though the question was not on the official agenda,

Many expressed disapproval of the manned program, along with the wish that the monies going to Apollo might be diverted to space science. Some expressed concern that not only was Apollo going to proceed but that NASA would even seek to justify the program on the basis of science, and this the scientists strongly objected to.[59]

Newell's recollection refers to those who, though they may have been critical of space research conducted by astronauts, still by their participation in Iowa certainly were sympathetic to science in space. There were others, however, less sympathetic to the entire enterprise, who when invited to Iowa made clear their deeper concerns. As we saw, Merle A. Tuve, Director of the Carnegie Institution of Washington's Department of Terrestrial Magnetism, reacted negatively to the invitation to attend the Iowa conference, feeling that his opinions about the inevitable distortions created by the SSB in endorsing manned space research might be too "distasteful to the leaders of this seminar."[60] He was willing to come for a few days, if they wanted him, but in the end, Merle Tuve played no visible part in the Iowa conference, although his sentiments were felt, and reflected by others.

At the Summer Study, held on the campus of the University of Iowa and therefore called the Iowa Summer Study, Berkner charged all the working groups to maximize scientific return in their planning, and to plan in concert with NASA's man in space program. Newell and other NASA officials at Iowa argued that Apollo was going ahead, not for science, but for "important national objectives other than science." They wanted scientists to take part and "take advantage of the opportunity before them and to help ensure that the science done in Apollo was the best possible." Indeed, Newell's stated position dovetailed nicely with Lloyd Berkner's agenda: both wished to maximize the science returned by Apollo. Newell needed the endorsement of the scientific community, but he needed the right kind of endorsement. To some extent, he received it from Berkner, who used the closing plenary sessions, in which the working groups came together to present preliminary findings, as a platform for his views. According to Newell, Berkner "stated that man in space was a good thing and that exploration was science."[61] Berkner's written commentary was somewhat more sophisticated than Newell's recounting. Arguing in his closing remarks on August 9 that "exploration of space *is* truly science in space," Berkner went beyond the usual appeal

FIGURE 11.5. Formal introductions at the Iowa Summer Study. Left to right, from the spotlight: Harry Hess, Lloyd V. Berkner, Hugh Dryden, Fred Seitz, university president Virgil M. Hancher, James Webb, Hugh T. Odishaw, James Van Allen, chairman of the Summer Study, in front of a blackboard depicting expected funding patterns for space science underneath an overall envelope describing Apollo funding. James A. Van Allen Collection.

to "national prestige" and "adventure" to remind readers of "man's questing spirit" which, embodied in the exploration of space "*basically* means the acquisition of information and a description of what is there ... This *is* science; and, thus, science is an integral and central objective of our man in space program." Berkner wanted those attending the Summer Study, and the broader scientific audience who would read his remarks, to understand that they had an obligation to make Apollo work to their own advantage: "I believe it is our responsibility to emphasize this concept and to point out that the present period in the development of technical capabilities is simply a necessary step which will lead inevitably to a more sophisticated contribution to *science in space*."[62]

Berkner took the long view, but knew well that scientists were not of one mind and had widely varying opinions. He knew that younger scientists concerned about tenure, or the intellectual health of their newly established graduate training programs, thought in terms of rapid payback, and that it would be very difficult to get them to think in terms of national purpose. He also relaized that the needs of students of lunar and planetary science were just too different from those working in solar, stellar, and extra-galactic areas. And all were distinct from those devoted to human physiology. Indeed, the Iowa Summer Study consisted of more than 20 working groups addressing a wide range of questions, from the degree of support required from NASA for ground-based research facilities of all types, to administrative and policy matters, international cooperation, science in the manned spaceflight program, and man as a scientific observer. If a consensus were to come out of this, it would require a firm hand. He almost got what he wanted.

At the end of the summer, Iowa conferees admitted that "there is considerable confusion about the Apollo mission and its proper justifications."[63] The summary report, edited in part by the Summer Study host, James Van Allen, tried to clarify matters. All agreed that Apollo was driven by national prestige and national security, if not by "man's innate drive to explore unknown regions." But some participants argued that "there are important scientific objectives in the Apollo program, and it is in terms of science and scientific opportunities that the program appears to have been most widely misunderstood."[64] Further, others contended that "the current program is primarily a technological and engineering effort, and this fact ought to be generally recognized." Once the engineering phase was finished, however, and Apollo became functional, then "a strong scientific validity follows." Reminiscent of William Swann's conservatism for automata, some summer study participants believed that man certainly would "contribute critical capacities for scientific judgement . . . which can never be accomplished by his instruments, no matter how sophisticated they become." Echoing Berkner's sentiments at the closing plenary session, the summary report proclaimed:

Hence, manned exploration of space *is* science in space, for man will go with the instruments that he has designed to supplement his capacities—to observe what is there, and to measure and describe phenomena in terms that his scientifically trained colleagues will clearly understand.[65]

The bartering point established in Iowa during the summer of 1962 was that once scientists understood Apollo in this context, and NASA was willing to present it as such, then the scientific community would accept and endorse the program. In effect, if NASA agreed to some of the major conditions identified by Summer Study participants as essential to the scientific robustness of Apollo, then NASA might find scientists more sanguine about working with the man-in-space program.

The conditions for playing ball were outlined by the Working Group on Man as a Scientist in Space Exploration, which keynoted Apollo: "Manned exploration of space promises great scientific return, and Apollo can be a fruitful first step in this effort."[66] The working group wanted the first step to be not only the development of the engineering capability to send man to the moon and bring him back, but also "the first step in the manned scientific study of the Moon and planets." To achieve the latter required that a scientist "fully trained as an astronaut" be a member of the crew.[67] The working group, composed chiefly of those interested in aspects of aerospace medicine and astronautical engineering, included Malcolm D. Ross, now retired from the Navy and working at General Motors. They diplomatically crafted their rhetoric to argue as strongly as possible for including scientists in the astronaut corps, and argued that trained scientists should be among the first to travel into space and to step onto the surface of the moon. Dedicated to studying man in space, they supported Apollo, but wished to maximize scientific return by creating a parallel corps of scientist-astronauts and by establishing an "institute for space sciences" in Houston.

The working group's own Subgroup on Scientific Undertakings for Man in Space presented an even harder sell. Cochaired by the planetary astronomer Gerard Kuiper and the astrogeologist Eugene Shoemaker, this group felt that science justified the entire effort:

The scientific exploration of space has been proclaimed a national goal, and manned lunar landing is a major step in the implementation of this program. Man's opportunities for scientific exploration of the moon are practically unlimited. The lack of an adequate scientific endeavor could invite serious criticism of the program, while the impact of a successful scientific mission by means of a lunar landing will enormously enhance the importance of the Apollo program in the eyes of the world.[68]

The trump card played in Iowa and held by the Space Science Board was scientific legitimacy for Apollo. Scientists knew that NASA wanted it, wanted as well to keep scientists from complaining to their congressmen, and might well play for it as the Apollo program was further defined. Some geologists and biologists were happy enough at first to provide that legitimacy, such as Eugene Shoemaker and Harry Hess, who were initially enthusiastic about what Apollo might mean for selenology and solar system studies, and Robert Galambos, who was particularly interested in human physiology in space. Other scientists felt that since Apollo was inevitable, the most politic thing to do was to throw in with the program, hoping it would return as much science as possible. After all, this was a period of expansion and a great deal of money was still around.[69] Least committed to man in space were representatives from the Iowa workshops on particles and fields, planetary atmospheres, and astronomy, who "concluded that manned units would contribute little or nothing at this time."[70] On the other hand, some members of the astronomy working group realized, quite correctly, that the technology of the day was not yet advanced enough to permit highly sophisticated experiments to work in orbit or on the surface of the moon without occasional human presence. In Iowa, many learned for the first time about NASA's plans to develop a capability to rendezvous and dock craft in orbit. "And right away," Leo Goldberg recalls, "we all began to think: 'gee, this is how you build a big telescope in space, or a big laboratory.' "[71] There was much informal talk at Iowa about a large imaging space telescope in orbit that extended the astronomer's traditional tools into space.[72] Large telescopes and detector arrays probably had to be assembled in orbit by astronauts, and photography remained the only reliable high resolution spatial detector available, which required physical return of data. Thus, much of what the scientific community hoped to accomplish in the near future would have to be manned, or at least "man-serviced."[73]

Berkner, Newell, and Webb in concert asked scientists in Iowa to buy into the man-in-space program in order to strengthen Apollo's image as a scientific venture. Unwilling to be railroaded, American scientists hoped that by setting out conditions, as strongly and positively as possible, under which they would participate, they might achieve "maximum scientific return." They also knew that Apollo was being sold as a scientific mission anyway, as Webb had made quite clear. And they knew that their most illustrious and visible body, the National Academy of Sciences and its Space Science Board, endorsed wholly what became Apollo in 1961 in a statement widely cited in newspaper articles and in official NASA and NAS arguments for manned spaceflight.[74] To paraphrase the recent musing of one participant, "We were manned if we did, and manned if we didn't."

The 1962 Iowa Summer Study summary report was intended to be a primer for NASA and Congress, pointing out the type of program Apollo must be in order for it to be endorsed by the world of science. But not everyone agreed with the cooptive tactics of Lloyd Berkner or Homer Newell, or had the hopeful attitude of the Working Group on Man as a Scientist in Space Exploration.[75] From the very beginning, NASA and other governmental bodies wishing to develop various avenues of activity in space alarmed the scientific community with their ideas and actions, which ranged from proposals for vast arrays of space mirrors to illuminate the landscape for tactical purposes to conducting projects Westford, Argus, and Starfish. *Westford* was created to fill the ionosphere with radio-reflecting needles. *Argus* and *Starfish* were both high-altitude atomic and nuclear explosions designed to improve DoD's understanding of differences "between a natural and an artificial population of geomagnetically trapped particles."[76] The low-yield Argus tests in the summer of 1958 were observed with Explorer satellites, but the higher-yield Starfish explosion temporarily poisoned the ionosphere, seriously damaging a number of satellites then in orbit.[77] In fact, a Starfish burst took place on July 9, 1962 when the summer study was in session, which made it a center of attention soon after. Thus although these issues were not on the agenda in Iowa, they were discussed in the halls along with many others of concern. Many scientists argued for increased NASA support for ground-based facilities, which were required to better define and articulate studies from space platforms, and worried over the high costs in terms of time, manpower, and the dedication required to enter into NASA's programs.[78] Others believed that science was weak and ill-served at NASA headquarters: There were not many good scientists there, and no scientist was among top NASA policymakers.[79] And still others were deeply uncomfortable about participating at all in the NASA adventure.

The range of sentiment expressed during the eight weeks of deliberations in Iowa was enormous. Verbal and written graffiti left its mark on many a participant. William E. Porter, a member of the journalism faculty at Iowa State, was invited by Van Allen and Berkner to participate in the Working Group on the Social Implications of the Space Program, but after attending this and other workshops week after week, he caught the sense of what was bothering the scientists. What Porter found was a large group of people far from comfortable with their perceived public responsibility as well as their self-image if they became absorbed into NASA's universe. First, among the more "conspicuously articulated states of mind" was concern over the bigness of the NASA enterprise. Most participants simply could not accept the huge budgets that were involved. "A few scientists opposed the whole NASA enterprise on this ground."[80] Porter reported that many felt it necessary "to keep a wary eye on future developments," as they sensed a degree of ineptitude in themselves and the NASA enterprise, summarized best by a blackboard sloganeer one day during the sessions: "If war is too big to be left to the generals, is science too big to be left to the scientists?"[81] There was "perhaps some sense of inability to do anything about this grand design even if one doesn't like it." We saw that Merle Tuve didn't like it, and did what he could to avert what he saw was a trend in science policy toward applied military research that might ultimately "reduce the future opportunity for devoted studies of the intricate mysteries of nature."[82] In 1959, Tuve expressed similar sentiments in a *Saturday Review* essay with essentially the same title scribbled on the University of Iowa classroom blackboard: "Is Science too Big for the Scientist?"

Some scientists did feel responsible for NASA, just as society held them responsible for the atomic bomb. The best they could hope for was to make clear a "distinction between stunts and science." Apollo, in particular, was held up as a prime example of "what they felt was a confusion between public relations and scientific investigation."[83] Anxiety over being drawn into such programs, something clearly desired by NASA and Berkner, gave rise to a collective paranoia. There was a "suspicion that the meeting had been convened for brainwashing, and that it was succeeding." Porter tried to explain the paranoia as a reflection of the scientists' sense of responsibility, but he was unable to get around his own problems with "big science." He also realized that many at Iowa that summer criticized what they thought was a strong "in-group" dominance in the proceedings, if not in the actual allocation of payload space and launch priorities. The elevated scientific status of instrument makers and providers, or of NASA scientists turned NASA bureaucrats, was debated throughout the summer. One cynic summed up the behavior of both the NASA and SSB administrators of science attending one working group: "The big wheels rolled in, rolled around, and rolled out."[84]

No doubt there were insiders and outsiders. Space science preceded NASA by over a decade. Those who had established successful contacts with the Office of Naval Research or the Air Force Cambridge Research Laboratory, had quickly moved along with NASA and had become familiar with the problems and complexities of launching instruments on balloons, rockets, and then early satellites. These pioneers naturally drew upon their extensive contacts and benefited from a deep social conditioning that dated literally from World War II. They knew that educated compromises with government bodies had to be made, but "tensions between the missions of the sponsors and the perceived needs of science" continued to be felt even though one of the goals of the planners of the summer study was to bring the two worlds closer together.[85]

On the positive side, Porter observed that the summer study was far from a "tangle of tensions" and some important work was being done to achieve a consensus about NASA's science programs. He felt that the scientists had learned more about NASA, and NASA learned more about the wishes and criticisms of scientists. "There also was a commitment to look at the NASA enterprise as science and thus, inevitably, at American science at large, and to help steer its development. It would be almost impossible to count the man hours which went into that painfully detailed process."[86]

The process did not end with Iowa; in fact, the Summer Study was a watershed for defining and articulating the problems the scientific community would have with NASA in following years. Even though a majority of its programs found general favor,[87] NASA's detailed design and execution of many ongoing ventures drew the criticism of scientists at Iowa. Driven by the fears and impressions reported by William Porter, this criticism kept the many issues addressed there very much alive in the next several years. Some of those most experienced with NASA and committed to its programs, such as Iowa Summer Study host James Van Allen, long appreciated that NASA was a "massive steamroller" that required diligent oversight by the scientific community.[88] Ever diligent was Philip Abelson, who kept his finger on the pulse of his readers' concerns and communicated them as editor of the influential American Association for the Advancement of Science (AAAS) publication, *Science*.

In the wake of the Iowa Summer study and the continuing skepticism over NASA's policies and programs, many congressmen became sufficiently curious about the rationale behind Apollo to convene special hearings before the Senate Committee on Aeronautical and Space Sciences in June 1963. Held as part of the appropriations process, they hoped to obtain scientists' testimony on the goals of the space program. Prior to Webb's NASA reorganization, some prominent scientists publicly supported the policies set forth by NASA's man-in-space program. But by the time of the June 1963 hearings, some among the most prominent argued that military advantage and national prestige would be preserved on the playing fields of space only if scientific goals were integral to the space program. Others argued as well that, although national prestige and a sense of global technological sovereignty would be achieved by winning a race to the moon, the meaning of that win would remain hollow unless the flights served the higher purpose of science. The most diplomatic among them pointed out that scientific goals could not justify Apollo, but nonetheless, Apollo should have them. Echoing Killian's first point, that man's "compelling urge to discover" was one factor used to justify space exploration, Princeton astronomer Martin Schwarzschild stated that the idea of a man landing on the moon would stimulate a new vitality sorely needed by a soft, lackluster American society that had lost the wonder and initiative that had made it great. Apollo was the most likely way to restore the pioneering spirit.[89]

At the outset of the June 1963 Senate hearings, "Scientists' Testimony on Space Goals," Senator Clinton B. Anderson charged each of the testifying savants to examine the veracity of the statement, "The expansion of knowledge and collection of scientific data are among the many reasons why this Nation has embarked upon a program designed to ensure preeminence in space." Beyond looking for confirmation of NASA's rhetoric, Anderson wished to hear the scientists' opinions on how "(1) to increase the amount of scientific return and (2) to insure better utilization of our scientific and technological resources."[90] In short, the scientists were called to Washington to discuss what Apollo could do for science and whether it was doing this properly and efficiently. "Human adventure" was not to be an issue, although some were happy to voice their opinions on that too in the Senate's public arena.[91]

In 1963, Philip Abelson was one of NASA's most vocal critics, and carried his concerns into the Senate appropriation hearings. He argued that the scientists who supported the NASA program were in large part financially connected to it. Indeed, Abelson was the only one to testify who was not somehow involved in some NASA-supported program or mission.[92] Harold Urey, Simon Ramo, Polykarp Kusch, Colin Pittendrigh, Fred Seitz, Lloyd Berkner, Lee DuBridge, Harry Hess, and Martin Schwarzschild all provided testimony from their different perspectives as representatives of the National Academy of Sciences Space Science Board, the President's Science Advisory Committee (PSAC), their home institutions, and the disciplines of geology, chemistry, physics, biology, and astronomy. Each was asked to compare the relative values to science of instrumented expeditions and manned expeditions, as well as provide answers to a number of directed questions from the senators.

Only Abelson argued that man was a poor scientific instrument, reflecting the opinion of a straw poll that drew 110 responses, which overwhelmingly argued against putting man in space for science. When questioned by Senator J. Howard Edmondson, Abelson admitted that he did not object to the manned space program per se, but to "the emphasis and glamour that has been attached to it," and the lack of emphasis on the unmanned scientific program.[93]

All of the others who testified admitted that man on the moon had important prestige value, though none argued that man was needed to conduct science on the moon. But there was a wide spectrum of opinion about man's value as an observer. Most of those testifying argued that instrumented landings had to precede manned landings. However, although Nobel laureate chemist Harold Urey admitted that science could not justify Apollo, he still felt that an astronaut would have the best chance of recovering the best samples.[94] Urey, in fact, chided his physicist colleagues who advocated only instrumented landings for being insensitive to the needs of chemists: "The chemists like to have their hands on matter."[95] Unlike the geologists, Urey did not feel that a trained geologist had to land on the moon to collect the best samples to test his theory of the origin of the moon. Since he believed that the lunar surface was a huge undifferentiated chunk of original material, any nice rock would probably do. Still, when no geologist was selected for the 1969 *Apollo 11* landing, Urey mused to a *Forbes* reporter: "I wish I could go rock hunting with the astronauts this month."[96]

To Urey, man on the moon was inevitable since exploration was an "innate" part of the human condition to do the impossible.[97] Columbia physicist Polykarp Kusch agreed, but added that exploration should not replace more pressing social needs. Kusch's call for deliberation and possibly a slower pace for manned flight was overshadowed by Colin Pittendrigh, professor of biology at Princeton, who countered that much could be learned about human physiology from these flights. He criticized Abelson's and Kusch's conservatism, which he thought would stifle the challenge, "without which science will dry up."[98]

The scientists' testimony covered a wide range of opinion, did little to convince or dissuade Congress that science was an important stimulus and rationale behind Apollo, and demonstrated once again, as in Iowa, that scientists on the Space Science Board are hardly of one mind. Geologists, geochemists, and biologists tended to appreciate the role of man in space as instrument, observer, and test object more than astronomers and physicists did, some of whom argued that man's presence would alter or obscure the very environment they hoped to study.[99] From Lloyd V. Berkner's argument that science was the only proper motivation for Apollo, to Martin Schwarzschild's separation of the two motives of scientific exploration and the uprising of the American spirit, to Abelson's criticisms and Kusch's social concerns, the senators had a great deal to digest if they cared to. Few really cared to, and the oversimplification of Apollo as a scientific mission, clearly the image sought by NASA, by most members of the committee, and by many of the scientists themselves, persisted.

As the 1960s wore on, many scientists lost what faith they initially had in Apollo as a scientific mission, or of their value as advisers. NASA's "scientist-astronaut" program is a good example of how NASA dealt with the recommendations of scientific panels. The 1962 Iowa Summer Study demanded such a program, and NASA obliged, hoping "to mollify some of the scientific grumblers and to strengthen its ties with the scientific community by emphasizing Apollo's potential contributions to science . . ."[100] The first generation of scientist-astronauts joined NASA in 1965, but were kept on the shelf. Brian T. O'Leary left the program in frustration, but others such as the astronomer Karl Henize and the geologist Harrison Schmitt persisted.[101] Schmitt was the only scientist-astronaut ever to reach the moon, on *Apollo 17*, the last of the landings in 1972. Henize had joined with O'Leary, but waited 18 years before he rode into space aboard *Challenger* to operate *Spacelab 2* in July 1985.[102]

In the past 25 years, the scientific community has become increasingly critical of NASA policy regarding how space science is to be conducted.[103] R. Cargill Hall recently observed, "During the course of Project Apollo, American scientists found themselves all but excluded from a major government-sponsored technological enterprise."[104] Yet scientists did little en masse in the early 1960s to dispel the myth of Apollo; in fact, some of them, especially those who found themselves working for NASA either through employment or aided by its contracts and grants, actively argued for the necessity of a human presence in space, even though, as in the ballooning era, few would have initiated a manned program themselves.

Apollo Science

When *Apollo 11* landed on the Sea of Tranquility on July 20, 1969, the front page of the *New York Times* talked of little else. Beneath the next day's headline "Men Land on Moon" and the photographs, framed by the poetry of Archibald Macleish and narrative on the flight by John Noble Wilford and above Walter Sullivan's byline, ran another large title: "Boulders May Prove Scientific Boons." The very rocks that the astronauts had tried so hard to avoid, for good reason, might in fact be clues to the origin of the moon and solar system, as well as to the differentiation of the lunar regolith. The rocks, thought Sullivan, might also provide shelter for the delicate seismometers the astronauts would leave on the lunar surface which might be damaged when the Lunar Module Ascent Stage blasted off for orbit and home.[105]

Enough science rode to the moon with Apollo to justify hearty enthusiasm in scientific circles. The $25 billion Apollo program included some

$63 million (as of 1978) for the analysis of lunar samples alone.[106] Stephen G. Brush, refining Ian Mitroff's 1974 study of lunar scientists, observed that continued Apollo data analysis increased knowledge of the moon significantly, and, along with other types of evidence (mainly from unmanned planetary probes), promoted greater consensus among scientists over theories for the lunar origin.[107] But, as in the balloon era, when Pfotzer and Regener's discovery of the cosmic-ray secondary maximum reduced Swann's Explorer efforts to the status of confirming observations, several scientists pointed out in the Apollo era that the manned flights mainly verified and refined the observations made possible by Surveyor's direct contact with and analysis of the lunar surface. Surveyor found that the lunar regolith was chemically homogeneous; variations were due to the "conscious efforts by the astronauts to bring back unusual samples." One significant variation, found by *Surveyor 5* and confirmed by *Apollo 11*, was that the Sea of Tranquility is rich in titanium. Apollo also confirmed that the lunar basalts were generally similar to terrestrial basalts. After 1970 the three unmanned Russian landers in the Luna series (starting with *Luna 16*) returned lunar samples by an automated system that further strengthened this view, which suggested that the basalts arose from magmatic processes, and hence were evidence that the surface of the moon was composed of chemically differentiated material, in opposition to Urey's prediction.[108]

An assessment of Apollo as a scientific program, however, requires not so much a recitation of results as an evaluation of David Knight's criteria (see Introduction) for determining if an expedition was scientific, or was designed to be useful to science. Were these standards satisfied by Apollo as they were by the National Geographic balloon flights? Yardsticks included the design of each flight, site for launch (or landing in the case of Apollo), the size and nature of the crew, the character of the scientific instrument array, the planned and actual flight profile, and the means used to organize, analyze, and publish the data.

We have already touched upon crew choice: American policy remained unchanged from the 1930s, only military test pilots on active duty flew in the early 1960s, and those who were sent into orbit and eventually to the moon, with the exception of Harrison Schmitt in 1972, were also military.[109] And even though there had been considerable attention early in the program to determining the most scientifically interesting spots to land on the moon, when the sites were selected, the primary criteria were crew safety and accessibility.[110] Favored were relatively uninteresting open spaces, and frowned upon were the geologically interesting lunar highlands. As in the 1930s, many of the instruments were automatic; some, such as the seismic probes and laser ranging experiments, were designed to be left on the moon by the astronauts for long-term observations.

There was, however, considerable support for data analysis and reduction, seminars, symposia, and post-flight research. Scientists wanted this support, and NASA at first provided it, but many became disappointed by NASA's reluctance to do more, or to continue adequate support levels for a satisfactory amount of time. Just the National Geographic Society, urged by Lyman Briggs, organized and published two long scientific and technical reports after Explorer, but failed to print enough copies for wide dissemination, NASA was criticized for spending an inappropriately small amount on data reduction and analysis over too few years compared with the overall cost of the program. Certainly the Lunar Receiving Laboratory served the interests of over 700 scientists, and many annual conferences were held at the Johnson Space Center to review results, but the short time-scale for this support and the abrupt cessation of most of the funding for data analysis when the flights were canceled left many scientists dissatisfied.[111]

Ultimately, however, the public spectacle of Apollo washed over the criticisms of scientists. There was much to be celebrated in the name of science. Possibly the most widely read articles on Apollo science appeared in the *National Geographic* magazine. To the *Geographic, Apollo 14* touched down "for the most scientifically challenging mission thus far," seeking clues to the origin of the earth and solar system.[112] *Apollo 15* returned even more provocative material: a wealth of full-color images from a dune buggy roving around craters and ejecta. The text burst with excitement for the mission and the scientific return, and clearly linked Apollo to science.[113] The impression the *Geographic* left, recalling the scenario long ago suggested by von Braun and Ley, was that scientists had struck a bonanza: "For a long time to come, scientists will be wresting valuable knowledge about the moon, the sun, and the earth from the splendid achievements of *Apollo 15* and its predecessors."[114] The impressions left by these well-crafted scenarios was that scientists stood ready to take advantage of the flood of information returned from the moon. Although many scientists indeed were "on tap" readying their laboratories for the rocks and numerical data, most remained diffident that science had not been "on top" of the process that brought those returns at such a high price.

Thus we return to the perpetual question of whether a human observer and operator are essential for scientific spaceflight, to look for parallels between the 1930s and 1960s. Just as Regener's balloonsonde payloads matched contemporary manned flight payloads in scientific return, Surveyor and Luna experiments rivalled those of Apollo. When Apollo was first planned, automata were not yet available that could do what man could do. Such was the case when Auguste Piccard first made his proposals in 1929. But in both eras, the automata soon became available. By the mid- to late 1960s, considerable improvements in robotics, computer control, attitude stability, pointing accuracy and two-dimensional electronic and solid state detection devices both promised and delivered instrumented systems that reduced the need for an on-site astronaut. Although NASA supported ever-larger and more sophisticated instrumented orbiting craft, planetary landers, and deep probes to study the sun, planets, their satellites, the earth from space, and the whole of the accessible universe, its top priority continued to be spectacular engineering feats making possible human adventure. Yet the selling of Apollo as a scientific enterprise, as with the selling of the manned flights of the 1930s, remained NASA's public relations strategy long after the last astronaut returned from the moon. Although in both eras an on-site human presence was never essential to conduct scientific observations, it was necessary to satisfy the demands created by patronage, politics, and national pride. And as far as the science was concerned, in both eras, it gave the crews something to do.

Notes

1. Auguste Piccard (1933d), p. 353.
2. Webb (1964), p. 3. Copy in Manned Spaceflight, Pros and Cons Folder, NHO.
3. Robert Gilruth oral history interview, 14 May 1986, GWS/NASM, pp. 83–84rd.
4. Maravelas (1980), p. 16.
5. J. F. Kennedy, "Special Message to the Congress on Urgent National Needs," in Senate Committee on Aeronautical and Space Sciences, "Documents on International Aspects of the Exploration and Use of Outer Space, 1954–1962," 88th Congress, First Session, 1963. p. 204. Quoted in Logsdon (1970), p. 2, and McDougall, (1985), pp. 302–303.
6. On national policy leading to Apollo, see: Logsdon (1970); and McDougall (1985).
7. Quoted in: McDougall (1985), p. 303.
8. Needell (1987b), pp. 108–109. The role of new observing techniques in astronomical discovery, or their absence, especially limitations on access to the full electromagnetic spectrum, has been examined in Harwit (1981), chap. 3.
9. Smith (1983), pp. 177–209; 193.
10. New York Times, 25 April 1986, quoted in Nelkin (1987), p. 137. On myth-making by NASA, see McDougall (1985), pp. 304–305.
11. Seamans (1961), p. 50.
12. Ibid., p. 51.
13. Ibid., pp. 57–58.
14. Odishaw (1962), p. xii.
15. On Berkner's success in overcoming the negativism of many board members, see McDougall (1985), pp. 315–316.
16. Needell (1987a), p. 264.
17. Needell (1987a), p. 284.
18. Odishaw (1962), p. 159. The SSB's "if and when" qualifier was usually dropped by the news media. Thus it appeared that the board had given its unqualified endorsement and that the only dissenters lay outside the academy's walls. See "Scientists Warn on Trips to Moon," New York Times, 7

August 1961. The *Washington Post* did include the qualifier, but its overall treatment was far more positive than that of the *Times*. See "Scientists Endorse Exploration of Space," *Washington Post* 7 August 1961. Space Science Board Folder 1961, NHO.

19. Quotation from Hall (1977), p. 113. See also McDougall (1985), pp. 315–316. On Webb's restatements of the SSB position, see "Manned Spaceflight Pro/Con" and NASA Public Affairs News Release folders, 1961–1962, NHO.

20. Merle A. Tuve to Detlev W. Bronk, February 26, 1962. Cabinet #3025 File 75 "NAS/NRC" Department of Terrestrial Magnetism.

21. Ibid.

22. Ibid., p. 2.

23. Homer E. Newell, "NASA and Space," in Odishaw (1962), p. 178.

24. Hall (1988), p. 6. A small but vocal segment of the scientific community was very concerned about the separation of the offices. A few perceived the move as an attempt to isolate science from the mainstream of NASA's programs, whereas others saw science being pitted against three far larger and more central engineering-dominated offices in NASA. See Lloyd V. Berkner Papers, SSB General Correspondence folder 1962, box 67. Impressions of Newell's discomfort can be found in Leo Goldberg oral history interview, 16 May 1978, SHMA/AIP, p. 143.

25. "Cost of Space Exploration," *Science 133* (30 June 1961), p. 2055.

26. Joseph Tatarewicz has studied how the community of planetary astronomers reacted to envoys from NASA at this time. See Tatarewicz (1984).

27. Newell (1963). See also NASA News Releases, 7 September 1962 (B. Holmes), 1 November 1962 (Newell), 28 March 1962 (Newell), Manned Spaceflight Pro/Con Folder, NHO.

28. Newell (1963), p. 5. See also NASA News Release, 5 November 1962, p. 3, Manned Spaceflight Pro/Con Folder, NHO.

29. Jastrow and Newell (1964), pp. 10–11.

30. James E. Webb to Robert Gilruth, 26 February 1963, Manned Spaceflight Pro/Con Folder, NHO.

31. Ibid.

32. See "Statement of Mr. James E. Webb . . ," 10 April 1961; "Press Conference on Russian Space Shot," 12 April 1961, No. 61-76; 5 May 1961, No. 61-97; "Remarks by James E. Webb . . ," 26 May 1961, No. 61-114; "Commencement Address . . ," 7 June 1961, No. 61-124; "Statement by James E. Webb . . ," 20 June 1961, No. 61-134; "Statement of James E. Webb . . ," 21 June 1961, p. 4. Public Affairs Releases file, 1961, NHO.

33. James Webb to the President, 30 November 1962, Kennedy Correspondence File, NHO.

34. "Commencement Address by James E. Webb . . ," 7 June 1961, NASA News Release 61-124, p. 8, Public Affairs Releases folder, May–June, 1961, NHO.

35. Nieburg (1966), p. 11.

36. Van Dyke (1964).

37. Logsdon (1970), pp. 167–168, quoting Alton Frye, "Politics—The First Dimension of Space," *Journal of Conflict Resolution 10* (March 1966), p. 105. To Logsdon, the "decision [to go to the moon] emerged from priorities assigning higher importance in the American decision-making process to the broad foreign policy goal of global preeminence than to the domestic

goal of improving the quality of our national life." On Webb's version of prestige, correctly described by Logsdon, see "Press Conference On Russian Space Shot," NASA News Release 61–76, p. 11; "Commencement Address by James E. Webb . . ," 7 June 1961, pp. 7–8, NASA News Release No. 61–124, NHO.

38. "Cost of Space Exploration," *Science 133*, (30 June 1961), p. 2055.

39. Nelkin (1987), pp. 90–91.

40. President's Science Advisory Committee, "Introduction to Outer Space," *Congressional Quarterly Weekly Review 16*, (January–June 1958) pp. 421–423. Quoted in Smith (1983), p. 198ff.

41. Clarke (1947), p. 6.

42. See Ley (1962), p. xiii. In a later edition, to introduce Ley's book the publishers reprinted excerpts from a review that appeared in the astronomical magazine *Sky & Telescope* that lauded the descriptions of spaceflight, and "the 1970 moon shot" by "a popular writer of science."

43. Pendray (1945), pp. 226–227.

44. Ley (1962), pp. 97, 116; von Braun, in Rabinowitch and Lewis (1969), p. 178.

45. Ernst Stuhlinger, "Apollo: A Pattern for Problem Solving," chap. 15 in Rabinowitch and Lewis (1969), p. 195.

46. See Hall (1977), chap. 7, pp. 112–123; McDougall (1985), pts. III, V.

47. Hall (1977), p. 113.

48. Hall (1988), p. 6.

49. Ivan Bekey, NASA Office of Space Flight, quoted in Ibid., p. 9.

50. Hall (1977), p. 128. One of those who advocated reducing the scientific agenda on Ranger after the first flights failed was the astronomer Martin Schwarzschild, who as a member of the PSAC panel looking at NASA programs, felt that until the Rangers were working properly, the emphasis should be on getting them to the moon. He recalls that this position required considerable "civil courage." Martin Schwarzschild oral history interview, session IV, 19 July 1979, SHMA/AIP, pp. 176–177.

51. Hall (1977), p. 128.

52. Ibid., p. 114.

53. Ibid., pp. 282–284.

54. Van Dyke (1964), p. 92.

55. Leo Goldberg oral history interview, 22 February 1983, SAOHP/NASM, pp. 70–71rd. On Sugar Grove, see McClain (1960); and Needell (1987a), p. 281. The result of this effort has been a series of 10-year plans in astronomy coordinated by the National Academy of Sciences.

56. Leo Goldberg, review of Newell (1980), in *Sky & Telescope 62*, no. 5 (November 1981), pp. 474–475; 475.

57. Eugene N. Cameron, quoted in Brooks, Grimwood, and Swenson (1979), p. 127.

58. See Lloyd V. Berkner to William E. Porter and others, 27 March 1962; Lloyd V. Berkner to Detlev Bronk, 5 January 1962, p. 3, SS-SSB General Correspondence, 1962 folder, box 67, Berkner Papers, LC.

59. Newell (1980), p. 208.

60. Merle A. Tuve to Detlev W. Bronk, February 26, 1962. Cabinet #3025 File 75 "NAS/NRC" Department of Terrestrial Magnetism.

61. Newell (1980), p. 209.
62. Lloyd V. Berkner, "Closing Remarks by Lloyd V. Berkner..," 9 August 1962, SSB-SS Folder, box 69, Berkner Papers, LC.
63. [National Academy of Sciences-National Research Council] (1962), "Introduction and Summary," p. 1.21.
64. Ibid., p. 1.22.
65. Ibid., p. 1.22.
66. Ibid., chap. 11, p. 11.2.
67. Ibid., chap. 11, p. 11.3.
68. Ibid., chap. 11, p. 11.3.
69. James Van Allen oral history interview no. 7, 28 July 1981, SAOHP/NASM, pp. 313–316.
70. [Working Group 8/19], "Report on Man as a Scientist in Space Exploration," 3 August 1962, p. 2, SSB-SS "Man as a Scientist in Space Exploration" Folder, box 68, Berkner Papers, LC. Reprinted in part in chap. 11 of [National Academy of Sciences-National Research Council] (1962), pp. 11.1–11.2.
71. Leo Goldberg oral history interview, 16 May 1978, SHMA/AIP, p. 126.
72. See [National Academy of Sciences-National Research Council] (1962), pp. 2.11–2.12; and Nancy Roman oral history interview, 19 August 1980, pp. 49–51, and John Naugle oral history interview, 20 August 1980, pp. 136–139, SHMA/AIP; James Van Allen oral history interview no. 7, 28 July 1981, SAOHP/NASM, pp. 317–321.
73. Ibid. This attitude persisted through the 1960s, see Martin Schwarzschild oral history interview session 4, SHMA/AIP, 19 July 1979, p. 181.
74. L. V. Berkner to James Webb, 31 March 1961; "Man's Role in the National Space Program," 7 August 1961, p. 1, quoted in Newell (1980), pp. 208–209.
75. For general impressions of the growing critical attitude of scientists toward NASA's man in space program, see "Manned Space Flight Pro/Con" Folder, NHO; and Newell (1980), chap. 12. For continuing problems planning for the post-Apollo Skylab program, see Leo Goldberg oral history interview, 22 February 1983, SAOHP/NASM, pp. 79–83rd.; and Leo Goldberg, review of Newell (1980), in *Sky & Telescope 62*, no. 5 (November 1981), pp. 474–476.
76. Van Allen (1983), p. 74. In chap. 8, Van Allen discusses how the *Argus* tests improved knowledge of the magnetosphere itself.
77. On astronomers' reactions to Westford, see Leo Goldberg oral history interview, 16 May 1978, SHMA/AIP, p. 89. On Argus, see Van Allen (1983), chap. 8. Among other satellites, the British *Ariel 1* was temporarily blinded by the radiation effects from Starfish, but its solar panels and power supply were permanently damaged. See Massey and Robins (1986), pp. 87–88.
78. On demands for better ground-based facilities, see Tatarewicz (1984). Ron Doel is also examining scientists' perceptions and demands during this period.
79. See SSB General Correspondence and Executive Committee Folders for 1962 in box 67, Berkner Papers, LC; see also Leo Goldberg oral history interview, 16 May 1978, SHMA/AIP, pp. 141–144.
80. William E. Porter, "The Context of the Space Science Summer Study," Special Report, 4 August 1962, SSB-SS "Agenda and Final Briefings" Folder, box 68, Berkner Papers, LC.

350   11. Anticipating Apollo

81. Quoted in Ibid., p. 2.
82. Merle A. Tuve to Detlev W. Bronk, February 26, 1962, p. 2. Cabinet #3025 File 75 "NAS/NRC" Department of Terrestrial Magnetism.
83. Porter, op. cit., p. 2.
84. Ibid., p. 3.
85. Needell (1987b), p. 95; pp. 105–107.
86. William E. Porter, "The Context of the Space Science Summer Study," Special Report, 4 August 1962, p. 4, SSB-SS "Agenda and Final Briefings" Folder, box 68, Berkner Papers, LC.
87. One of the decisions reached at the Summer Study was that the group would ask either NASA or the NAS to respond formally to its conclusions and suggestions. In May 1963, J. F. Clark prepared "NASA Comments on 'A Review of Space Research'," pointing out that although there was not "complete agreement" with NASA's "official position," some 63 percent "of the recommendations in the report are included in NASA's planning and thinking." But overall, Clark concluded that there was "general satisfaction with NASA's space policy and program." See J. F. Clark, "NASA Comments on 'A Review of Space Research' draft," 15 May 1963, p. 2, "Space Science Board 1962–1963" Folder, NHO.
88. James Van Allen oral history interview no. 7, 28 July 1981, SAOHP/NASM, pp. 300–301.
89. Comments of Martin Schwarzschild, in [Committee on Aeronautical and Space Sciences] (1963), p. 167.
90. Remarks of Clinton B. Anderson, in [Committee on Aeronautical and Space Sciences] (1963), p. 1.
91. Vernon Van Dyke, reflecting upon the congressional testimony of the many prominent scientists called to Washington, chastised those who failed to appreciate the value of Apollo as a human adventure. Van Dyke chose not to repeat the Senate's charge to its testifiers, nor did his criticism reflect the fact that a number of the scientists did highlight the importance of the human adventure. See Van Dyke (1964), pp. 90–91.
92. [Committee on Aeronautical and Space Sciences] (1963), pp. 136–137. The Princeton astronomer Martin Schwarzschild, for example, was on the PSAC space panel, but also was principal investigator on Project Stratoscope, co-sponsored by NASA, ONR, and NSF, and was in the midst of making the huge 36-inch balloon-borne telescope work properly. He recalls being asked by NASA to testify as an "official witness." See Martin Schwarzschild oral history interview, 18 June 1982, SAOHP/NASM, p. 67rd.
93. [Committee on Aeronautical and Space Sciences] (1963), p. 19.
94. Ibid., p. 57.
95. Ibid., pp. 57–58.
96. Harold Urey, quoted in Brush (1982a), pp. 891–898; 891. On Urey's theories, see also Brush (1982b). Urey's belief in an undifferentiated moon consequently weakened arguments by others, notably Gerard Kuiper and Eugene Shoemaker, that a professional geologist trained as an astronaut had to be among the first crews to land on what they thought was a highly differentiated moon. When scientist–astronauts continued to remain in the background, these scientists became vocal critics of the program, as historian Ron Doel is showing in his continuing studies of the astrogeological research community.

97. [Committee on Aeronautical and Space Sciences] (1963), pp. 50–51.
98. Ibid., p. 75.
99. On the reluctance of astronomers to become involved in space activities, see Tatarewicz (1984), pp. 98, 117.
100. Brooks, Grimwood, and Swenson (1979), p. 179. A somewhat different perspective on the development of the scientist–astronaut corps in given in Atkinson and Shafritz (1985), chap. 4.
101. O'Leary (1970), pp. 12–13.
102. Besides admitting to a "Buck Rogers" complex, Henize was driven by his conviction that "Space is ultimately for scientists." [Henize] (1986), p. 446.
103. See, for instance, O'Leary (1970); Newell (1980), pp. 290–294. More recent examples are Van Allen (1986), pp. 32–39; Waldrop (1987), pp. 426–429; Roland (1987), pp. 104–111; Van Allen (1987), pp. 183–186.
104. Hall (1988), p. 6, rough draft version.
105. Walter Sullivan, "Boulders May Prove Scientific Boons," *New York Times* 21 July 1969, p. 1.
106. [Science Policy Research Division, CRS] (1981), pp. 385, 394. These costs include estimates for ALSEPs (Apollo Lunar Surface Experiment Packages) and the experiments performed in lunar orbit, but do not explicitly include that portion of the NASA fellowship program or NASA support for basic science at universities that may have been associated with lunar science. In contrast, the $2.5 billion Skylab program, a beneficiary of the development costs of Apollo and the result of its Applications Program in the early 1970s, included over $200 million in contracts and services directly related to the chief solar experiments. See Compton and Benson (1983), p. 361; "Appendix B," pp. 377–378.
107. Brush (1988), pp. 239–240; Mitroff (1974), pp. 146–160.
108. Turkevich (1973), pp. 27–34, p. 33.
109. On the astronauts' backgrounds and on-duty status, see [Congressional Research Service] *Astronauts and Cosmonauts Biographical and Statistical Data.* (U.S. Government Printing Office, December 1985).
110. Brooks, Grimwood, and Swenson (1979), p. 363.
111. Newell (1980), p. 294.
112. Hall (1971), p. 136.
113. Weaver (1972), pp. 230, 234.
114. Ibid., p. 264.

CHAPTER 12

Conclusions

Perceived and Created Needs

Although more than 30 years separate the balloon flights of the 1930s from Apollo, they are of one mold in terms of perceived and created needs. When Auguste Piccard first flew in Europe, Erich Regener was just demonstrating to the world of science that unmanned sondes could be made to perform the same observations at altitudes no manned flight could hope to achieve. And when *Apollo 11* touched down on the lunar surface, *Surveyor 3* and *5* had already provided the basic picture of what it would find there.

Yet, both Auguste Piccard and Albert Stevens, as well as Malcolm Ross and advocates in the Apollo era believed sincerely in the scientific value of piloted flight. In Stevens's view, Explorer made it possible to send full-size laboratory instruments into the stratosphere, "to insure the greatest attainable accuracy."[1] His participating scientists bought into this rationale. With little modification, such as adding automatic timers, temperature insulation, and wide-angle light collectors, the same devices used in laboratories or on mountaintops flew on Explorer. They could have flown on unmanned sondes if sufficiently large balloons had been available to loft them. Even Stevens admitted as much when he told Millikan that the huge Explorer balloon might well be available after the first successful piloted flights. Clearly, it was because of patronage, not technological limitations, that piloted missions were the exclusive domain of large instruments. The degree of technical ability required to automate scientific tasks was for the most part available, as Regener had amply demonstrated in the 1930s, and both the Luna and Surveyor missions in the 1960s and 1970s demonstrated in the Apollo era. Although a human presence for in-flight operation and servicing may well have been required at the outset of both the ballooning and Apollo eras, during the actual flights a human presence was required only to gain and maintain large-scale patronage. Unfortunately for the instruments and experiments involved, many scientists argued, priorities also shifted from doing science in the stratosphere to flying people in the stratosphere. As one astronomer remarked during the Apollo era, the demands of human-attended flight made man "more of a handicap technically than an asset. Suddenly the whole project becomes how to send a man up, not how to send a telescope up."[2] Millikan sensed this too in the 1930s when he worried about the added vibration and shadowing that would take place in a human-tended gondola.

We can therefore appreciate why the flight capabilities created by the advocates of manned ballooning or manned spaceflight were not readily embraced by the scientific community. Constructed in the name of science and proclaimed as the servant of science, manned craft did not fulfill a need but created one: They needed scientific patronage to justify their creators' efforts. The most telling indicator for the 1930s was that, even though Auguste Piccard had originally argued that one could trust a laboratory instrument controlled directly by a human observer, and Swann certainly agreed with this assessment later, the majority of the instruments flown in the American piloted ascents had to be made automatic so that the pilots would be free to handle the balloon. When the crew on *Explorer II* was reduced from three to two in order to reduce weight, the scientific instrumentation was cut in half, and experiments that required the greatest amount of onboard attention were eliminated.

Ultimately, simplified and miniaturized versions of Compton's and Millikan's instruments, and even O'Brien's complete spectrograph flew in unmanned sondes to good purpose. Thus one is led to conclude, in retrospect, that the stated scientific purposes of the manned flights were, at the very least, sincere self-delusions on the part of determined advocates. Large laboratory-style instruments could be flown in the manned systems, but the compromises that were required to make the flights possible reduced their value in comparison with simplified single purpose sondes. Of course, the self-delusion of service to science was stimulated solely to justify the funding of manned flight. Advocates of manned flight deftly exploited the mystery of the cosmic ray, as well as the riddle of the origin of the moon as symbols that would capture and hold public attention to what were narrowly focused feats of engineering and daring heroics in search of a mission.[3]

Since none of the piloted ventures in the 1930s and 1940s was to be the start of a continuing program (with the exception of *Helios*), the ambitions for each flight became too great. In their zeal for patronage, flight promoters signed on too many experiments. Thus Millikan became frustrated with Jean Piccard and others who were so eager to carry out what appeared at first to be redundant experiments. As "all-up" ventures, each flight had too many goals: establishing new altitude records; following flight profiles most suited to scientific observation; or, as in *Helios*, testing military systems. These created complex flight-profile problems that were never reconciled. The housekeeping needs of human observers required flight profiles that in many respects were immiscible with profiles suited for scientific observation. The flight agendas thus created left in doubt why and for whom the stratospheric ascents were being made. This lack of a clear mission confused patron and participant alike, who therefore fell back on generalities and images of service to science that were hard to pin down, even by the scientists who allowed themselves to be exploited to rationalize spending a great deal of money, whether for political reasons like Millikan, or out of a sincere fascination with the enterprise, like Swann.

The simple fact that each succeeding manned flight, from *A Century of Progress* through the two Explorers, became more useful and productive seemed to be lost on all but Albert Stevens, who advocated a third Explorer. But the money and energy were not there. Even Swann, the most ardent supporter of manned flight among the physicists, turned away from manned ballooning after Explorer. He admitted that the "normal" mode of doing science by manned flight—carrying his familiar laboratory environment into the stratosphere—had to be put aside for a while. Unmanned sondes and aircraft-borne experiments, in his mind less reliable or less capable, had their advantages in times when little real money was spent on science.

Thus the selling of manned ballooning in the 1930s and 1940s was mission-oriented and short sighted. Each manned ballooning flight was an "all-up" effort to set a record; most patrons, save for some of the National Geographic Society officials, had little appreciation of the highly complex and experimental nature of these flights. When immediate results did not meet expectations, or when, as in the case of *Helios*, the entire system became too cumbersome, support simply vanished. Yet *Helios* can be considered a watershed in the sense that it taught the Navy an important lesson about how to conduct large-scale development programs. Skyhook was born out of *Helios* as a development program that, although it soon became operational for single scientific investigations, remained for the Navy a means of improving its capability for high-altitude long-range flight. Although the pattern developed for Apollo was developmental as well, and provided for ever-increasing sophistication on succeeding flights, the program itself became an "all-up" engineering effort to meet Kennedy's deadline.[4]

Selling the Image of Scientific Exploration

The intellectual backers of "A Century of Progress" equated the power and universal applicability of the scientific method with the image of conquering new territory. They were followed by the National Bureau of Standards seeking to restore its image and retain its best staff by underwriting the interests of the National Geographic Society and the Army

Air Corps. Just as the Apollo program was a way to pull America together in the lingering Cold War, and to restore its "sense of mission,"[5] the A Century of Progress and Explorer flights were ways to signal that better days lay ahead, and they could be reached by nurturing and exploiting "science." In both eras, indeed, throughout recent history, institutional survival depended upon successful myth-making, both to create an acceptable character profile and to assume a safe niche in the social fabric of America. NASA was no more an exception than was the National Geographic Society, the National Bureau of Standards, or the National Academy of Sciences. All had to demonstrate how they were vital to the national interest.

Auguste Piccard and the intrepid flyers that followed him into the stratosphere enjoyed a mystique not unlike that surrounding NASA's Mercury, Gemini, and Apollo astronauts. Whether they flew under explosive hydrogen-filled balloons, or sat on top of precipitous rockets, they were risking their lives for sport, country, and science. Though both the aeronaut and astronaut were honored as national heroes, their historical roles were quite different. More like the Powells, Byrds, or Lindberghs than the Shepards, aeronauts such as the Piccards, Stevens, Settle, or in the post-war era Moore and Ross, were the planners, designers, and entrepreneurs: they were the ones who raised the funds, planned the flight scenarios, and did what they could to convince scientists to hop aboard. They were the intrepid explorers of the romantic past.

Explorers found their way into the stratosphere just at the time that the world's last open frontier was vanishing, and the limits of the terrestrial world were becoming known.[6] The individuals and the institutions examined here were nonetheless fired by the romance of exploration, and their identities depended on performing feats of discovery. Scientific exploration by humans riding stratospheric balloons fit in with the American mystique surrounding exploration. To maintain that mystique, however, people, and not instruments, had to make the journey of discovery. Romantic America held onto its frontier status as tenaciously as it sought to become a new world leader. Thus the explorer as romantic hero had to have a mission worthy of a mature society. Discovering nature's secrets and wonders would bring glory, prestige, and advantage, and provide a legitimate reason for exploration in the 1930s as it would in the 1960s.

Extending boundaries was the explorer's romantic task and the responsibility of enlightened institutions and nations, as so faithfully captured in popular synthetic works that have fascinated the reading public.[7] Scientific substance added to spectacle proved to be a powerful formula for prospective patrons of scientific ballooning. Therefore, although reliable high-altitude unmanned balloons carrying solar, meteorological, and cosmic-ray measuring devices became a reality by the 1930s, it was the race of nations to send man into the stratosphere in the name of science that captured the popular imagination and both corporate and military support.

We have seen that these forces worked for and against one another. As in the Apollo era, the romantic drive to explore, teamed up with the need to reestablish national pride, fostered the idea that serving science was service to the nation. But in both eras, the question of whether man needed to be involved created difficulties and compromises, and at the same time revealed priorities: The reduction of the crew on *Explorer II* and the decision to let Settle fly solo reflect tensions that reappeared in the Apollo era: the goal was to beat the world's record, and to do it safely, quickly and efficiently. Similarly, priorities in both periods were revealed by what happened once the primary goal was obtained: The programs were canceled. Stevens's pleas for an *Explorer III*, especially his argument that those involved should now capitalize on the success of *Explorer II* by mounting a purely scientific mission, fell on deaf ears. Reactions were much the same when geologist Eugene Shoemaker, echoing the concerns of many of his colleagues, criticized NASA for ending Project Apollo prematurely.[8] At least Shoemaker enjoyed the support of a large and vocal scientific community, whereas Stevens was very much alone. Unlike Apollo, which in fact had planned more missions but then canceled them, Explorer was never expected to last beyond one or two flights, and no scientist or community of scientific opinion ever thought that pushing the society or the Army made any sense. As Stevens's closest ally Lyman Briggs well knew, the scientific community had little real influence at a time when the U.S. government, military or civilian, was not a leader among nations in the funding of science.[9]

Scientific Patronage

Without doubt, the prospect of a human presence in an alien world, whether it be the high atmosphere or the surface of the moon, both required and made possible a level of patronage that scientific pursuit alone could not have achieved. Thus two very different but nevertheless interdependent worlds came together in an uneasy alliance that reflected a centuries-old pattern of behavior. Ever since naturalists were placed on military frigates to ply the seven seas, voyages of exploration returned with intelligence of foreign lands, informed by trained scientific eyes. But unlike all past voyages, which had civilian scientists as passengers (with the exception of the Piccard flights in Belgium and the U.S.), the crews and passengers of the 1930s manned balloon flights and those of Mercury, Gemini and Apollo were all active-duty military officers. Although this was not a concern in the 1930s, it became something of a publicity issue in the 1960s when Cold War rhetoric demanded that the NASA venture be wholly civilian.

There are striking similarities between the scientists involved in manned ballooning and in Apollo. In the balloon era, most scientists signed on willingly but few were satisfied with the results, and fewer still encouraged further manned flights. They did want to fly their instruments as high as possible, but became disillusioned when the ballast and lift requirements for maximum altitude caused the balloons to rise too quickly to take good observations en route. Compton's embarrassment after the crash of *A Century of Progress*, and Millikan's retreat from *Explorer II* when the going got rough indicate that both were not interested in developing manned flight for itself. Neither wished to commit intellectual support to a costly effort that put most of the energy and funds into preparing a vehicle for human habitation rather than for scientific experimentation, an attitude we have seen was consistently held in the Apollo era. In both eras, many pushed for instrumented alternatives and concentrated their own resources on them. In the early 1960s, some endorsed Apollo in the hope that this support would help align NASA policy with their own scientific priorities, just as Millikan agreed to participate in Explorer in the hope that he could influence the flight profiles one way or another, or at least garner points with military officers he felt he needed to patronize.

Just as Apollo was a response to Cold War tensions, balloon flight sponsors, and their scientific underwriters, were responding to pressures created by the Depression. They knew that in the depths of the Depression a "revolt against science" was in full swing, and to combat it, they orchestrated events that, if successful, might demonstrate that recovery was at hand. Exploration was a deeply ingrained symbol of the vitality of America; therefore just as Millikan and Compton were duty bound to defend science by working together in Chicago, scientists testified before Congress in the early 1960s that the symbolic value of a manned flight to the moon more than justified the effort and expense in the rebuilding of the character of the nation.

Military Participation

In the United States both the Navy and Army entered stratospheric ballooning quietly, initially as participant and then as patron. The military was always present or had some connection with the flights. At first the Navy was an interested supporter willing to supply manpower and some logistic support, partly for the experience, but soon it became directly involved and gained control in the 1933 flight from Akron. Explorer was distinguished both by its more ambitious battery of scientific experiments, and by the expanding role of the Army—in this case the Army Air Corps, which was eager to better the Navy record. The Army, spurred on by the Navy's accomplishment, entered as a full partner in the Explorer series and came away quite satisfied with its involvement, and, with the world's record safely in hand, saw no reason to pursue the activity.

After the war, the Navy once again took up a leadership role, now in the position of patron in the Helios program born of a very different world. The stratosphere turned from being a curiosity and a place to compete for sporting records to an operational realm ripe for exploitation. Humans would soon be flying and fighting in that realm, and had to be protected against the hazards of supersonic flight as well as be left free to fly and to perform complex tasks. To the military, the gondola became a terrestrial laboratory transformed to operate under stratospheric conditions to test the performance of man. Now interested in all aspects of the high atmosphere—from its radio-reflecting and transmission properties to upper-air wind patterns—and in developing unmanned balloon systems for tactical purposes and exploring the capabilities of long-range photoreconnaissance, the military learned how to be a useful participant and ultimately governing patron.

The Navy responded at first with a mixed and conflicting agenda, quickly learning lessons that gave birth to modern scientific ballooning. The Belgian rationale for establishing the FNRS—that enterprises such as Auguste Piccard's required support beyond that normally assumed by a university—re-emerged full-blown but far from mature in postwar America. The Navy knew it was important to support large-scale ventures in scientific research and engineering development, not only to maintain wartime momentum but to keep intact and in active reserve the best scientific and technical talent. Thus the patronage of large-scale projects that were beyond the scope of normal university funding became an ONR mission, but as yet it and its service counterparts and interservice oversight panels were unable to manage such programs properly; specfically how to separate what were engineering efforts from those professing an immediate scientific return. Helios provided an important and valuable lesson for ONR in the organization and management of civilian scientific research.

The Necessity for Compromise

Throughout this discussion, we have chronicled the many compromises reached over payload capacity, instrument choice, crew safety, launch and landing sites, and the nature of the flight profile. Compromises between the interests of patrons and those hoping to maximize the scientific return became essential factors in the design of all flights, from the Piccard–Compton attempt to the last Apollo landing. Millikan and Compton insisted that flights should take place from a new range of geomagnetic latitude, while scientists participating in experiments aboard both Ranger and Apollo hoped that they might land in selenographically interesting places. In both cases scientists had to back down, first for the publicity value and gate receipts of a launch from Chicago, and second to ensure a safe landing on the moon.

There was also compromise on how best to use available payload capacity. Aerial photography was naturally interesting to the National Geographic Society just as missile drops, reconnaissance photography, and tactical experience in the stratosphere were naturally interesting to the military. A new perspective of the earth—from balloons in the 1930s, from the moon during Apollo, or from large format motion picture cameras flown on the Space Shuttle—provided the payback that was needed to turn public criticism into praise. Along with aerial photography, the results of which could only be appreciated later, live communications with the stratospheric explorers provided an immediacy appreciated by the sponsors of all manned missions. Live radio conversations were promoted, even when they affected the scientific agenda. Although these activities were never pursued by scientists, and tended to interfere with the scientific agendas of many missions, those most enlightened in the world of science realized that, given the true motivations for manned flight in America, these types of activities were actually the most appropriate. Those able to take a populist view knew that the legacy of world's records and landing first on the moon resided ultimately with how the people of a nation felt about itself. More in the realm of the Olympics than of science, manned flight in space indeed had a place in the collective psyche of the nation, if only all parties concerned accepted openly that it was so.

Unfortunately, this has not happened in the United States. Therefore, the most difficult issue to reach a compromise on was, and remains, manned versus unmanned flight. The issue first arose when Auguste Piccard and Erich Regener sought access to the stratosphere in very different ways, and it was first articulated in Millikan's blunt criticisms. The issue was raised again within ONR, was revived by early Space Science Board deliberations, and continues unabated today. Those who saw the need for compromise clearest in the balloon era were either unabashed advocates such as Swann or those who were closely tied to the institutions that stood to benefit from stratospheric spectacles, such as Arthur Holly Compton or Lyman Briggs. Those who hoped that Kennedy would equate space with science were by and large statesmen-scientists like Lloyd Berkner. By their endorsement they hoped to move Kennedy to change space policy to maximize their oversight and control. In addition, Berkner charted an activist role for the SSB as a leader in the salvation of the nation through science. Although his sentiments were echoed in the reports and testimony from the Iowa Summer Study and the 1963 Senate hearings, the many scientists who labored over the question—painfully equating patronage, politics, and national pride with what they really hoped could be accomplished for science by landing on the moon—never reached a clear consensus that influenced, or at the very least, informed national policy.

Coming of age in a nation that had spent its first century supporting the charting and taming of the western wilderness, much of American science turned from expeditions and exploration in the field to the confines of the laboratory as the preferred site for controlled experiments and research.[10] Even though Auguste Piccard and his followers found a means of carrying the laboratory once again into the field, few scientists were anxious to follow this path because it had just barely been broken. Obstacles along the way were high cost, compromise, delay, and failure. In the 1930s and even in the 1960s, large team efforts were still rare in most scientific circles and too consuming and deadening for all but the most ambitious or daring to tackle. Only those most dependent on data from heavy shielded counters, sample returns, or physiological performance, or those comfortable in the corridors of power who hoped to influence the profile of a centralized national system of research required by the space program, remained advocates of manned stratospheric or space flight. Some, like Berkner in the latter group, also saw the space program slipping away from scientific oversight after Webb's reorganization, and for them compromise became a necessity.

Technical compromises were closely related to the issue of human safety, but they had to be made if the ultimate goals of the patrons were threatened. Helium was substituted for explosive hydrogen only with reluctance at first. Thus crews were reduced in size to regain altitude. But these decisions could and did run into trouble when the smaller crews tried to handle the scientific agenda and control the balloon. Thus there were, are, and always will be compromises. Lessons learned in Belgium and Chicago improved the chances for success in Akron and Rapid City, somewhat for the science but mainly for the safety of the passengers. One lesson not learned quickly was that a flight agenda could be overloaded. In manned flight—Explorer, *Helios*, Gemini, or Apollo—the crew could do only so much in the time available. Yet, knowing that there were likely to be only a few flights, mission entrepreneurs such as Albert Stevens, aided and abetted by enthusiastic scientists such as Swann or O'Brien, loaded on as much equipment as possible partly to show how desirable such a platform was for science. As a result, the flights became incredibly complex to design, logistics were a constant headache, and the experiments competed for space and attention. In the end, few returned fully satisfactory data when they did fly.

One specific lesson learned by the Navy, as we noted before, was that manned flight was too problematic to attempt before the new plastic balloon skins were fully understood, or their behavior in clusters was found to be acceptable. Compromises had to be made, and when they were too late in coming, the flight suffered, or in the case of *Helios*, was abandoned for years and not reconsidered until the technology was ready and the scientific agenda rationalized with, or separated from, military priorities. Out of this the Navy learned a critical lesson. Mission-oriented manned flight was not feasible until a thorough development program took the time to provide a reliable means of transportation. The program the Navy developed was sound. Unlike Webb's reorganization of NASA, which was done to meet Kennedy's schedule for a manned landing on the moon, the Navy threw out all schedules for manned flight until they knew what they were doing. Thus although there was nothing intrinsically wrong with the separation of manned and unmanned flight, the reasons for the separation, and where the power centers lay after the separation, defined the fault lines along which the fracture had to take place.

The Science that Was Done

For the most part, the actual data retrieved from the manned flights of the 1930s, although acknowledged to be useful, were not in any way revolutionary and did not solve any of the major problems of the day. There was, of course, no reason to expect that they would, unless one believed the entrepreneurial rhetoric for manned flight that naturally implied that fundamental questions were going to be attacked with instruments far more sensitive, more discriminating, and more reliable than any available automata.

Most of the cosmic-ray physicists used their data to add points to already existing curves, or quietly shelved their data because something had not gone right. Swann's Stösse chamber, removed from *Explorer I*, was taken to Pike's Peak for a productive summer of studying large burst phenomena, but returned nothing when it was taken along on the Piccards' Detroit flight or on *Explorer II*. None of the major discoveries in cosmic-ray physics came from manned flights; and although some came from unmanned sondes, most came from traditional ground-based observations. The latitude and east–west effects were detected on the ground, as were the major classes of new particles. Intensity-versus-altitude curves came mainly from unmanned sonde flights, which also held the most promise for ultimately solving the question of the identity of cosmic-ray primaries. But one should not bluntly compare cosmic-ray results from the six manned flights discussed here to the hundreds of unmanned flights during the same period or to the extended observations made on mountains or on shipboard around the world. The detection of the cosmic processes each searched for required long periods of observation. The trade-off between larger and more reliable instruments and shorter look times simply did not pay off. Each was a valid mode in which to conduct experimentation. One was simply far more expensive, and certainly more frustrating, than the other, and therefore was used far less.

When we consider observations of continuously acting phenomena, such as the physical conditions in the upper atmosphere itself, its temperature, pressure, humidity, and composition, we find, at least in the case of Explorer, a greater degree of success. Again, although the data amounted to additional points on a curve, the accuracy of those points was greater than was previously available. Yet even here, the height restriction of manned flight and the existence of seasonal variables and local contamination by the gondola and the balloon compromised some results such as O'Brien's ozone profiles. O'Brien's results did show that the indirect ground-based methods employing the Umkehr effect yielded very different results. But O'Brien's efforts did no more than support the unmanned observations of Erich Regener since they also differed from the indirect techniques of Götz and his colleagues. Instruments for air temperature, pressure, and composition, either sampling or spectroscopic, and certainly the static biological studies, could all have been flown as unmanned ventures, as many of those who participated in Explorer indicated. Only the experiments directly involving man, such as studies of human physiology during flight or voice communications from the stratosphere, required the presence of man.

Regener's cosmic-ray measurements and his later ozone studies based on the ultraviolet solar spectra he obtained both have remained oftencited milestones in the history of such work, along with the unmanned studies of his associates, such as Pfotzer and the various groups that formed around Compton, Millikan, Swann, and the Minnesota Physics Department. Observational results from balloonsonde flights began to appear in the scientific literature in the 1930s alongside observations taken from sea level and mountaintops. They became regular additions to the database and were regarded as a normal mode of doing science in the 1950s. In contrast, results from manned flights of that era, which were seldom cited, were usually restricted in subsequent review literature to the fact that they were daring, expensive and somewhat dangerous curiosities conducted in the name of science.

Jean Piccard's Dream

And what of Jean Piccard? Although it can be said that his exploits first in Chicago and then at Minneapolis fostered both an awareness of and enthusiasm for high flight in the United States, which was his ultimate legacy, Jean Piccard always lived in the shadow of his brother. His collaborations with Thomas Johnson, John D. Akerman and then with Otto Winzen produced useful techniques and procedures for building and handling unmanned sondes, in keeping with trends at the time, but this normal mode of activity for a scientist was not sufficient for a Piccard. He never gave up his dream of stratospheric flight and prepared more extensive proposals for 31-kilometer flights in the 1950s. Even when he retired at the age of 68, Jean Piccard hoped that somehow he and his wife would be the ones to break the record established by Stevens and

Anderson in 1935. Jean Piccard was as uncompromising in his dream as he was with his hoped-for patrons. Unwavering as both he and his wife were that their flights would be for science, even to the most casual eye it was obvious that "to Adventurer Piccard, no gondola probing the unexplored purple twilight of the stratosphere would be complete without him and his wife in it."[11]

Notes

1. Stevens (1934), p. 398.
2. Martin Schwarzschild oral history interview, 20 April 1983, SAOHP/NASM, p. 29rd.
3. On NASA's engineering proclivity to create complex experimental systems in search of applications, and their efforts to encourage the application of what they develop, see Mack (1983), pp. 8–19; 304–310; Mack (1981), pp. 135–148; Chapman (1967).
4. Brooks, Grimwood, and Swenson (1979), pp. 129–131.
5. Smith (1983), p. 192.
6. Goetzmann (1979), p. 33. See also Dupree (1957), p. 290.
7. The latest and best example of the latter is Boorstein (1983).
8. Newell (1980), p. 293.
9. See Kuznick (1987), p. 128.
10. Dupree (1957), p. 290; Dupree (1965), p. 3.
11. Herbert Morton, "20 Miles Into Stratosphere," newspaper fragment, n.d. (circa 1952), p. 13; Jean Piccard, "Report on the Project for a 100,000 Foot Balloon Flight," manuscript 31 July 1947; Jean Piccard, *Physics of the Stratosphere* (manuscript draft textbook, n.d.), PFP/LC.

Published References

Abbot, C. G. (1939a). "Report on Solar Radiation," in J. Oort, ed., *Transactions of the IAUVI*. (Cambridge, 1939), p. 62.

Abbot, C. G. (1939b). "Report of the Astrophysical Observatory," in *Annual Report of the Smithsonian Institution for 1938*. (Smithsonian, 1939), appendix 8, pp. 99–103.

Abramson, Howard S. (1987). *National Geographic: Behind America's Lens on the World*. (Crown, 1987).

Akerman, John D., and Jean F. Piccard (1937). "Upper Air Study by Means of Balloons and the Radio Meteorograph," *Journal of the Aeronautical Sciences 4*, no. 8 (1937), pp. 332–337.

Alfvén, H. (1985). "Recollection of Early Cosmic Ray Research," in Y. Sekido and H. Elliot, eds., *Early History of Cosmic Ray Studies*. (Reidel, 1985), pp. 427–431.

Allen, Hugh (1949). *The House of Goodyear*. (Hugh Allen, 1949).

Allen, James Albert (1968). *Studies in Innovation in the Steel and Chemical Industries*. (Kelley, 1968).

Alvarez, Luis W. (1987). *Alvarez, Adventures of a Physicist*. (Basic Books, 1987).

Armstrong, Harry G. (1939). *Principles and Practice of Aviation Medicine*. (Williams and Wilkins, 1939).

Arnstein, Karl (1935). "The Design of the Stratosphere Balloon 'Explorer'," in *National Geographic Society–U.S. Army Air Corps Stratosphere Flight of 1934 in the Balloon "Explorer."* (National Geographic Society, 1935), pp. 95–109.

Arnstein, Karl, and F. D. Swann (1936). "The Design of the Stratosphere Balloon, 'Explorer II'," in *The National Geographic Society–U. S. Army Air Corps Stratosphere Flight of 1935 in the Balloon "Explorer II."* (National Geographic Society, 1936), pp. 240–245.

Atkinson, Joseph D., and Jay M. Shafritz (1985). *The Real Stuff*. (Praeger, 1985).

Auger, Pierre (1945). *What Are Cosmic Rays?* Trans. M. M. Shapiro, (Chicago, 1945).

Bailes, Kendall E. (1976). "Technology and Legitimacy: Soviet Aviation and Stalinism in the 1930s," *Technology and Culture 17*, no. 1 (January 1976), pp. 55–81.

Barr, N. L. (1957). *Aero Medical Aspects of the Strato-Lab Program*. (unpublished, May 1957).

Bates, Charles C., and John F. Fuller (1986). *America's Weather Warriors 1814–1985.* (Texas A & M, 1986).

Baxter, James Phinney (1946). *Scientists Against Time.* (Little, Brown, 1946).

Beghin, Pierre. (1938). *Le Fonds National de la Reserche Scientifique et l'Industrie.* (FNRS, 1938).

Benade, J. M., and R. L. Doan (1935). "Apparatus for Transmitting Cosmic Ray Data from Stratosphere," (Abstract.) *Physical Review 47,* (1935), p. 198.

Benn, A. W. (1906). *A History of English Rationalism in the Nineteenth Century I.* (Longmans, 1906).

Bernson, Reysa (1934). "L'Ascension et la catastrophe de l'"Ossoaviakhim-1,' " *l'Aeronautique,* (June 1934), pp. 131–134. Translated as NACA Miscellaneous Paper 39, August 1934.

Beson, Everett E. (1959). "The Balloon-Borne Capsule as a Space Flight Trainer," in *Applications of Manned Stratospheric Laboratories.* (Vitro Laboratories, 1959), pp. 125–129.

Biehl, A. T., R. A. Montgomery, H. V. Neher, W. H. Pickering, and W. C. Roesch (1948). "A New Cosmic Ray Telescope for High Altitudes," *Reviews of Modern Physics 30,* no. 1 (January 1948), pp. 353–359.

Bilstein, Roger E. (1984). *Flight in America 1900–1983.* (Johns Hopkins 1984).

Bonham, L. D. (1936). "Air Conditioning of the Stratosphere Gondola," in *The National Geographic Society–U. S. Army Air Corps Stratosphere Flight of 1935 in the Balloon "Explorer II."* (National Geographic Society, 1936), pp. 259–267.

Boorstein, Daniel (1983). *The Discoverers.* (Random House, 1983; reprint, Vintage, 1985).

Bowen, I. S., and R. A. Millikan (1933). *Physical Review 43,* (1933), p. 695.

Bowen, I. S., R. A. Millikan, and H. V. Neher (1934). "A Very High Altitude Survey of the Effect of Latitude upon Cosmic-Ray Intensities And an Attempt at a General Interpretation of Cosmic-Ray Phenomena," *Physical Review 46,* (1934), pp. 641–652.

Bowen, I. S., R. A. Millikan, and H. V. Neher (1937). "The Influence of the Earth's Magnetic Field on Cosmic-Ray Intensities up to the Top of the Atmosphere," *Physical Review 52,* (July 15, 1937), pp. 80–88.

Braddick, H. J. J. (1939). *Cosmic Rays and Mesotrons.* (Cambridge, 1939).

Bradt, H. L., and B. Peters (1948). "Investigation of the Primary Cosmic Radiation with Nuclear Photographic Emulsions," *Physical Review 74,* no. 12 (December 15, 1948), pp. 1828–1837.

[Briggs, Lyman J.] (1935a). "Report of the Joint Board of Review Appointed to Investigate the Causes of the Accident to the Balloon 'Explorer'," in *National Geographic Society–U.S. Army Air Corps Stratosphere Flight of 1934 in the Balloon "Explorer."* (National Geographic Society, 1935), pp. 71–82.

Briggs, Lyman J. (1935b). "Laboratories in the Stratosphere," *Scientific Monthly 40,* (April 1935), pp. 295–306.

Briggs, Lyman J. (1936a). "Summary of the Results of the Stratosphere Flight of the 'Explorer II'," in *The National Geographic Society–U. S. Army Air Corps Stratosphere Flight of 1935 in the Balloon "Explorer II."* (National Geographic Society, 1936), pp. 5–12.

[Briggs, Lyman J.] (1936b). *Bulletin of the American Physical Society 11,* no. 2 (April 13, 1936), pp. 23–24.

Brooks, Courtney C., James M. Grimwood, and Loyd W. Swenson, Jr. (1979). *Chariots for Apollo.* (NASA, 1979).

Brush, Stephen G. (1982a). "Nickel for Your Thoughts: Urey and the Origin of the Moon," *Science 217,* (3 September 1982), pp. 891–898.

Brush, Stephen G. (1982b). "Harold Urey and the Origin of the Moon: The Interaction of Science and the Apollo Program," *Twentieth Goddard Memorial Symposium.* (NASA/Goddard Space Flight Center, March 17–19, 1982).

Brush, Stephen G. (1988). "A History of Modern Selenogony: Theoretical Origins of the Moon, From Capture to Crash 1955–1984," *Space Science Reviews 47,* (1988), pp. 211–273.

Burgess, Robert F. (1975). *Ships Beneath the Sea.* (McGraw Hill, 1975).

Burnham, John C. ed. (1971). *Science in America: Historical Selections.* (Holt, Rinehart and Winston, 1971).

Burrows, William (1986). *Deep Black: Space Espionage and National Security.* (Random House, 1986).

Capshew, James (1986). "Engineering a Technology of Behavior: B. F. Skinner's Kamikaze Pigeons in World War II," manuscript draft.

Carmichael, Hugh (1985). "Edinburgh, Cambridge and Baffin Bay," in Sekido and Elliot (1985), pp. 99–113.

Chapman, R. L. (1967). *A Case Study of the U.S. Weather Satellite Program: The Interaction of Science and Politics.* (Syracuse University Ph.D. dissertation, 1967).

Chapman, Sydney (1943). "The Photochemistry of Atmospheric Oxygen," in W. B. Mann, ed., *Reports on Progress in Physics 9,* (1943), pp. 92–100.

Chapman, Sydney, and J. Bartels (1940). *Geomagnetism.* (Oxford, 1940), Chaps. 21–22.

Clarke, Arthur C. (ca. 1947). *Interplanetary Flight.* (Harper, ca. 1947).

Clarke, E. T., and S. A. Korff (1941). "The Radiosonde: The Stratosphere Laboratory," *Journal of the Franklin Institute 232,* (1941), pp. 217–238, 339–355.

Coblentz, W. W., and R. Stair (May 1939). "Distribution of Ozone in the Atmosphere," *Journal of the National Bureau of Standards 22,* (May 1939), paper 1207, pp. 573–606.

Cochrane, Rexmond C. (1966). *Measures for Progress: A History of the National Bureau of Standards.* (Department of Commerce, 1966).

[Committee on Aeronautical and Space Sciences, U.S. Senate] (1963). *Scientists' Testimony on Space Goals: Hearings before the Committee on Aeronautical and Space Sciences.* 88th Cong., 1st sess., June 10, 11, 1963.

Compton, A. H. (1932a). "Progress of Cosmic Ray Survey," *Physical Review 41,* (1932), pp. 681–682.

Compton, A. H. (1932b). "Studies of Cosmic Rays," *Carnegie Institution of Washington Yearbook 31,* (1932), pp. 331–333.

Compton, A. H. (1933). "A Geographical Study of Cosmic Rays," *Physical Review 43,* no. 6 (March 15, 1933), pp. 387–403.

[Compton, A. H.] (1934a). "Pilotless Balloons to Explore the Stratosphere," *Science—Supplement 80,* (August 31, 1934), p. 8.

Compton, A. H. (1934b). "Scientific Work in the 'Century of Progress' Stratospheric Balloon," *Proceedings of the National Academy of Sciences 20,* (January 1934), pp. 79–81.

Compton, A. H. (1936). "Cosmic Rays as Electrical Particles," *Physical Review 50,* (December 1936), p. 1130.

Compton, A. H. (1967). *The Cosmos of Arthur Holly Compton.* (Knopf, 1967).

Compton, A. H., and R. J. Stephenson (1934). "Cosmic-Ray Ionization in a Heavy Walled Chamber at High Altitudes," Abstract. *Physical Review 45,* no. 8 (April 15, 1934), p. 564

Compton, W. David, and Charles D. Benson (1983). *Living and Working in Space: A History of Skylab.* (NASA, 1983).

Cornell, Thomas D. (1986). *Merle A. Tuve and his Program of Nuclear Studies at the Department of Terrestrial Magnetism: The Early Career of a Modern American Physicist.* (The Johns Hopkins University Ph.D. dissertation, 1986).

Cosyns, Max (1939). *Bull Techn. Ass. Ing. Ec. Polyt. Bruxelles 35,* (1939), p. 73.

Cosyns, Max, Paul Kipfer, and Auguste Piccard (1933). *Étude du rayonnement cosmique.* (Brussels, 1933).

Crawford, Elisabeth, J. L. Heilbron, and Rebecca Ullrich (1987). *The Nobel Population.* (Berkeley, Calif.: Office for History of Science and Technology, 1987).

Crouch, Tom D. (1983). *The Eagle Aloft.* (Smithsonian, 1983).

Curtiss, L. F., and A. V. Astin (1935). "A Practical System for Radiometeorography," *Journal of the Aeronautical Sciences 3,* (November 1935), pp. 35–39.

Curtiss, L. F., and A. V. Astin (May 1936). "High Altitude Stratosphere Observations," *Science 83,* (May 1, 1936), pp. 411–412.

Curtiss, L. F., and A. V. Astin (1938a). "Cosmic Ray Observations in the Stratosphere," *Physical Review 53,* (1938), p. 23.

Curtiss, L. F., and A. V. Astin (1938b). "An Improved Radiograph on the Olland Principle," NBS Research Paper 1169, (1938).

Davis, Lance E., and Daniel J. Kevles (April 1974). "The National Research Fund: A Case Study in the Industrial Support of Academic Science," *Minerva 12,* no. 2 (April 1974), pp. 207–220.

De Latil, Pierre, and Jean Rivoire (1962). *Le professor Auguste Piccard.* (Seghers, 1962).

De Maria, Michelangelo, and Arturo Russo (1987). "Cosmic Ray Romancing: The Discovery of the Latitude Effect and the Compton–Millikan Controversy," (unpublished draft, April 1987).

DeVorkin, David H. (1980). "An Astronomer Responds to War: Otto Struve and the Yerkes Observatory During World War II." *Minerva 18,* no. 4 (Winter 1980), pp. 595–623.

DeVorkin, David H. (1985). "Electronics in Astronomy: Early Applications of the Photoelectric Cell and Photomultiplier for Studies of Point Source Celestial Phenomena," *Proceedings of the IEEE 73,* no. 7 (July 1985), pp. 1205–1220.

DeVorkin, David H. (1987a). "Organizing for Space Research: The V-2 Panel," *Historical Studies in the Physical and Biological Sciences 18,* no. 1 (1987), pp. 1–24.

DeVorkin, D. H. (1987b). "John Strong's First Aluminized Mirror," *Rittenhouse 2,* no. 1 (November 1987), pp. 1–10.

DeVorkin, D. H. *Origins of Space Science.* (in progress).

Dollfus, Audouin (1950). *Notice Sur Les Travaux Scientifiques.* (Meudon, 1950).

Dollfus, Audouin (1983). "Les Debuts De L'Observation Astronomique en Ballon," *L'Astronomie,* (November 1983), pp. 471–473.

Dupree, A. Hunter (1957). *Science in the Federal Government.* (Harvard, 1957; reprint, Johns Hopkins, 1986).

Dupree, A. Hunter (1960). "Influence of the Past: An Interpretation of Recent Development in the Context of 200 Years of History," *Annals of the American Academy of Political and Social Science 327*, (1960), pp. 19–26.

Dupree, A. Hunter (1965). "Paths to the Sixties," in D. L. Arm, ed., *Science in the Sixties.* (University of New Mexico Press, 1965), pp. 1–9.

Dupree, A. Hunter (1972). "The Great Instauration of 1940: The Organization of Scientific Research for War," in Gerald Holton, ed., *The Twentieth Century Sciences: Studies in the Biography of Ideas.* (Norton, 1972), pp. 443–467.

Dwyer, James F. (1979). "Zero Pressure Balloon Shapes, Past Present and Future," in W. Riedler, ed., *Scientific Ballooning.* (Pergamon, 1979), pp. 9–19.

Ehmert, A. (1953). "The Measurement of Cosmic Ray Time Variations with Ion Chambers and Counter Telescopes," in Gerard Kuiper, ed., *The Sun.* (Chicago, 1953), pp. 711–714.

Emme, E. (1961). *Aeronautics and Astronautics, 1915–1960.* (NASA, 1961).

Encyclopedia Britannica 14 (1929). "Belgium," *Volume 3* pp. 353ff.

Fassim, Gustave, and H. F. Kurtz (1935). "The Stratosphere Spectrographs," in *National Geographic Society–U.S. Army Air Corps Stratosphere Flight of 1934 in the Balloon "Explorer."* (National Geographic Society, 1935), pp. 112–118.

Field, Adelaide (1969). *Auguste Piccard, Captain of Space, Admiral of the Abyss.* (Houghton Mifflin, 1969).

FNRS (1953). *Au Service de la Science 1927–1952.* (Bruxelles, FNRS, 1953).

Forman, Paul (1988). "Social Niche and Self-Image of the American Physicist," (unpublished draft prepared for the 19–23 September 1988 Rome conference, *The Restructuring of Physical Sciences in Europe and the United States 1945–1960.*).

Freier, Phyllis, E. J. Lofgren, E. P. Ney, and F. R. Oppenheimer (University of Minnesota), and H. L. Bradt and B. Peters (University of Rochester) (1948a). "Evidence for Heavy Nuclei in the Primary Cosmic Radiation," *Physical Review 74*, no. 2 (July 15, 1948), pp. 213–217.

Freier, Phyllis, E. J. Lofgren, E. P. Ney, and F. R. Oppenheimer (1948b). "The Heavy Component of Primary Cosmic Rays," *Physical Review 74*, no. 12 (December 15, 1948), pp. 1818–1827.

Friedman, Herbert (1981). "Rocket Astronomy—An Overview," in P. Hanle and V. Del Chamberlain, eds., *Space Science Comes of Age.* (Smithsonian, 1981), pp. 40–41.

Galison, Peter (1987). *How Experiments End.* (Chicago, 1987).

Garber, Sheldon, and Felicia Holton (1960). "Voyage of the 'Valley Forge'," *University of Chicago Reports 10*, (January–February 1960), no. 4; no. 5.

Geiger, Hans, and W. Müller (1928). "Elektronenzählrohr zur Messung Schwächster Aktivitäten," *Naturwissenschaften 16*, (1928), p. 617.

Gish, O. H. (1939). "Atmospheric Electricity," in J. A. Fleming, ed., *Physics of the Earth VIII.* (McGraw Hill, 1939), pp. 149–230.

Gish, O. H., and K. L. Sherman (1936). "Electrical Conductivity of Air to an Altitude of 22 Kilometers," in *The National Geographic Society–U. S. Army Air Corps Stratosphere Flight of 1935 in the Balloon "Explorer II."* (National Geographic Society, 1936), pp. 94–116.

Goetzmann, William H. (1979). "Paradigm Lost," in N. Reingold, ed., *The Sciences in the American Context.* (Smithsonian, 1979), pp. 21–34.

Goetzmann, William H. (1986). *New Lands, New Men.* (Viking, 1986).

Goody, R. M. (1954). *The Physics of the Stratosphere*. (Cambridge, 1954).

Götz, F. W. Paul, A. R. Meetham, and G. M. B. Dobson (1934). "The Vertical Distribution of Ozone in the Atmosphere," *Proceedings of the Royal Society (London), Series A 145*, (July 1934), pp. 416–425.

Gray, James (1954). *Business Without Boundary: The Story of General Mills*. (Minnesota: University of Minnesota Press, 1954).

Grosvenor, Gilbert (1935). "The Society Announces New Flight Into the Stratosphere," *National Geographic, 65* (February 1935), pp. 265–272.

Gutbier, Rolf (1955). "Ansprachen bei der akademischen Trauerfreier für Professor Dr. Erich Regener am 11 März 1955," *Technische Hochschule Stuttgart Reden und Aufsätze 21*, (1955), pp. 37–48.

Hagan, Raymond D. (1986). *The Office of Naval Research: Windows to the Origins*. (ONR, [1986]).

Hall, Alice J. (1971). "The Climb Up Crater Cone," *National Geographic 140*, no. 1 (July 1971), pp. 136–148.

Hall, R. Cargill (1977). *Lunar Impact: A History of Project Ranger*. (NASA, 1977).

Hall, R. Cargill (1988). "Thirty Years Into the Mission: NASA at the Crossroads," (manuscript, 1988). Published as: "NASA: Thirty Years of Space Flight," *Aerospace America 26*, no. 12 (December 1988), pp. 6–9.

Harwit, Martin (1981). *Cosmic Discovery: The Search, Scope, and Heritage of Astronomy*. (Basic Books, 1981).

[Henize, Karl] (1986). "Conversations with an Astronomer–Astronaut," *Sky & Telescope 72*, (November 1986), p. 446–449.

Hess, Victor (1940). *Thought, The Fordham University Quarterly 15*, (1940), p. 225.

Hirsh, Richard F. (1983). *Glimpsing an Invisible Universe, The Emergence of X-ray Astronomy*. (Cambridge, 1983).

Hollinger, David A. (1984). "Inquiry and Uplift: Late Nineteenth-Century American Academics and the Moral Efficacy of Scientific Practice," in Thomas Haskell, ed., *The Authority of Experts*. (University of Indiana Press, 1984), Chap. 5.

Honour, A. (1957). *Ten Miles High, Two Miles Deep*. (McGraw Hill, 1957).

Hulburt, E. O. (1939). "The Upper Atmosphere," in J. A. Fleming, ed., *Physics of the Earth VIII*. (McGraw Hill, 1939), pp. 492–572.

Ingrao, Hector C., and Donald H. Menzel (1964). "The Technique of Stratospheric Balloons and Space Research," in T. M. Tabanera, ed., *Advances in Space Research*. (Pergamon, 1964), pp. 207–229.

Jastrow, Robert, and Homer E. Newell (1964). *Why Land on the Moon?* (NASA, 1964).

Jánossy, L. (1948). *Cosmic Rays*. (Clarendon Press, 1948).

Johnson, T. H. (1932). "Cosmic Rays—Theory and Observation," *Journal of the Franklin Institute 214*, (December 1932), pp. 665–689.

Johnson, Thomas H. (1937). "Radio Transmission of Coincidence Counter Cosmic Ray Measurements in the Stratosphere," *Journal of the Franklin Institute 223*, (March 1937), pp. 339–354.

Kargon, Robert H. (1981). "Birth Cries of the Elements: Theory and Experiment Along Millikan's Route to Cosmic Rays," in Harry Woolf, ed., *The Analytic Spirit*. (Cornell, 1981), pp. 309–325.

Kargon, Robert H. (1982). *The Rise of Robert Millikan: Portrait of a Life in American Science*. (Cornell, 1982).

Kargon, Robert H., and Elizabeth Hodes (1985). "Karl Compton, Isaiah Bowman, and the Politics of Science in the Great Depression," *Isis 76*, (1985), pp. 301–318.

Keen, C. D. (1937). "High Altitude Test of Radio-Equipped Cosmic Ray Meter," *Journal of the Franklin Institute 223*, (March 1937), pp. 355–373.

Kepner, William E., Jr. (1935). "Report of the Commanding Officer of the National Geographic Society-Army Air Corps Stratosphere Flight to Chief of the Air Corps, United States Army, 1 August 1934," in *The National Geographic Society–U.S. Army Air Corps Stratosphere Flight of 1934 in the Balloon "Explorer."* (National Geographic Society, 1935), pp. 26–32.

Kepner, William E., Jr., and John H. Scrivner, Jr. (1971). "The Saga of Explorer I: Man's Pioneer Attempts to Reach Space," *Aerospace Historian 18*, (September 1971), pp. 123–128.

Kevles, Daniel (1974). "Robert A. Millikan," *Dictionary of Scientific Biography 9*, (Scribners, 1974), p. 399.

Kevles, Daniel J. (1975). "Scientists, the Military, and the Control of Postwar Defense Research: The Case of the Research Board for National Security, 1944–1946," *Technology and Culture 16*, no. 1 (1975), pp. 20–47.

Kevles, Daniel J. (1978). *The Physicists: The History of a Scientific Community in Modern America.* (Knopf, 1978; reprinted, Vintage, 1979).

Knight, David (1986). *The Age of Science.* (Blackwell, 1986).

Konecci, E. B. (1959). "Manned Space Cabin Systems," in F. I. Ordway III, ed., *Advances in Space Science 1.* (Academic Press, 1959), p. 175.

Korff, S. A. (1985). "Cosmic Ray Neutrons, 1934–1959," in Sekido and Elliot (1985), pp. 255–261.

Kresser, Theodore O. J. (1957). *Polyethylene.* (Reinhold, 1957).

Krolak, Leo J. (1952). *The Vertical Distribution of Ozone in the Upper Atmosphere.* (Master's thesis, Institute of Optics, University of Rochester, 1952).

Kuznick, Peter J. (1987). *Beyond the Laboratory.* (Chicago, 1987).

LePrince-Ringuet, Louis (1950). *Cosmic Rays.* Trans. Fay Ajzenberg. (Prentice-Hall, 1950).

Ley, Willy (1962). *Satellites, Rockets and Outer Space.* (Signet, revised 1962). 1st ed. 1957/58.

Liebowitz, Ruth P. (1985). *Chronology, From the Cambridge Field Station to the Air Force Geophysics Laboratory 1945–1985.* Special Report no. 252. (AFGL, Hanscom AFB, 1985).

Lincoln, Ben (1933). "Piccard's Ocean Diving Balloon," *Modern Mechanix and Inventions*, (April 1933), pp. 34–37.

Logsdon, John M. (1970). *The Decision to Go To The Moon.* (MIT, 1970).

Lutz, F. W. (1936). "The Aerial Camera Equipment Used on the Flight of 'Explorer II'," in *The National Geographic Society–U. S. Army Air Corps Stratosphere Flight of 1935 in the Balloon "Explorer II."* (National Geographic Society, 1936), pp. 276–277.

McClain, Edward F., Jr. (1960). "The 600-Foot Radio Telescope," *Scientific American 202*, (January 1960), pp. 45–51.

McDougall, Walter A. (1985). *The Heavens and The Earth, A Political History of the Space Age.* (Basic Books, 1985).

McFarland, Marvin W. (1974). "Auguste Piccard," *Dictionary of Scientific Biography 10*, (Scribners, 1974), pp. 597–598.

Mack, Pamela E. (1981). "Space Science for Applications: The History of Landsat," in P. Hanle and V. Del Chamberlain, eds., *Space Science Comes of Age*. (Smithsonian, 1981), pp. 135–148.

Mack, Pamela E. (1983). *The Politics of Technological Change: A History of Landsat*. (University of Pennsylvania Ph.D. dissertation, 1983).

Maravelas, Paul (1980). "Jeannette Piccard Interviewed," *Ballooning*, (July–August 1980), pp. 15–19.

Massey, Harrie, and M. O. Robins (1986). *History of British Space Science*. (Cambridge, 1986).

Mayo-Wells, Wilfred J. (1963). "The Origins of Space Telemetry," *Technology and Culture 4*, no. 4 (1963), pp. 499–514.

Medaris, Maj. Gen. J. B., and Arthur Gordon (1960). *Countdown for Decision*. (Putnam, 1960).

Menzel, Donald H. (1959). "Introduction," *Applications of Manned Stratospheric Laboratories*. (Vitro, 1959), pp. 1–6.

Middleton, W. E. Knowles (1969a). *Invention of the Meteorological Instruments*. (Johns Hopkins, 1969).

Middleton, W. E. Knowles (1969b). *Catalog of Meteorological Instruments in the Museum of History and Technology*. (Smithsonian, 1969).

Miller, Howard (1970). *Dollars for Research*. (University of Washington Press, 1970).

Millikan, Robert A. (1928). "Available Energy," *Science 63*, (1928), pp. 279–284.

Millikan, Robert A. (1935). "The Flight of the 'Explorer' From the Standpoint of Cosmic Ray Intensities," in *The National Geographic Society–U.S. Army Air Corps Stratosphere Flight of 1934 in the Balloon "Explorer."* (National Geographic Society, 1935), pp. 24–25.

Mitra, S. K. (1948). *The Upper Atmosphere*. (Calcutta, 1948).

Mitroff, Ian I. (1974). *The Subjective Side of Science*. (Elsevier, 1974).

Moore, Charles B. (1952). "Plastic Balloons: A Platform for Experiments in the Upper Atmosphere," in Clayton S. White and Otis O. Benson, eds., *Physics and Medicine of the Upper Atmosphere: A Study of the Aeropause*. (University of New Mexico Press, 1952), pp. 395–404.

Moore, Charles B., J. R. Smith, W. D. Murray, et al. (1950). *Technical Reports 93.01–93*. Balloon Group Research Division, New York University, 1948–1950.

Morrill, Claire, and Philip E. Rich, (1933). "Up in the Air," *United Effort 13*, no. 8 (August 1933), pp. 3–4.

[National Academy of Sciences–National Research Council] (1962). *A Review of Space Research*. (NAS/NRC Publication no. 1079, 1962).

Naumann, J. W., and M. Hohenester, eds., (1931). *Professor Piccard's Forschungsflug in die Stratosphäre*. (Haas und Grabherr, 1931).

Needell, Allan A. (1987a). "Lloyd Berkner, Merle Tuve, and the Federal Role in Radio Astronomy," *Osiris, 2nd series 3*, (1987), pp. 261–288.

Needell, Allan A. (1987b). "Preparing for the Space Age: University-Based Research, 1946–1957," *Historical Studies in the Physical and Biological Sciences 18*, no. 1 (1987), pp. 89–109.

Neher, H. Victor (1985). "Some of the Problems and Difficulties Encountered in the Early Years of Cosmic-Ray Research," in Sekido and Elliot (1985), p. 91.

Nelkin, Dorothy (1987). *Selling Science*. (W. H. Freeman, 1987).

Newell, Homer E. (1963). *The Mission of Man in Space*. (NASA, 1963).

Newell, Homer E. (1980). *Beyond the Atmosphere.* (NASA, 1980).

Nieburg, H. L. (1966). *In the Name of Science.* (Quadrangle, 1966).

O'Brien, Brian J. (1925). "A Programme of Research on the Action of Light on Living Matter," *American Review of Tuberculosis 11,* (August 1925), pp. 486–491.

O'Brien, Brian J. (1930). "Energy Distribution in the Ultraviolet Spectrum of Skylight," Abstract, *Physical Review 36,* (1930), p. 381.

O'Brien, Brian J. (1931). "Intermittent Exposure in Photographic Spectrophotometry," *Physical Review 37,* (1931), p. 471.

O'Brien, Brian J. (1936). "Vertical Distribution of Ozone in the Atmosphere," in *The National Geographic Society–U. S. Army Air Corps Stratosphere Flight of 1935 in the Balloon "Explorer II."* (National Geographic Society, 1936), pp. 49–70.

O'Brien, Brian J., Fred L. Mohler, and H. S. Stewart, Jr. (1936). "Spectrographic Results of 1935 Flight," in *The National Geographic Society–U. S. Army Air Corps Stratosphere Flight of 1935 in the Balloon "Explorer II,"* (National Geographic Society, 1936), pp. 71–93.

O'Brien, Brian J., L. T. Steadman, and H. S. Stewart, Jr. (1937). "Measurement of Solar Radiation from High Altitude Sounding Balloons," Abstract, *Science 86,* no. 2237 (November 12, 1937), p. 445.

Odishaw, Hugh T. ed. (1962). *The Challenges of Space.* (Chicago, 1962).

Oehser, Paul H. comp. (1975). *National Geographic Society Research Reports 1890–1954.* (National Geographic Society, 1975).

O'Leary, Brian T. (1970). *The Making of an Ex-Astronaut.* (Houghton-Mifflin, 1970. Repr. Pocket Books, 1971).

Owens, Larry (1985). "Pure and Sound Government: Laboratories, Playing Fields, and Gymnasia in the Nineteenth-Century Search for Order," *Isis 76,* (1985), pp. 182–194.

Paetzold, H. K. (1985). "Erich Regener: A Pioneer of Geophysical Research," in Wilfred Schröder, ed., *Historical Events and People in Geosciences.* (Peter Lang, 1985), pp. 59–63.

Pauly, Philip (1979). "The World and All That is in It: The National Geographic Society, 1888–1918," *American Quarterly 31,* no. 4 (1979), pp. 517–532.

Pendray, G. Edward (1945). *The Coming Age of Rocket Power.* (Harper, 1945).

Pfotzer, Georg (1935). *Zeitschrift für Technischen Physik 16,* (1935), p. 400.

Pfotzer, Georg (1936). "Dreifachkoinzidenzen der Ultrastrahlung aus vertikaler Richtung in der Stratosphäre," *Zeitschrift für Physik 102,* (1936), p. 23.

Pfotzer, Georg (1974). "History of the Use of Balloons in Scientific Experiments," *Space Science Reviews 13,* (1974), pp. 199–242.

Pfotzer, Georg (1985). "On Erich Regener's Cosmic Ray Work in Stuttgart and Related Subjects," in Sekido and Elliot (1985), pp. 75–89.

Philp, Charles G. (1937). *The Conquest of the Stratosphere.* (Pitman, 1937).

Piccard, Auguste (1933a). *Auf 16,000 Meter.* (Schweizer Aero-Revue, 1933).

Piccard, Auguste (1933b). *Au-Dessus Des Nuages.* (Bernard Grasset, 1933).

Piccard, Auguste (1933c). "The Stratosphere," *Proceedings of the Founder's Meeting.* (New York: Institute of the Aeronautical Sciences, 1933).

Piccard, Auguste (1933d). "Ballooning in the Stratosphere," *National Geographic Society 63,* (March 1933), pp. 353–384.

[Piccard, Auguste] (1938). In *J.C. Poggendorff's Handwörterbuch VI.* (Berlin, 1938), p. 2007.

Piccard, Auguste (1950). *Between Earth and Sky*. Trans. Claude Apcher. (Falcon, 1950).

Piccard, Auguste, and Max Cosyns (1932). "Étude du rayonnement cosmique en grande altitude," *Comptes Rendus 195*, (1932), pp. 604–606.

Piccard, Auguste, E. Stahel, und Paul Kipfer (1932). "Messung der Ultrastrahlung in 16,000 M Höhe," *Naturwissenschaften 20*, (1932), pp. 592–593.

Piccard, Jean (1933). "Construction of Welded Gondolas for Stratospheric Balloons," *The Welding Engineer 18*, no. 6 (June 1933), pp. 30–31.

Piccard, Jean (1934). "Stratosphere—Superhighway of the Air," *Rotarian* (February 1934), n.p.

Piccard, Jean (1937). "Why We Explore the Stratosphere," *Minnesota Techno-Log* (November 1937), p. 31.

[Piccard, Jean] (1938). In *J.C. Poggendorff's Handwörterbuch VI*. (Berlin, 1938), p. 2008.

Piccard, Jean, and Harold Larsen (1939). "Improvement in Radio-Sounding Balloons: A Short Cycle Radiosonde," *Review of Scientific Instruments 10*, (November 1939), pp. 352–355.

Piccard, Jean, Harold Larsen, and John Blomstrand (1954). "Thin-Wire Thermometer for Radiosondes," *Review of Scientific Instruments 24*, (October 1954), pp. 959–963.

Primera Ascensión Estratosférica en el Hemispherio Sur. n.d. (circa August 1939). NASM Technical Files "Balloons—Science and Technology," file A2003600.

Proceedings of the Founder's Meeting, Institute of the Aeronautical Sciences. (1933) New York.

Pyne, Steven J. (1986). *The Ice: A Journey to Antarctica*. (University of Iowa Press, 1986).

Rabinowitch, Eugene, and Richard S. Lewis, eds. (1969). *Man on the Moon*. (Harper & Row, 1969).

Raff, R. A. V., and J. B. Allison (1956). *Polyethylene*. (Interscience, 1956).

Regener, Erich (1932a). "Messung der Ultrastrahlung in der Stratosphäre," *Naturwissenschaften 20*, (1932), p. 695.

Regener, Erich (1932b). *Physikalische Zeitschrift 34*, (1932), p. 306.

Regener, Erich (1932c). *Nature 130*, (1932), p. 364.

Regener, Erich, and Victor H. Regener (1934a). "Aufnahmen des ultravioletten Sonnenspektrums in der Stratosphäre und vertikale Ozonverteilung," *Physikalische Zeitschrift 35*, (October 1, 1934), pp. 788–793.

Regener, Erich, and Victor H. Regener (1934b). "Ultra-Violet Solar Spectrum and Ozone in the Stratosphere," *Nature 134*, (September 8, 1934), p. 380.

Regener, Erich (1935). *Nature 136*, (1935), p. 718.

Riedler, W. ed. (1979). *Scientific Ballooning: Cospar Advances in Space Exploration, Vol. 5*. (Pergamon, 1979).

Roland, Alex (1985). *Model Research: The National Advisory Committee for Aeronautics 1915–1958*. (NASA 1985), 2 Vols.

Roland, Alex (1987). "Priorities in Space for the USA," *Space Policy*, (May 1987), pp. 104–111.

Ross, Malcolm D. (1958). "Plastic Balloons for Planetary Research," *The Journal of the Astronautical Sciences 5*, (Spring 1958), pp. 5–10.

Ross, Malcolm D., and M. Lee Lewis (1958). "The Role of Manned Balloons in the Exploration of Space," *Aerospace Engineering 17*, no. 8 (August 1958), pp. 45–52.

Rossi, Bruno (1953). "Where Do Cosmic Rays Come From?" *Scientific American* *189*, (September 1953), pp. 64–70.

Rossi, Bruno (1964). *CosmicRays*. (McGraw Hill, 1964).

Rossi, Bruno (1981). "Early Days in Cosmic Rays," *Physics Today 34*, (October 1981), pp. 34–41.

Rossi, Bruno (1985). "Arcetri, 1928–1932," in Sekido and Elliot (1985), pp. 53–73.

[R.T.P.W.] (1938). "Observations Made in the Highest Stratosphere Flight," *Nature 141*, (February 12, 1938), pp. 270–274.

Rumbaugh, L. H., and G. L. Locher (1936). "Neutrons and Other Heavy Particles in Cosmic Radiation of the Stratosphere," in *The National Geographic Society–U. S. Army Air Corps Stratosphere Flight of 1935 in the Balloon "Explorer II,"* (National Geographic Society, 1936), pp. 32–36. Reprinted from *Physical Review 49*, no. 11 (June 1, 1936).

Rydell, Robert W. (1984). *All the World's a Fair: Visions of Empire at American International Expositions, 1876–1916*. (Chicago, 1984).

Rydell, Robert W. (1985). "The Fan Dance of Science," *Isis 76*, no. 284 (December 1985), pp. 525–542.

Sagan, Carl (1987). "Why We Must Continue to be Explorers," *Parade Magazine* (November 22, 1987), pp. 4–6.

Sandström, A. E. (1965). *Cosmic Ray Physics*. (Wiley, 1965).

[Science Policy Research Division, Congressional Research Service] (1981). *United States Civilian Space Programs 1958–1978. Vol. 1*. (Govt. Printing Office, January 1981).

Seamans, Robert C. Jr. (1961). "The Challenge of Space Exploration," *Smithsonian Report for 1961*. Updated 1965, and reprinted in W. P. True ed., *Smithsonian Treasury of 20th Century Science*. (Simon and Schuster, 1966).

Seidel, Robert (1978). *Physics Research In California: The Rise of a Leading Sector in American Physics*. (Ph.D. dissertation, University of California, Berkeley, 1978).

Sekido, Yataro, and Harry Elliot, eds. (1985). *Early History of Cosmic Ray Studies*. (Reidel, 1985).

Seymour, Raymond B., and Tai Cheng, eds. (1986). *History of Polyolefins*. (Reidel, 1986).

Shepherd, Martin (1936). "The Composition of the Atmosphere at Approximately 21.5 Kilometers," in *The National Geographic Society–U. S. Army Air Corps Stratosphere Flight of 1935 in the Balloon "Explorer II."* (National Geographic Society, 1936), pp. 117–133.

Simpson, John A. (1948). *Physical Review 73*, (1948), p. 1389.

Simpson, John A. (1951). *Physical Review 83*, (1951), p. 1175.

Simpson, John A. (1953). "Evidence for a Solar Cosmic Ray Component," in Gerard Kuiper, ed., *The Sun*. (Chicago, 1953), pp. 715–721.

Simpson, John A. (1985). "Cosmic Ray Astrophysics at Chicago (1947–1960)," in Sekido and Elliot (1985), pp. 385–409.

Simpson, John A. (1986). *To Explore and Discover*. (Ryerson Lecture, University of Chicago, 1986).

Smith, Michael L. (1983). "Selling the Moon," in R. W. Fox and T. J. Jackson Lears, eds., *The Culture of Consumption*. (New York, 1983), pp. 177–209.

Snyder, Wilbert F., and Charles L. Bragaw (1986). *Achievement in Radio: Seventy Years of Radio Science, Technology, Standards, and Measurement at the Na-*

tional Bureau of Standards. National Bureau of Standards Special Publication no. 555. (Department of Commerce, 1986).

"Soviet Balloons for Exploring the Stratosphere," *Science-Supplement 81,* (17 May 1935), p. 6.

Spilhaus, Athelstan F., C. S. Schneider, and C. B. Moore (1948). "Controlled Altitude Free Balloons," *Journal of Meteorology 5,* (August 1948), pp. 130–137.

Stair, R., and W. W. Coblentz (1938). "Radiometric Measurements of Ultraviolet Solar Intensities in the Stratosphere," *Journal of Research of the NBS 20,* Paper RP1075, (February 1938), pp. 185–215.

Stehling, Kurt R., and William Beller (1962). *Skyhooks.* (Doubleday, 1962).

Steinmaurer, Rudolf (1985). "Erinnerungen an V.F. Hess," in Sekido and Elliot (1985), p. 17.

Stevens, Albert W. (1934). "Exploring the Stratosphere," in *The National Geographic Society–U.S. Army Air Corps Stratosphere Flight of 1934 in the Balloon "Explorer."* (National Geographic Society, 1935). Reprinted from the *National Geographic 66,* no. 4 (October 1934), pp. 397–434.

Stevens, Albert W. (1936). "Report of the Commanding Officer of the National Geographic Society–Army Air Corps Stratospheric Flight of 1935 to Chief of the Air Corps, United States Army," in *The National Geographic Society–U. S. Army Air Corps Stratosphere Flight of 1935 in the Balloon "Explorer II."* (National Geographic Society, 1936), pp. 158–163.

Stewart, Irwin (1948). *Organizing Scientific Research for War. The Administrative History of the Office of Scientific Research and Development.* (Atlantic–Little, Brown, 1948).

Street J. C., and T. H. Johnson (1932). "Concerning the Production of Secondaries by Cosmic Radiation," *Physical Review 42,* (1932), pp. 142–144.

Stuewer, Roger H. ed. (1979). *Nuclear Physics in Retrospect.* (Minnesota, 1979).

Swann, W. F. G. (1932). "Electrons as Cosmic Rays," *Physical Review 41,* (1932), pp. 540–541.

Swann, W. F. G. (1933a). "A Mechanism for Acquirement of Cosmic-Ray Energy by Electrons," *Physical Review 43,* (1933), p. 217.

Swann, W. F. G. (1933b). "On the Nature of the Primary Cosmic Radiation," *Physical Review 43,* (1933), pp. 945–946.

Swann, W. F. G. (1933c). "Hoffmann Stösse and the Origin of Cosmic-Ray Ionization," *Physical Review 44,* (December 1933), pp. 1025–1027.

Swann, W. F. G. (1934). "The Relation of the Primary Cosmic Radiation to the Phenomena Observed," *Physical Review 46,* (November 1934), pp. 828–829.

Swann, W. F. G. (1935a). "Cosmic Rays," *The Book of the Hayden Planetarium.* (New York, 1935), pp. 255–261.

Swann, W. F. G. (1935b). "The Nature of the Cosmic Radiation," *Physical Review 47,* (April 1935), pp. 575–577.

Swann, W. F. G. (1936a). "Report on the Work of the Bartol Research Foundation, 1935–1936," *Journal of the Franklin Institute 222,* no. 6 (December 1936), pp. 646–714.

Swann, W. F. G. (1936b). "Report on the Work of the Bartol Research Foundation, 1934–1935," *Journal of the Franklin Institute 222,* no. 1 (July 1936), pp. 1–81.

Swann, W. F. G. (1937). "Report on the Work of the Bartol Research Foundation, 1936–1937," *Journal of the Franklin Institute 224,* no. 4 (October 1937), pp. 415–473.

Swann, W. F. G. (1955). *The Story of Cosmic Rays*. (Sky Publishing Corporation, 1955).

Swann, W. F. G., and G. L. Locher (1935). "The Variation of Cosmic Ray Intensity with Direction in the Stratosphere," in *The National Geographic Society–U.S. Army Air Corps Stratosphere Flight of 1934 in the Balloon "Explorer."* (National Geographic Society, 1935), pp. 7–14.

Swann, W. F. G., G. L. Locher, and W. E. Danforth (1936). "Measurement of Cosmic Ray Intensities by Geiger–Müller Counter Telescopes," in *The National Geographic Society–U. S. Army Air Corps Stratosphere Flight of 1935 in the Balloon "Explorer II."* (National Geographic Society 1936), pp. 16–25.

Swann, W. F. G., C. G. Montgomery, and D. D. Montgomery (1936). "Observations of the Production of Cosmic Ray Burst of Large Size," in *The National Geographic Society–U. S. Army Air Corps Stratosphere Flight of 1935 in the Balloon "Explorer II,"* (National Geographic Society, 1936), pp. 26–29.

Talley, B. B. (1934). *Study of the Vertical Aerial Photographs taken on the National Geographic Society–U.S. Army Air Corps Stratospheric Flight 1934.* (Corps of Engineers, Wright Field, 1934), Stevens Folder, Swann Papers, APS.

Tatarewicz, Joseph N. (1984). *"Where Are the People Who Know What They Are Doing?" Space Technology and Planetary Astronomy, 1958–1975.* (Ph.D. dissertation, Indiana University 1984).

"Ten Miles Up," *Aeroplane 40*, (June 3, 1931), p. 1026.

[The Bird Dogs] (1961). "The Evolution of the Office of Naval Research," *Physics Today*, (August 1961). Reprinted in Spencer R. Weart and Melba Phillips, eds., *History of Physics.* (American Institute of Physics, 1985), pp. 94–99.

Titterton, George F. (1937). *Aircraft Materials and Processes.* (Pitman, 1937. Rev. 1941), pp. 200–204.

Turkevich, Anthony L. (1973). "The Chemical Analysis of the Lunar Surface on Surveyor," *Bulletin of the Atomic Scientists 29*, (December 1973), pp. 27–34.

Turner, Frank M. (1978). "The Victorian Conflict between Science and Religion: A Professional Dimension," *Isis 69*, (1978), pp. 356–378.

Vaeth, J. Gordon (August 1963). "When the Race for Space Began," *U.S. Naval Institute Proceedings*, (August 1963), pp. 69–78.

Van Allen, James A. (1983). *Origins of Magnetospheric Physics.* (Smithsonian, 1983).

Van Allen, James A. (1986). "Space Science, Space Technology and the Space Station," *Scientific American 254*, (January 1986), pp. 32–39.

Van Allen, James A. (1987). "Space Station: The Next Logical Step to National Distress," *International Security*, (Spring 1987), pp. 183–186.

Van Dyke, Vernon (1964). *Pride and Power: The Rationale of the Space Program.* (University of Illinois, 1964).

Van Orman, Ward T. (1978). *The Wizard of the Winds.* (North Star Press, 1978).

Vernov, Sergei N. (1934). "On the Study of Cosmic Rays at Great Altitude," *Physical Review 46*, (1934), p. 822.

Vernov, Sergei N. (1935). "Radio Transmission of Cosmic Ray Data from the Stratosphere," *Nature 135*, (June 29, 1935), pp. 1072–1073.

Vernov, Sergei N., Naum L. Grigorov, et al. (1985). "From Balloons to Space Stations," in Sekido and Elliot (1985), p. 357.

Waldrop, M. Mitchell (1987). "A Crisis in Space Research," *Science 235*, (23 January 1987), pp. 426–429.

Weaver, Kenneth F. (1972). "To The Mountains of the Moon," *National Geographic 141*, no. 2 (February 1972), pp. 230–265.

Webb, James E. (1964). *Man Must Take Environment into Space—Project Gemini.* (NASA, 1964).

Weiner, Charles (1970). "Physics in the Great Depression," *Physics Today*, (October 1970), pp. 31–37. Reprinted in Spencer R. Weart and Melba Phillips, eds., *History of Physics.* (American Institute of Physics, 1985), pp. 115–121.

Weisz, Paul (1941). "The Geiger–Müller Tube: An Electronic Instrument," *Electronics*, (December 1941), pp. 18–21.

White, Clayton S., and Otis O. Benson, eds. (1952). *Physics and Medicine of the Upper Atmosphere: A Study of the Aeropause.* (University of New Mexico, 1952).

Wilkins, T. R. (1936). "Cosmic Ray Tracks in Photographic Emulsions," in *The National Geographic Society–U. S. Army Air Corps Stratosphere Flight of 1935 in the Balloon "Explorer II."* (National Geographic Society, 1936), pp. 37–48.

Wilkins T. R., and H. St. Helens (1935). *Physical Review 49*, (1935), p. 855.

Winckler J. R. and D. J. Hoffman (n.d.). *Cosmic Rays.* ("Resource Letter CR-1 on Cosmic Rays,"). Reprint. American Association of Physics Teachers.

Winston, A. W. (1935). "The Design and Construction of the Gondola for the 'Explorer'," in *The National Geographic Society–U.S. Army Air Corps Stratosphere Flight of 1934 in the Balloon "Explorer."* (National Geographic Society, 1935), pp. 110–111.

Winston, A. W. (1936). "The Design and Construction of the Gondola for 'Explorer II'," in *The National Geographic Society–U. S. Army Air Corps Stratosphere Flight of 1935 in the Balloon "Explorer II."* (National Geographic Society, 1936), pp. 248–250.

Winzen, Otto C. (1958). "10 Years of Plastic Balloons," in Friedrich Hecht, ed., *Proceedings, VIIIth International Astronautical Congress, Barcelona, 1957.* (Springer-Verlag, 1958), pp. 436–459.

Winzen, Otto C. (1959). "From Balloon Capsules to Space Cabins," in Friedrich Hecht, ed., *Proceedings, IXth International Astronautical Congress, Amsterdam, 1958 Vol. II.* (Springer-Verlag, 1959), pp. 562–569.

Ziegler, Charles A. (1986). "Ballooning and the Birth of Cosmic Ray Physics," *National Air and Space Museum Research Report 1986.* (Smithsonian, 1986), pp. 69–92.

Ziegler, Charles A. (1988). "Waiting for Joe-1: Decisions Leading to the Detection of Russia's First Atomic Bomb Test," *Social Studies of Science 18*, (May 1988), pp. 197–229.

Index